The Role of Toll-Like Receptors (TLR) in Infection and Inflammation

The Role of Toll-Like Receptors (TLR) in Infection and Inflammation

Editors

Ralf Kircheis
Oliver Planz

MDPI • Basel • Beijing • Wuhan • Barcelona • Belgrade • Manchester • Tokyo • Cluj • Tianjin

Editors
Ralf Kircheis
Research & Development
Syntacoll GmbH
Saal a.d. Donau
Germany

Oliver Planz
Institute of Cell Biology and
Immunology
Eberhard Karls University
Tübingen
Germany

Editorial Office
MDPI
St. Alban-Anlage 66
4052 Basel, Switzerland

This is a reprint of articles from the Special Issue published online in the open access journal *International Journal of Molecular Sciences* (ISSN 1422-0067) (available at: www.mdpi.com/journal/ijms/special_issues/TLR_role).

For citation purposes, cite each article independently as indicated on the article page online and as indicated below:

LastName, A.A.; LastName, B.B.; LastName, C.C. Article Title. *Journal Name* **Year**, *Volume Number*, Page Range.

ISBN 978-3-0365-7615-2 (Hbk)
ISBN 978-3-0365-7614-5 (PDF)

Contents

About the Editors . . . . **vii**

Preface to "The Role of Toll-Like Receptors (TLR) in Infection and Inflammation" . . . . **ix**

Ralf Kircheis and Oliver Planz
The Role of Toll-like Receptors (TLRs) and Their Related Signaling Pathways in Viral Infection and Inflammation
Reprinted from: *Int. J. Mol. Sci.* **2023**, *24*, 6701, doi:10.3390/ijms24076701 . . . . **1**

Ralf Kircheis and Oliver Planz
Could a Lower Toll-like Receptor (TLR) and NF-B Activation Due to a Changed Charge Distribution in the Spike Protein Be the Reason for the Lower Pathogenicity of Omicron?
Reprinted from: *Int. J. Mol. Sci.* **2022**, *23*, 5966, doi:10.3390/ijms23115966 . . . . **9**

Abdul Manan, Rameez Hassan Pirzada, Muhammad Haseeb and Sangdun Choi
Toll-like Receptor Mediation in SARS-CoV-2: A Therapeutic Approach
Reprinted from: *Int. J. Mol. Sci.* **2022**, *23*, 10716, doi:10.3390/ijms231810716 . . . . **45**

Mohammad Enamul Hoque Kayesh, Michinori Kohara and Kyoko Tsukiyama-Kohara
Toll-like Receptor Response to Hepatitis C Virus Infection: A Recent Overview
Reprinted from: *Int. J. Mol. Sci.* **2022**, *23*, 5475, doi:10.3390/ijms23105475 . . . . **71**

Aleksandra Dondalska, Sandra Axberg Pålsson and Anna-Lena Spetz
Is There a Role for Immunoregulatory and Antiviral Oligonucleotides Acting in the Extracellular Space? A Review and Hypothesis
Reprinted from: *Int. J. Mol. Sci.* **2022**, *23*, 14593, doi:10.3390/ijms232314593 . . . . **85**

Mohammed Y. Behairy, Ali A. Abdelrahman, Eman A. Toraih, Emad El-Deen A. Ibrahim, Marwa M. Azab and Anwar A. Sayed et al.
Investigation of TLR2 and TLR4 Polymorphisms and Sepsis Susceptibility: Computational and Experimental Approaches
Reprinted from: *Int. J. Mol. Sci.* **2022**, *23*, 10982, doi:10.3390/ijms231810982 . . . . **105**

Longfei Yan, Yanran Li, Tianyu Tan, Jiancheng Qi, Jing Fang and Hongrui Guo et al.
RAGE–TLR4 Crosstalk Is the Key Mechanism by Which High Glucose Enhances the Lipopolysaccharide-Induced Inflammatory Response in Primary Bovine Alveolar Macrophages
Reprinted from: *Int. J. Mol. Sci.* **2023**, *24*, 7007, doi:10.3390/ijms24087007 . . . . **129**

Finn Jung, Raphaela Staltner, Anja Baumann, Katharina Burger, Emina Halilbasic and Claus Hellerbrand et al.
A Xanthohumol-Rich Hop Extract Diminishes Endotoxin-Induced Activation of TLR4 Signaling in Human Peripheral Blood Mononuclear Cells: A Study in Healthy Women
Reprinted from: *Int. J. Mol. Sci.* **2022**, *23*, 12702, doi:10.3390/ijms232012702 . . . . **145**

Yonghong Luo, Rawipan Uaratanawong, Vivek Choudhary, Mary Hardin, Catherine Zhang and Samuel Melnyk et al.
Advanced Glycation End Products and Activation of Toll-like Receptor-2 and -4 Induced Changes in Aquaporin-3 Expression in Mouse Keratinocytes
Reprinted from: *Int. J. Mol. Sci.* **2023**, *24*, 1376, doi:10.3390/ijms24021376 . . . . **159**

Charlotte E. Roth, Rogerio B. Craveiro, Christian Niederau, Hanna Malyaran, Sabine Neuss and Joachim Jankowski et al.
Mechanical Compression by Simulating Orthodontic Tooth Movement in an In Vitro Model Modulates Phosphorylation of AKT and MAPKs via TLR4 in Human Periodontal Ligament Cells
Reprinted from: *Int. J. Mol. Sci.* **2022**, *23*, 8062, doi:10.3390/ijms23158062 . . . . **175**

Marina Nicolai, Julia Steinberg, Hannah-Lena Obermann, Francisco Venegas Solis, Eva Bartok and Stefan Bauer et al.
Identification of an Optimal TLR8 Ligand by Alternating the Position of 2′-O-Ribose Methylation
Reprinted from: *Int. J. Mol. Sci.* **2022**, *23*, 11139, doi:10.3390/ijms231911139 . . . . **193**

Shuquan Zhang, Yu Liu, Ji Zhou, Jiaxin Wang, Guangyi Jin and Xiaodong Wang
Breast Cancer Vaccine Containing a Novel Toll-like Receptor 7 Agonist and an Aluminum Adjuvant Exerts Antitumor Effects
Reprinted from: *Int. J. Mol. Sci.* **2022**, *23*, 15130, doi:10.3390/ijms232315130 . . . . **207**

About the Editors

Ralf Kircheis

Ralf Kircheis has more than 30 years of experience in translational research in the fields of immunology, gene therapy, cancer therapy, virology, pharmacology, and drug delivery. He has published more than 60 original papers and review articles in leading international journals as well as several book chapters, and is co-inventor of various patent applications in the fields of cancer immune therapy, virology, gene therapy, cancer therapy, and drug delivery. His major topics of interest comprise targeted DNA/RNA delivery and drug delivery, immunotherapy of cancer, involvement of TLR, and related signal transduction pathways in viral infection and inflammation.

Oliver Planz

Oliver Planz is Professor for Immunology and Cell Biology and heads the group "Translational Immunology of Infection" in the Department of Immunology at the University of Tuebingen. He is an expert in antiviral research and translational preclinical R&D up to clinical studies and functioned as CSO for Atriva Therapeutics GmbH.

Preface to "The Role of Toll-Like Receptors (TLR) in Infection and Inflammation"

Toll-like receptors (TLRs) represent a powerful system for the recognition and elimination of pathogen-associated molecular patterns (PAMPs) from bacteria, viruses, and other pathogens and damage-associated molecular patterns (DAMPs) released from dying cells. TLRs are expressed on immune cells but can also be present on other cell populations. Typical PAMPs include bacterial cell wall components, viral pathogens, or pathogenic nucleic acids, including viral RNA and DNA. Activation of TLRs leads to the production of proinflammatory cytokines and type I interferons which are important for the induction of the host immune response against bacterial and viral infections. However, dysregulation and overstimulation can be detrimental, leading to hyper-inflammation, sepsis, and loss of tissue integrity. The involvement of TLRs in inflammation and bacterial infection has been widely recognized. TLRs are involved in the pathogenesis of acute viral infections, including COVID-19. Consequently, TLRs are promising targets for pharmacological intervention and treatment.

The involvement of TLRs in inflammation and bacterial infection has been recognized for a long time. There is an increasing number of reports demonstrating the involvement of TLR activation in a variety of viral infections, associated with protective immunity, but also immune hyper activation and even viral replication. Recent data show the involvement of TLR activation in various acute respiratory viral infections, including SARS-CoV-2 and indicate an essential role in COVID-19 pathology. The present Special Issue aimed to gather newest data and hypotheses regarding molecular and cellular mechanisms of TLR triggering and activation by viral infections and inflammation. Downstream signaling pathways of TLRs, and their correlation to immunology and pathophysiology of the associated diseases, were also discussed. Finally, this Special Issue facilitated on translational research resulting in new targets for the treatment of viral infectious diseases including COVID-19.

Ralf Kircheis and Oliver Planz

Editors

International Journal of
Molecular Sciences

MDPI

Editorial

The Role of Toll-like Receptors (TLRs) and Their Related Signaling Pathways in Viral Infection and Inflammation

Ralf Kircheis [1,*] and Oliver Planz [2]

1 Syntacoll GmbH, 93342 Saal an der Donau, Germany
2 Institute of Cell Biology and Immunology, Eberhard Karls University Tuebingen, 72076 Tuebingen, Germany; oliver.planz@uni-tuebingen.de
* Correspondence: rkircheis@syntacoll.de; Tel.: +49-15116790606

1. Introduction

Toll-like receptors (TLRs) belong to a powerful system for the recognition and elimination of pathogen-associated molecular patterns (PAMPs) from bacteria, viruses, and other pathogens. They are also involved in recognizing and eliminating self-derived, damage-associated molecular patterns (DAMPs) released from dying or lytic cells. Typical PAMPs are nucleic acids, including viral RNA and DNA, but they also include surface-exposed glycoproteins, lipoproteins, and various membrane components.

TLRs are highly expressed in immune cells, such as dendritic cells (DCs) and macrophages, as well as in non-immune cells, such as fibroblasts and endothelial cells. Activation of TLRs leads to the production of proinflammatory cytokines and type I interferons, which are important for induction of the host immune response against bacterial, fungal, and viral infections and malaria. However, dysregulation and overstimulation can be detrimental, leading to hyperinflammation, sepsis, and loss of tissue integrity. TLRs are involved in the pathogenesis of acute viral infections.

The common theme of this Special Issue is the role of toll-like receptors (TLRs) and their related signaling pathways in viral infection and inflammation, including sterile inflammation and inflammation in response to mechanical stress, tissue remodeling, and cancer. Four reviews and six experimental articles are published in this Special Issue. The reviews focus on the topics of viral infection-induced inflammation, such as SARS-CoV-2 [1,2] and hepatitis C virus (HCV) infection [3], related to TLR-induced pathological processes; they also discuss the protective role of TLRs against viral infection and the strategies used by viruses to inhibit TLR activation. Furthermore, the role of immunoregulatory and antiviral oligonucleotides is addressed [4]. The six experimental research papers cover topics ranging from TLR2 and TLR4 polymorphisms and susceptibility to bacterial sepsis [5]; decrease in endotoxin-induced activation of TLR4 signaling by xanthohumol-rich hop extract [6], advanced glycosylation end-products and TLR- induced changes in aquaporin-3 expression in mouse keratinocytes [7]; TLR4 signaling in periodontal ligament cells in response to mechanical compression [8]; identification of optimal TLR8 ligand by altering the position of 2′-O-ribose methylation [9]; to a breast cancer vaccine containing a novel TLR7 agonist showing antitumor effects [10].

The two facets of TLR activation in viral infection are well illustrated by two review articles focusing on the detrimental effects of massive activation of TLR and the antiviral protective effects of TLR activation in the context of SARS-CoV-2 infection [1,2]. Patients in critical stage show acute respiratory distress syndrome (ARDS) of the lungs; coagulopathies; multi-organ dysfunction correlating with massive cytokine and chemokine release; and distortion of the complement and coagulation system, which is mostly triggered by hyperactivation of innate immune receptors and their related pathways, such as NF-κB, JAK/STAT, and MAPK [1,2].

Citation: Kircheis, R.; Planz, O. The Role of Toll-like Receptors (TLRs) and Their Related Signaling Pathways in Viral Infection and Inflammation. *Int. J. Mol. Sci.* **2023**, *24*, 6701. https://doi.org/10.3390/ijms24076701

Received: 26 March 2023
Accepted: 31 March 2023
Published: 4 April 2023

The involvement of cell types that differ greatly in their TLR patterns and their pro-inflammatory potential will finally determine the level of pro-inflammatory activation. Whereas alveolar cells that highly express TMPRSS2 in the lungs are the main drivers of viral replication, the major pathophysiological effects are derived from over-activation of innate immune cells, such as macrophages and endothelial cells. In this context, the lower pathology (despite higher infectivity) of the omicron variant compared to previous VOCs has been hypothesized to correlate with a lower TLR- and NF-κB activation due to a changed charge distribution in the spike protein (Figure 1) [1].

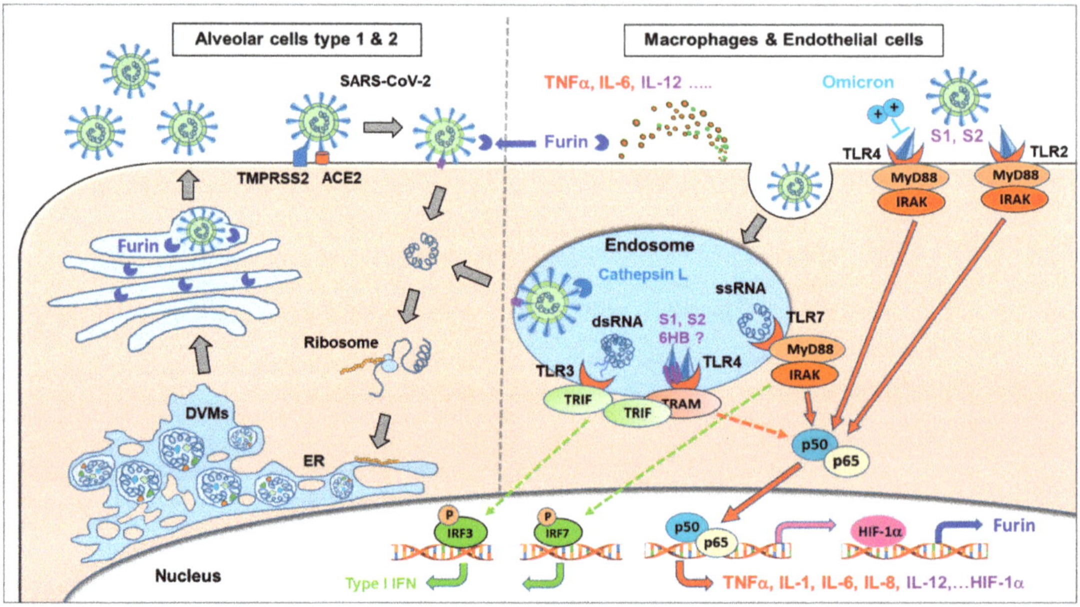

Figure 1. SARS-CoV-2 binds via the spike protein to ACE2, followed by fusion with the host cell membrane. Viral RNA transcription and translation of viral proteins occur in double-membrane vesicles (DMVs). The newly assembled virus particles leave the cells via the Golgi apparatus where the spike protein undergoes proteolytic cleavage by furin. This process occurs preferably in virus-producing cells with a high TMPRSS2 expression, such as alveolar cells (**left side**). Alternatively, the virus can be taken up into endosomes, predominantly in cathepsin L-rich cells, such as innate immune cells and endothelial cells (**right side**). Several components of the SARS-CoV-2 virus act as PAMPs by activating various TLRs, resulting in massive activation of the NF-kB (p50/p65) pathway. Different cells differ regarding their expression of TLRs, with high levels of TLR4 and TLR2 expressed by macrophages and endothelial cells, but low levels by alveolar lung cells. Adapted from [1].

The second article discusses how the outcome of SARS-CoV-2 infection is dependent on the balance between induced antiviral immunity and tissue damage. Surface (TLR2 and 4) and intracellular (TRL3, 7/8, and 9) factors are involved in the recognition of SARS-CoV-2 infection by the immune system. Adaptors such as MyD88 and TRIF are recruited by TLRs to initiate the signaling pathways (Figure 2).

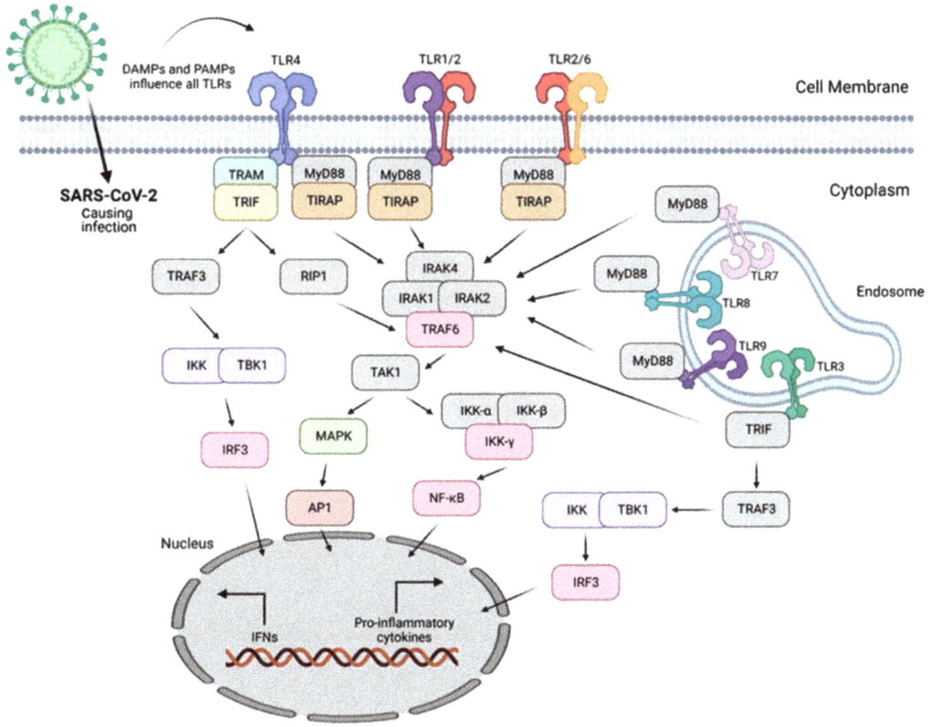

Figure 2. SARS-CoV-2 infection causes DAMPs and PAMPs that are recognized by a broad variety of toll-like receptors (TLRs), leading to the expression of pro-inflammatory cytokines and interferons. Reproduced from [2].

The nature of the ligand and downstream adaptor molecules direct the TLR signaling cascade into two distinct pathways, i.e., MyD88-dependent and -independent pathways. The MyD88-dependent pathway employed by all TLRs (except TLR3) triggers the activation of pro-inflammatory signal transduction pathways, such as NF-κB and mitogen-activated protein kinases (MAPKs), resulting in the release of inflammatory cytokines. In contrast, the TRIF-dependent pathway involved in TLR3 and TLR4 signaling leads to the stimulation of IRF3 expression of IFN-I, which primes the cells for antiviral activities. Expression of another important interferon regulatory factor, IRF7, is triggered by TLR7–9. The production of type I and type III IFNs by TLRs is a significant antiviral feature essential for systemic viral control and viral clearance [2].

The dual role of TLR activation has also been shown for HCV infection. Protective effects have been shown for TLR3/4/7/8/9, whereas adverse effects have been demonstrated, e.g., for TLR4 and TLR8. The protective effects of TLR activation may potentially outbalance the adverse side effects in the case of chronic HCV infection. Accordingly, HCV has developed multiple strategies to inhibit the innate immune response and dispose the host toward chronic infection, largely via virus-derived TLR inhibitors (Figure 3).

A clear understanding of TLR interactions in HCV infection is critical for developing new therapeutic approaches to fight the disease, including TLR agonist-adjuvanted HCV vaccines [3].

The fourth review discusses the role of immune regulatory and antiviral oligonucleotides regarding TLR-3, 7, 8, and 9, which are located within endosomes and sense oligonucleotides that are taken up into the endosomes. There seems to be an abundant pool of 30–40 oligonucleotides derived from different RNAs, such as tRNA, rRNA, and mRNA. This 30–40 oligonucleotide pool may confer a "buffering" system to regulate the uptake of endocytic cargo into cells. Some single-stranded oligonucleotides (ssONs) may have the capacity to inhibit endocytic pathways and have immunoregulatory functions. In addition, this pool of 30–40 oligonucleotide RNAs may also prevent the entry of certain viruses (Figure 4) [4].

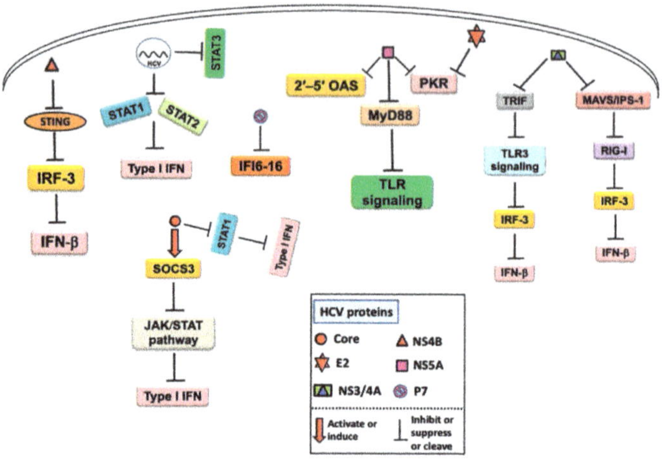

Figure 3. An overview of the mechanism of host innate immune response inhibition by HCV and its proteins. Reproduced from [5].

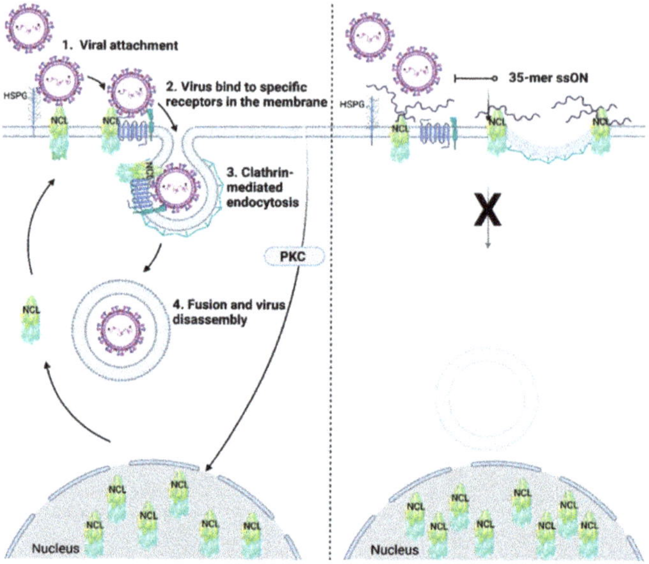

Figure 4. Schematic illustration of nucleolin (NCL) as a receptor for viruses and its participation in clathrin-mediated endocytosis (**left**), which can be inhibited by 35-mer ssON (**right**). Viruses attach to the cellular membrane using, e.g., heparan sulfate proteoglycan and/or NCL, followed by clathrin-mediated endocytosis. 35-mer ssON, but not 15-mer ssON, can inhibit viral attachment by shielding NCL. 35-mer ssON is hypothesized to confer steric hindrance for endocytosis. Reproduced from [4].

Toll-like receptors (TLRs) play an eminent role in immune responses to bacterial pathogens, and are involved in the development of sepsis. TLR genetic variants might influence individual susceptibility to develop sepsis. In an experimental study, the association of genetic polymorphisms of TLR2 and TLR4 with the risk of developing sepsis was investigated based on single nucleotide polymorphisms (SNPs). The DNA samples of patients in intensive care unit were genotyped using RT-PCR technology. Significant associations between TLR2 Arg753Gln polymorphisms and sepsis and between TLR4 Asp299Gly polymorphisms and *Acinetobacter baumannii* infection were found. This study concludes that the TLR2 genotype may be a risk factor for sepsis in adult patients [5].

Infections with Gram-negative bacteria are among the leading causes of infection-related deaths. To find new therapeutic approaches to control bacterial infection-triggered sepsis, the effect of xanthohumol-rich hop extract on endotoxin-induced activation of TLR-4 signaling in human peripheral blood mononuclear cells was investigated. Previous studies had suggested that chalcone xanthohumol (XN) found in hop has anti-inflammatory effects. A placebo-controlled, randomized cross-over design study assessed if the oral intake of a single low dose of XN could affect lipopolysaccharide (LPS)-induced immune responses in peripheral blood mononuclear cells (PBMCs) ex vivo in normal-weight healthy women. LPS-dependent activation of hTLR4-transfected cells was significantly and dose-dependently suppressed by the XN-rich hop extract, which was attenuated when cells were co-challenged with sCD14. These results suggest that even low doses of XN consumed in a XN-rich hop extract can suppress LPS-dependent stimulation of PBMCs due to the interaction of the hop compound with the CD14/TLR4 signaling cascade [6].

The present Special Issue also covers the roles of TLRs in wound healing, after mechanical stress, and in cancer. Usually, inflammation is finally exchanged by processes restoring physiological homeostasis, such as fibrinolysis, wound healing, and re-epithelization. Prolonged inflammation and impaired re-epithelization are major contributing factors to chronic non-healing diabetic wounds. Advanced glycation end-products (AGEs) and the activation of toll-like receptors (TLRs) can trigger inflammatory responses. Aquaporin-3 (AQP3) plays an essential role in keratinocyte function and skin wound re-epithelialization, regeneration, and hydration. Suberanilohydroxamic acid (SAHA), a histone deacetylase inhibitor, mimics the increased acetylation observed in diabetes. The effects of TLR2/TLR4 activators and AGEs on keratinocyte AQP3 expression in the presence and absence of SAHA was studied in primary mouse keratinocytes. The results indicate that TLR2 activation and AGEs may be beneficial for wound healing and skin hydration under normal conditions via AQP3 upregulation, but these pathways are likely deleterious in diabetes [7].

The involvement of TLRs in sterile inflammation induced by mechanical stress on tissues was studied in another study. Mechanical compression by simulating orthodontic tooth movement in in vitro models induced pro-inflammatory cytokine expression in periodontal ligament (PDL) cells. Primary PDL cells were studied for cell signaling downstream of key molecules involved in the process of sterile inflammation via TLR4. The TLR4 monoclonal blocking antibody was found to reverse the upregulation of phospho-AKT caused by compressive force. Overall, this study provides evidence that TLR4 is involved in the modulation of sterile inflammation during mechanical stress, such as during orthodontic therapy and periodontal remodeling [8].

Recognition of RNA by TLRs is regulated by various posttranslational modifications. Different single 2′-O-ribose (2′-O-) methylations have been shown to convert TLR7/TLR8 ligands into specific TLR8 ligands. A study investigated whether the position of 2′-O-methylation is crucial for its function. An 18S rRNA-derived TLR7/8 ligand, RNA63, was found to be differentially digested as a result of 2′-O-methylation, leading to variations in TLR8 and TLR7 inhibition. The suitability of certain 2′-O-methylated RNA63 derivatives as TLR8 agonists was further supported by the fact that other RNA sequences were only weak TLR8 agonists. Specific 2′-O-methylated RNA derivatives were identified as optimal TLR8 ligands [9].

The last paper of this Special Issue addresses the use of TLR agonists as adjuvants for anti-cancer vaccination against mucin 1 (MUC1), a tumor-associated antigen that is highly expressed in breast cancer. The authors constructed a novel tumor vaccine (SZU251 + MUC1 + Al) containing MUC1 and two types of adjuvants, a TLR7 agonist (SZU251) and an aluminum adjuvant (Al). Immunostimulatory responses were first verified in vitro, showing that the vaccine promoted the release of cytokines and the expression of costimulatory molecules in mouse bone marrow dendritic cells and spleen lymphocytes. Importantly, SZU251 + MUC1 + Al was effective and safe against tumors expressing the MUC1 antigen in both prophylactic and therapeutic schedules in vivo. The immune responses in vivo were attributed to the increase in specific humoral and cellular immunity, including antibody titers, $CD4^+$, $CD8^+$, and activated $CD8^+$ T cells. The results indicate that TLR agonists can be successfully used as new vaccine adjuvant candidates for the prevention and treatment of breast cancer [10].

2. Discussion

Overall, this Special Issue illustrates the two sides of TLRs as drivers of pathogenesis of acute bacterial and viral infections, including COVID-19, but also as potent players for antiviral, antibacterial, and anti-cancer immune activation. Dependent on the direction of the disbalance in TLR activation in different pathologies, TLR agonists [9,10] or antagonists [2,4,6] may be of interest. For stimulating innate immune responses against bacterial, viral, or parasitic invasions or for use as an adjuvant in anti-cancer vaccination, one single selected TLR agonist may be sufficient to trigger an effective innate immune response. In contrast, inhibiting the hyperactivation of the innate system can be much more complex due to the involvement of multiple and diverse TLRs in various pathological conditions [1,3]. Therefore, it will be difficult to choose one TLR antagonist to inhibit all TLRs that may be involved in the induced systemic hyperactivation. Moreover, secondarily induced cytokines, such as TNFα, IL-1, and IL-6, will activate additional receptors, amplifying the immune stimulatory signaling [1,11].

In order to inhibit the entire immune stimulatory signaling cascades induced, e.g., by life-threatening acute viral infections (such as Ebola, dengue fever, SARS-CoV-2, and RSV) [1,2,12] or by bacterial sepsis [5], inhibition of **(i)** specific downstream adaptor molecules or of **(ii)** central signaling pathways may be more specific to obtain the desired effect or to provide a more generalized modulation of signaling, respectively. Regarding downstream adaptor molecules, the MyD88-dependent pathway (involved in the signaling of all TLRs, except for TLR3) primarily triggers the activation of pro-inflammatory signal transduction pathways, such as NF-κB and mitogen-activated protein kinases (MAPKs) [1–3]. In contrast, the TRIF-dependent pathway involved in TLR3 and TLR4 signaling leads to the stimulation of IRF expression of IFN-I, which primes the cells for antiviral activities [2]. Specific inhibition of Myd88 signaling without affecting TRIF-mediated signaling may inhibit the expression of pro-inflammatory cytokines, with no (or only marginal) effect on IRF expression. Furthermore, secondary adaptor molecules, such as TRAF3 and TRAF6, may provide additional levels for modulating the ratio between the induction of IRF and pro-inflammatory signaling pathways after TLR activation [13].

To attenuate TLR signaling, the inhibition of whole signal transduction pathways, such as NF-κB [1] or JAK/STAT [11], may provide a powerful modality to control the multiple additive, synergistic, triggering, and amplifying signaling cascades that have been induced during life-threatening acute viral infections [1,11].

3. Conclusions

TLRs are the drivers of the pathogenesis of acute bacterial and viral infections, but they are also essential for antiviral, antibacterial, and anti-cancer immune activation. Accordingly, TLRs, including their molecular adaptors and related signaling transduction pathways, are promising targets for pharmacological intervention and treatment.

Author Contributions: Conceptualization, R.K. and O.P.; methodology, R.K.; validation, O.P.; formal analysis, R.K. and O.P.; investigation, R.K.; data curation, R.K.; writing—original draft preparation, R.K.; writing—review and editing, O.P. and R.K.; visualization, R.K. All authors have read and agreed to the published version of the manuscript.

Conflicts of Interest: The authors declare no conflict of interest.

References

1. Kircheis, R.; Planz, O. Could a Lower Toll-like Receptor (TLR) and NF-κB Activation Due to a Changed Charge Distribution in the Spike Protein Be the Reason for the Lower Pathogenicity of Omicron? *Int. J. Mol. Sci.* **2022**, *23*, 5966. [CrossRef] [PubMed]
2. Manan, A.; Pirzada, R.H.; Haseeb, M.; Choi, S. Toll-like Receptor Mediation in SARS-CoV-2: A Therapeutic Approach. *Int. J. Mol. Sci.* **2022**, *23*, 10716. [CrossRef] [PubMed]
3. Kayesh, M.E.H.; Kohara, M.; Tsukiyama-Kohara, K. Toll-like Receptor Response to Hepatitis C Virus Infection: A Recent Overview. *Int. J. Mol. Sci.* **2022**, *23*, 5475. [CrossRef] [PubMed]
4. Dondalska, A.; Axberg Pålsson, S.; Spetz, A.-L. Is There a Role for Immunoregulatory and Antiviral Oligonucleotides Acting in the Extracellular Space? A Review and Hypothesis. *Int. J. Mol. Sci.* **2022**, *23*, 14593. [CrossRef] [PubMed]
5. Behairy, M.Y.; Abdelrahman, A.A.; Toraih, E.A.; Ibrahim, E.E.-D.A.; Azab, M.M.; Sayed, A.A.; Hashem, H.R. Investigation of TLR2 and TLR4 Polymorphisms and Sepsis Susceptibility: Computational and Experimental Approaches. *Int. J. Mol. Sci.* **2022**, *23*, 10982. [CrossRef] [PubMed]
6. Jung, F.; Staltner, R.; Baumann, A.; Burger, K.; Halilbasic, E.; Hellerbrand, C.; Bergheim, I. A Xanthohumol-Rich Hop Extract Diminishes Endotoxin-Induced Activation of TLR4 Signaling in Human Peripheral Blood Mononuclear Cells: A Study in Healthy Women. *Int. J. Mol. Sci.* **2022**, *23*, 12702. [CrossRef] [PubMed]
7. Luo, Y.; Uaratanawong, R.; Choudhary, V.; Hardin, M.; Zhang, C.; Melnyk, S.; Chen, X.; Bollag, W.B. Advanced Glycation End Products and Activation of Toll-like Receptor-2 and -4 Induced Changes in Aquaporin-3 Expression in Mouse Keratinocytes. *Int. J. Mol. Sci.* **2023**, *24*, 1376. [CrossRef] [PubMed]
8. Roth, C.E.; Craveiro, R.B.; Niederau, C.; Malyaran, H.; Neuss, S.; Jankowski, J.; Wolf, M. Mechanical Compression by Simulating Orthodontic Tooth Movement in an In Vitro Model Modulates Phosphorylation of AKT and MAPKs via TLR4 in Human Periodontal Ligament Cells. *Int. J. Mol. Sci.* **2022**, *23*, 8062. [CrossRef] [PubMed]
9. Nicolai, M.; Steinberg, J.; Obermann, H.-L.; Solis, F.V.; Bartok, E.; Bauer, S.; Jung, S. Identification of an Optimal TLR8 Ligand by Alternating the Position of 2′-O-Ribose Methylation. *Int. J. Mol. Sci.* **2022**, *23*, 11139. [CrossRef] [PubMed]
10. Zhang, S.; Liu, Y.; Zhou, J.; Wang, J.; Jin, G.; Wang, X. Breast Cancer Vaccine Containing a Novel Toll-like Receptor 7 Agonist and an Aluminum Adjuvant Exerts Antitumor Effects. *Int. J. Mol. Sci.* **2022**, *23*, 15130. [CrossRef] [PubMed]
11. Hojyo, S.; Uchida, M.; Tanaka, K.; Hasebe, R.; Tanaka, Y.; Murakami, M.; Hirano, T. How COVID-19 induces cytokine storm with high mortality. *Inflamm. Regener.* **2020**, *40*, 37. [CrossRef] [PubMed]
12. Olejnik, J.; Hume, A.J.; Mühlberger, E. Toll-like receptor 4 in acute viral infection: Too much of a good thing. *PLoS Pathog.* **2018**, *14*, e1007390. [CrossRef] [PubMed]
13. Häcker, H.; Redecke, V.; Blagoev, B.; Kratchmarova, I.; Hsu, L.C.; Wang, G.G.; Kamps, M.P.; Raz, E.; Wagner, H.; Häcker, G.; et al. Specificity in Toll-like receptor signalling through distinct effector functions of TRAF3 and TRAF6. *Nature* **2006**, *439*, 204–207. [CrossRef] [PubMed]

International Journal of *Molecular Sciences*

MDPI

Review

Could a Lower Toll-like Receptor (TLR) and NF-κB Activation Due to a Changed Charge Distribution in the Spike Protein Be the Reason for the Lower Pathogenicity of Omicron?

Ralf Kircheis [1,*] and Oliver Planz [2]

1 Syntacoll GmbH, 93342 Saal an der Donau, Germany
2 Interfaculty Institute for Cell Biology, Department of Immunology, Eberhard Karls University Tuebingen, 72076 Tübingen, Germany; oliver.planz@uni-tuebingen.de
* Correspondence: rkircheis@syntacoll.de; Tel.: +49-151-167-90606

Abstract: The novel SARS-CoV-2 Omicron variant B.1.1.529, which emerged in late 2021, is currently active worldwide, replacing other variants, including the Delta variant, due to an enormously increased infectivity. Multiple substitutions and deletions in the N-terminal domain (NTD) and the receptor binding domain (RBD) in the spike protein collaborate with the observed increased infectivity and evasion from therapeutic monoclonal antibodies and vaccine-induced neutralizing antibodies after primary/secondary immunization. In contrast, although three mutations near the S1/S2 furin cleavage site were predicted to favor cleavage, observed cleavage efficacy is substantially lower than in the Delta variant and also lower compared to the wild-type virus correlating with significantly lower TMPRSS2-dependent replication in the lungs, and lower cellular syncytium formation. In contrast, the Omicron variant shows high TMPRSS2-independent replication in the upper airway organs, but lower pathogenicity in animal studies and clinics. Based on recent data, we present here a hypothesis proposing that the changed charge distribution in the Omicron's spike protein could lead to lower activation of Toll-like receptors (TLRs) in innate immune cells, resulting in lower NF-κB activation, furin expression, and viral replication in the lungs, and lower immune hyper-activation.

Keywords: Omicron; spike protein; SARS-CoV-2; COVID-19; cytokine storm; NF-kappaB; Toll-like receptor (TLR)

Citation: Kircheis, R.; Planz, O. Could a Lower Toll-like Receptor (TLR) and NF-κB Activation Due to a Changed Charge Distribution in the Spike Protein Be the Reason for the Lower Pathogenicity of Omicron? *Int. J. Mol. Sci.* **2022**, *23*, 5966. https://doi.org/10.3390/ijms23115966

Academic Editor: Dominic De Nardo

Received: 2 May 2022
Accepted: 23 May 2022
Published: 25 May 2022

1. Introduction

The novel SARS-CoV-2 Omicron variant of concern (VoC) B.1.1.529 was first detected in the context of exceptionally high infection numbers in South Africa and Botswana in November 2021, spread worldwide within a few weeks, and is now replacing all other variants worldwide, including the previously dominant Delta variant [1]. The emergence of Omicron has been associated with a dramatic increase in infection case numbers, with doubling times of few days, currently counting for far more than 90% of the cases in the USA and in most European countries, with the fastest growth reported for UK, Denmark, and France, reaching up to more than 400,000 infections per day.

In addition to the dramatically increased infectivity, Omicron has been found to largely evade therapeutic monoclonal antibodies and vaccine-induced polyclonal neutralizing antibodies after primary/secondary immunization. Mutations in the N-terminal domain (NTD) and receptor binding domain (RBD) domains of the spike protein targeted by therapeutic monoclonal antibodies correlate with significantly decreased binding and neutralization in experimental studies and in serum after primary and secondary vaccination with all tested vaccine types, including mRNA-based vaccines, BNT162b2 (BioNTech/Pfizer) and mRNA-1273 (Moderna), and even a more dramatic decrease in neutralizing activity in serums following vector-based SARS-CoV-2 vaccines, such as ChAdOx1 nCoV-19 (AZD1222, AstraZeneca), Ad26.COV2. S (Johnson & Johnson/Janssen), Gam-COVID-19-Vac ("Sputnik

V", Gamaleya National Centre of Epidemiology and Microbiology, Moscow, Russia), or the inactivated vaccine Coronavac (Sinovac, Sinovac Biotech, Beijing, China) [2–4]. This had been expected very soon after the sequence of the Omicron VoC became available because of the highly unusual genetic profile of Omicron with 50 genetic changes, including, exclusively, spike protein 30 amino acid substitutions, one insertion of three amino acids, and several small deletions when compared to the original Wuhan strain. These changes comprise an unprecedented sampling of mutations from earlier VoC, i.e., Alpha, Beta, Gamma, and Delta, together with other substitutions not found in any of the previous VoCs [3,5,6].

The spike protein is responsible for both the adherence of the virus to the host cells and the invasion of the virus into the host cell making the spike protein most critical for viral transmission. All vaccines available today target the spike protein and are based on the original strain first detected in Wuhan, China, at the end of 2019. Changes in the amino acid sequence in the spike protein are expected to affect both the transmissibility of the virus and the ability of the virus to evade neutralizing antibodies. Many of the substitutions and deletions in the NTD and in the RBD can meanwhile be correlated with the observed increased infectivity and transmissibility, and with the high evasion potential from therapeutic monoclonal antibodies and vaccine-induced polyclonal neutralizing antibodies. More enigmatic are the three mutations near the S1/S2 furin cleavage site which were expected to favor cleavage. Contrary to what had been expected, the observed cleavage efficacy has been found to be substantially lower in the Omicron variant compared to the Delta variant and the Wuhan wild-type virus, correlating with significantly lower TMPRSS2-dependent replication in the lungs [3,7–9].

A lower virus replication in lungs, together with a faster replication in the upper respiratory system, such as nasopharyngeal and bronchi, can largely explain Omicron's greater ability for transmission between people while apparently causing less frequently acute respiratory distress syndrome (ARDS) of the lungs and systemic symptoms of COVID-19. However, the molecular mechanisms responsible for these reciprocal changes in cellular tropism of the Omicron variant regarding the upper and lower respiratory system are not sufficiently defined at this moment. Based on the data available so far, we present here a new hypothesis proposing that the changed distribution of charged amino acids in the spike protein of the Omicron variant compared to all other VoCs may disturb the recognition by innate Pattern-recognition receptors (PRRs), in particular of certain Toll-like receptors (TLR), resulting in lower activation of the NF-κB pathway and related signaling pathways, and also resulting in lower furin expression, lower viral replication in the lungs, and lower systemic immune hyper-activation.

2. SARS-CoV-2 and COVID-19—The Virus and the Disease

SARS-CoV-2 belongs to enveloped positive-sense, single-stranded RNA viruses, similar to the two other highly pathogenic coronaviruses, SARS-CoV and Middle East respiratory syndrome (MERS-CoV) [10]. SARS-CoV-2 binds primarily to the angiotensin-converting enzyme-related carboxypeptidase-2 (ACE2) receptor on target cells by its spike (S) protein. In addition to ACE2, additional cellular co-receptors have been identified as potential binding targets for SARS-CoV-2, including integrins, CD147, heparane sulfate, sialic acid, and neutropilin-1 [11,12]. There are two principal cellular entry routes for the virus: the highly efficient plasma membrane route and the cathepsin L-dependent endosomal entry route, dependent on whether TMPRSS2 is co-expressed with ACE2 on the host cell or not [13–16].

The S protein of SARS-CoV-2 is a class I viral membrane fusion protein that exists as a trimer, covered with 22 predicted N-glycosylation sites [17,18]. The spike glycoprotein consists of a large ectodomain, a single-pass transmembrane anchor, and a short C-terminal intracellular tail. The ectodomain contains the receptor binding S1 subunit and the S2 subunit responsible for membrane fusion. The RBD of S1 binds to ACE2 on the target cells with high affinity, whereas S2 mediates fusion between the viral and host cell membranes.

The S2 subunit includes two heptad repeat (HR) regions (HR1 and HR2), the proteolytic site (S2) for the TMPRSS2 serine protease, a hydrophobic fusion peptide, and the transmembrane (TM) domain [16,19]. Between the S1 and S2 subunits, there is a polybasic PRRAR furin-like cleavage site which is unique to the S protein of SARS-CoV-2 and may, together with the particularly high-binding affinity to the target receptor ACE2 and the peculiarity of a long symptom-free but nevertheless highly infectious time period between infection and appearance of first symptoms or asymptomatic transmission [20], be responsible for the particularly efficient spread of SARS-CoV-2 compared to previous pathogenic hCoVs. The ACE2 receptor is widely expressed in pulmonary and cardiovascular tissues, which may explain the broad range of pulmonary and extrapulmonary effects of SARS-CoV-2 infection on the cardiac system, gastrointestinal organs, and kidneys [21–23]. Most individuals infected with SARS-CoV-2 show mild-to-moderate symptoms, and up to 20% of infections may be asymptomatic. This ratio, however, can differ for different virus variants. Symptomatic patients show a wide spectrum of clinical manifestations ranging from mild febrile illness and cough up to acute respiratory distress syndrome (ARDS), multiple organ failure, and death. Whereas for the Omicron variant, a generally lower severity is found compared to the original Wuhan strain and previous VoCs, the clinical picture of severe cases of COVID-19 in general is rather similar to that seen in SARS-CoV-1- and MERS-CoV-infected patients [10]. Whereas younger individuals show predominantly mild-to-moderate clinical symptoms, elderly individuals frequently exhibit severe clinical manifestations [24–27]. Pre-existing comorbidities, including diabetes, respiratory and cardiovascular diseases, renal failure and sepsis, older age and male sex, are associated with more severe disease and higher mortality [28–31].

Regarding the pathophysiological manifestations, a diffuse alveolar disease with capillary congestion, cell necrosis, interstitial edema, platelet-fibrin thrombi, and infiltrates of macrophages and lymphocytes are typical for critical and fatal COVID-19 cases [32]. Furthermore, induction of endotheliitis in various organs (including lungs, heart, kidney, and intestine) by the SARS-CoV-2 infection as a direct consequence of viral involvement and of the host inflammatory response has been demonstrated [22,23].

The molecular and cellular mechanisms for the morbidity and mortality of SARS-CoV-2 are getting increasingly clarified. Virus-induced cytopathic effects and viral evasion of the host immune response, and a dysregulated host IFN type I response by SARS-CoV-2 [33], seem to play a role in disease severity. Furthermore, clinical data from patients, in particular those with severe clinical manifestations, show that highly dysregulated exuberant inflammatory and immune responses correlate with the severity of disease and lethality [25,34–36]. In particularly, upregulated cytokine and chemokine levels, also termed "cytokine storm", have been demonstrated to play a central role in the severity and lethality of SARS-CoV-2 infections. Elevated plasma levels of IL-1β, IL-7, IL-8, IL-9, IL-10, G-CSF, GM-CSF, IFNγ, IP-10, MCP-1, MIP-1α, MIP-1β, PDGF, TNFα, and VEGF have been measured in both ICU (intensive care unit) patients and non-ICU patients, with significantly higher plasma levels of IL-2, IL-7, IL-10, G-CSF, IP-10, MCP-1, MIP-1α, and TNFα found in patients with severe pneumonia developing ARDS and requiring ICU admission and oxygen therapy compared to non-ICU patients showing pneumonia without ADRS [25,37]. Several studies have shown that highly stimulated epithelial-immune cell interactions escalate into exuberant dysregulated inflammatory responses with significantly (topically and systemically) elevated cytokine and chemokine release [38,39].

Regarding the underlying signaling pathways, several reports indicate the NF-κB pathway as one of the critical signaling pathway for the SARS-CoV-2 infection-induced proinflammatory cytokine/chemokine response, playing a central role in the severity and lethality of COVID-19, probably in the context of related pathways such as the IL-6/STAT pathway [40–46]. Notably, this NF-κB-triggered proinflammatory response in acute COVID-19 is shared with other acute respiratory viral infections caused by highly pathogenic influenza A virus of H1N1 (e.g., Spanish flu) and H5N1 (avian flu origin), as well as SARS-CoV-1 and MERS-CoV [10,45].

Excessive activation of exuberant inflammatory responses with involvement of endothelial cells, epithelial cells, and immune cells may lead to further disturbances of other integrated systems, such as the complement system, coagulation, and bradikinine systems, leading to increased coagulopathies and positive signaling feedback loops accelerating COVID-19-associated inflammatory processes [47–53]. Furthermore, vascular occlusion by neutrophil extracellular traps (NETs) [54,55] and disturbances of coagulation including thromboses and multiple microthromboses seem to be another (beside cytokine storm) hallmark of the COVID-19 disease. The development of coagulopathies is one of the key features associated with poor outcome, with elevated D-dimer levels, prolonged prothrombin time, thrombocytopenia, and low fibrinogen (indicating fibrinogen consumption) found as prognostic indicators for poor outcome [56–59]. Lung histopathology often reveals fibrin-based blockages in the small blood vessels of patients who succumb to COVID-19 [32]. Furthermore, various types of antiphospholipid (aPL) antibodies targeting phospholipids and phospholipid-binding proteins were found in half of the serum samples from patients hospitalized with COVID-19. Higher titers of aPL antibodies were associated with neutrophil hyperactivity, including the release of neutrophil extracellular traps (NETs), higher platelet counts, and more severe respiratory disease [60]. High rates of thrombotic-related complications have been reported in adult patients with severe COVID-19 as well as in children developing COVID-19 or multisystem inflammatory syndrome (MIS-C). Studies in adults have invoked thrombotic microangiopathy (TMA) from endothelial cell damage to small blood vessels, leading to hemolytic anemia, thrombocytopenia, and organ damage [61–65].

3. Molecular Changes in the Omicron's Spike Protein and Their Impact on Transmissibility, Immune Escape, and Pathogenicity

Omicron has gathered more than 30 amino acid substitutions, one insertion of three amino acids, and three small deletions in the spike protein compared to the original Wuhan strain, with various mutations shared with previous VoCs and 26 unique modifications. The mutations in the Omicron spike protein can be grouped more or less by four distinct parts of the molecule: the NTD, the RBD, near the S1/S2 cleavage site, and in the S2 subunit with accumulation in the HR1 region [3]. The structure of the spike protein and the mutations found in the Omicron variant are shown in Figure 1:

The highest number of the mutations in the spike protein are in the S1 region, whereas the S2 is relatively conserved in the Omicron variant and actually harbors only six unique mutations i.e., N764K, D796Y, N856K, Q954H, N969K, and L981F, which are not detected in other variants of concern. Remarkable is the increase in positively charged amino acids in four out of these six mutations, plus one change leading to the loss of a negative charge (D796Y). Additionally, mutations in the RBD region led to a significant increase in positively charged residues (N440K, T478K, Q493R, Q498R, Y505H, and T547K), and the S1/S2 cleavage site amino acid substitutions lead to positively charged amino acids (N679K, P681H). In contrast, there is a negative charge accumulation on the NTD surface derived from G142D and EPE insertion after R214 [3].

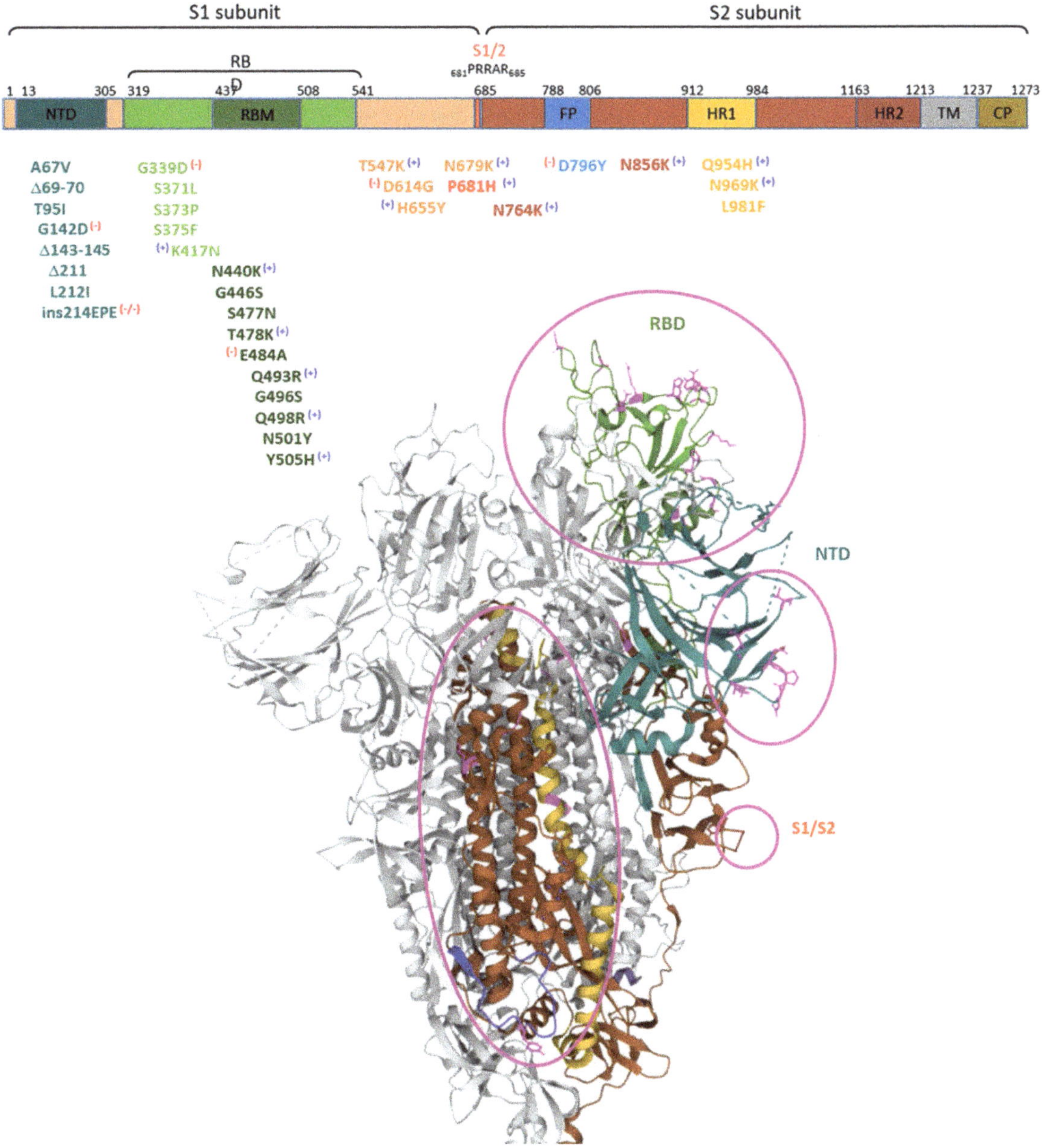

Figure 1. Structure of the SARS-CoV-2 spike protein and mutations in the Omicron variant and spatial illustration of mutated sites based on the published cryo-structure PDB 7TEI SARS-CoV-2 Omicron 1-RBD up Spike Protein Trimer. PDB DOI: 10.2210/pdb7TEI/pdbEM Map EMD-25846: EMDB EMDataResource [66]. Protomer 1—brown, except RBD—light green, NTD—dark turquoise (ride side), HP1—yellow, FP—blue, Omicron mutations—pink, Protomers 2 & 3—grey. Regions of the Omicron spike protein with higher number of mutations are circled in pink.

3.1. Mutations in the Omicron Spike Protein RBD Region Strengthen the Spike-ACE2 Interaction

Mutations in the RBD region have been shown to intensify the interface interaction with ACE2 (Figure 2).

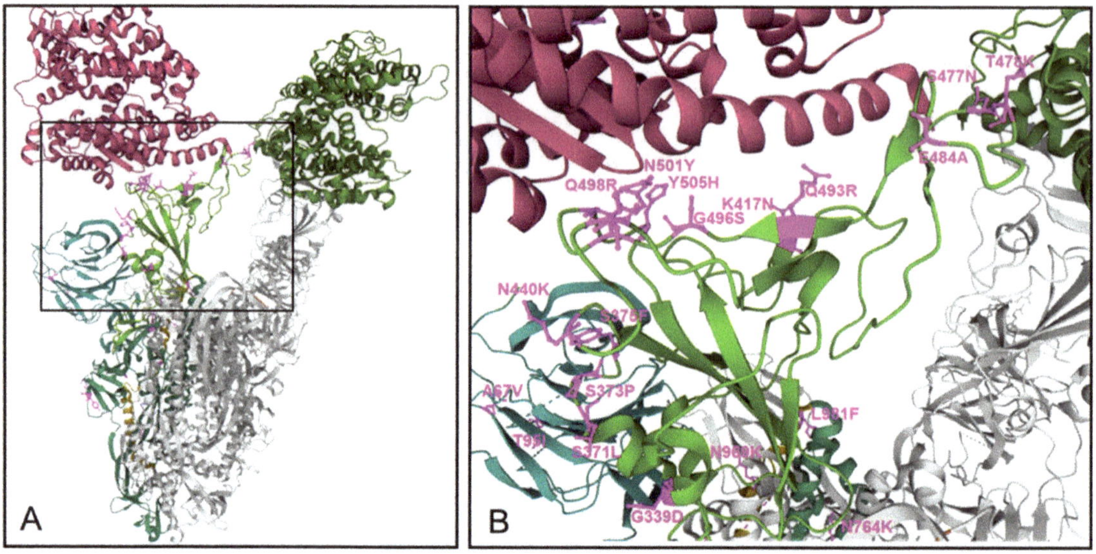

Figure 2. The Omicron spike protein trimer, binding to two ACE2 molecules (**A**), and the RBD—ACE2 binding interface (**B**) are shown, derived from the PDB **7T9K** [67]. Protomer 1—dark green, except RBD—light green, binding to one ACE2 molecule (dark red, **upper left corner**), NTD—dark turquoise (**left side**), HP1—yellow, Omicron mutations—pink, Protomers 2 and 3—grey, with one of them binding to another ACE2 (middle green, **upper right corner** (**A**)).

Computational mutagenesis and binding free energy analyses confirmed that Omicron spike protein binds ACE2 are stronger than wild-type SARS-CoV-2. Notably, three substitutions to positively charged amino acids in the RBD, i.e., T478K, Q493R, and Q498R, significantly contribute to the binding energies and doubled electrostatic potential of the RBD^{Omic}-ACE2 complex, suggesting that the Omicron binds ACE2 with greater affinity, enhancing its infectivity and transmissibility [6]. Other recently published Cryo-EM structural analyses of the Omicron variant spike protein in complex with human ACE2 revealed new salt bridges and hydrogen bonds formed by the mutated residues Q493R, G496S, and Q498R in the RBD with ACE2 [67,68]. Furthermore, the N501Y and S477N (distinct to Omicron) mutations enhance transmission primarily by enhancing binding [69] with the N501Y mutation, which is common to all VoCs except the Delta variant, enhancing binding to ACE2 receptor by a factor of 10 compared to the Wuhan strain spike protein [70].

Apart from the significant increase in binding affinity to ACE2 by amino acid substitutions in the Omicron spike RBD region, the increase in positive charge in various regions of the spike protein may increase also binding to some of the various proposed co-receptors for SARS-CoV-2, in particular those with high negative charge such as heparane sulfate and sialic acid [11].

3.2. Enhanced Escape from Therapeutic Antibodies and Immune Sera by Mutations in RBD and NTD

Mutations in the NTD and RBD region have been shown to affect binding of therapeutic monoclonal antibodies and immune sera from vaccinated individuals. Instead of E484K substitution that helped the neutralization escape of Beta, Gamma, and Mu variants, the Omicron variant harbors the E484A substitution. Together, T478K, Q493R,

Q498R, and E484A substitutions contribute to a significant drop in the electrostatic potential energies between RBD^{Omic}-mAbs, in six out of seven tested therapeutic antibodies: Etesevimab (AbCellera&Eli Lilly), Bamlanivimab (AbCellera&Eli Lilly), CTp59 (Celltrion), Imdevimab/REGN10987 (Regeneron), Casirivimab/REGN10933) (Regeneron), and a moderate drop for AZD9995 (Astrazeneca). Regarding the question of which mutations are particularly involved in weakening the RBD^{Omic}-mAb interactions, calculated changes in energy indicated that, e.g., for Bamlanivimab and CT-p59, highly stable salt bridges were lost due to E484A mutation or due to the combination of E484A, Q493K, and Y505H, respectively, in RBD^{Omic}. These data suggest that mutations in the Omicron spike were precisely selected to utilize the same mutations to enhance receptor binding and resist antibody binding [6]. Another study provided further evidence that amino acid substitutions in the RBD, E484A, and Q493R impact interactions with Casivirimab (REGN 10933) and S375F and N501Y with Imdevimab (REGN 10897). Indeed, whereas the antibodies individually were partially effective and inhibited highly potent against VoC Delta in combination, there was a complete loss of neutralizing activity against Omicron [3]. Furthermore, testing the neutralizing activity of sera from serum samples derived from persons vaccinated two times with either BNT162b2, mRNA 1273, ChAdOx-1, or Coronovac showed more than a 10-fold decrease in neutralization activity for Omicron compared to Delta, with almost no neutralizing activity against Omicron found for ChAdOx-1 in serum samples after two immunizations, and only low activity against Delta and no activity against Omicron found for individuals immunized with Coronavac [3].

The Omicron spike contains some of the mutations also reported in previous VoC, particularly D614G, found in all VoCs, which has been shown to enhance the receptor-binding by increasing its "up/open" conformation necessary for binding of the RBM to ACE2 and to enhance the overall density of the spike protein at the virus' surface [71,72]. The Omicron unique insertion mutation, i.e., Ins214EPE, maps to the NTD distant from the known antibody binding sites. However, the loop with the insertion maps to known human T-cell epitope on SARS-CoV-2 [73].

For multiple mutations in the RBD and NTD regions of Omicron, the correlation with enhanced escape behavior has been found or suggested [74]. The deletion Δ143–145 is also found in the spike protein of the Alpha variant. The resistance of Alpha to most monoclonal antibodies in the NTD is largely conferred by this deletion. Mutations in this region abolish the binding of monoclonal antibody 4A8 [75,76]. The N440K mutation was observed in a virus isolated in India associated with patient re-infection and described as an "immune escape variant" [77] and emerged under selection pressure against the human monoclonal antibody C135 [78]. For G446S, other mutations at this position in different variants have conferred escape from multiple antibodies [79,80]. The S477N mutation in the spike protein is resistant to neutralization by multiple monoclonal antibodies and resulted in a degree of resistance across the entire panel of antibodies [80]. The T478K mutation is also present in Delta [81]. Other alterations at this position that have provided resistance to neutralizing antibodies confers resistance to monoclonal antibody 2B04 and 1B07 [80]. Q493R mutations escape neutralization by the monoclonal antibody cocktail LY-CoV555 + LY-CoV016 [82] and confers a greater than two log reduction in IC_{50} for the REGN10989/10934 pair of monoclonal antibodies compared to the protein [83]. The Q498R mutation confers escape against the COV2-2499 antibody non-mutated spike [79].

3.3. Mutations in S2

The mutations found in the S2 subunit are of particular interest in relation to another essential step in the virus infection process, i.e., the fusion of the virus with the host cell membrane. S2 is a typical viral class I fusion protein, which includes a hydrophobic fusion peptide (FP), two α-helical hydrophobic (heptad) repeats (HR1 and HR2), a long, linking loop region, and a transmembrane domain. HR2 is located close to the transmembrane anchor, and HR1 is close to the FP. Binding of the S1 receptor binding domain to the ACE2 receptor on the target cells triggers a series of conformational changes in the S2 subunit, resulting in the proteolytic cleavage between S1 and S2, and its transition from a prefusion metastable form to a postfusion stable form, with insertion of the putative fusion peptide into the lipid layer of the target cell membrane due to the abundance of hydrophobic residues. The precise localization of the fusion peptide (FP) in the S2 fusion protein in SARS-CoV-2 is still under controversion. For SARS-CoV-2, a stretch around the amino acid sequence $_{788}$IYKTPPIKDFGGFNFSQIL$_{806}$ [16,19] has been suggested to be involved in membrane fusion. The fusion peptide is characterized by its higher hydrophobicity, due to a high density of nonpolar amino acid residues, such as glycine (G), alanine (A), phenylalanine (F), leucine (L), Isoleucine (I), proline (P), and tyrosine (Y). This hydrophobic core plays an essential role in the interaction and penetration into the host membrane lipid. A computer-aided drug design study using FDA-approved small molecules docking to the fusion peptide hydrophobic pocket of S2 suggested that the potential binding site at the fusion peptide region is centralized amid the Lys790, Thr791, Lys795, and Asp808 residues (with some additional interactions also near Gln872) [84]. This is followed by a further conformational change, leading to the association of the two heptad repeat (HR) regions HR1 and HR2 domains to form a six-helix bundle fusion core structure (6HB) motif where the HR1 helices form a central coiled–coil fusion core surrounded by three HR2 helices in an anti-parallel arrangement (see Figure 3C) [85]. This brings the viral envelope and target cell membrane into close proximity enabling fusion. The six-helix bundle is linked by a beta-hairpin loop, which finally acts as a hinge end-to-end in-groove attachment of HR1 and HR2 [84]. Notably, the formation of 6HB in class I fusion proteins is a common step in viral entry and is used by various virus types, including HIV-1, Parainfluenza, Newcastle disease, Respiratory syncytial virus, Herpes simplex virus, Ebola virus, as well as members of the coronaviridae family, including HCoV-229E, MERS-CoV, and SARS-CoV-2. For insertion into the cellular membrane, the FP must be accessible, which is achieved through cleavage at the S2′ site by the transmembrane serine protease 2 (TMPRSS2). Alternatively, the pH-dependent enzyme cathepsin L can take over the processing function after endocytic uptake [86]. Notably, endocytic uptake and activating cleavage by cathepsin L may also overcome the requirement for furin-mediated priming of the S protein [14,87,88].

Regarding the Omicron-unique mutations in the S2 subunit, most of them have been suggested primarily to stabilize the spike trimer. The D796Y mutation replaces a charged surface-exposed acidic residue with tyrosine, containing an aromatic side chain allowing for potential carbohydrate-pi interactions with the N-linked glycan chain originating from N709 of the neighboring monomer chain, this way having a stabilizing effect for the spike trimer. In addition, for the N856K mutation, the longer side-chain of the lysine residue has been suggested to form new interactions with T572 from an adjacent monomer. For N764K, two new interactions of an amine head-group of K764 with Q314 and N317 from an adjacent monomer, are expected to stabilize the Spike trimer [89].

We used the PDB 7TEI and PDB 7T9K Cryo-EM structures of the Omicron spike protein file [66,67] to visualize the amino acid substitutions in the S2 subunit and in the interface area between S1 and S2. The substitutions Q954H, N969K, and L981F are located within the HR1 region and D796Y is located in the area of a putative fusion protein region. Notably, together with the spatially and proximally located N764K and N856K, these mutations represent four changes to positively charged amino acids and one exchange of the negatively charged glutamic acid by a neutral tyrosine (D796Y), with an increase at five sites of the positive electrostatic charge in the S2 region (see Figure 3A,B), which may have an impact on interaction with various innate receptors, as discussed below.

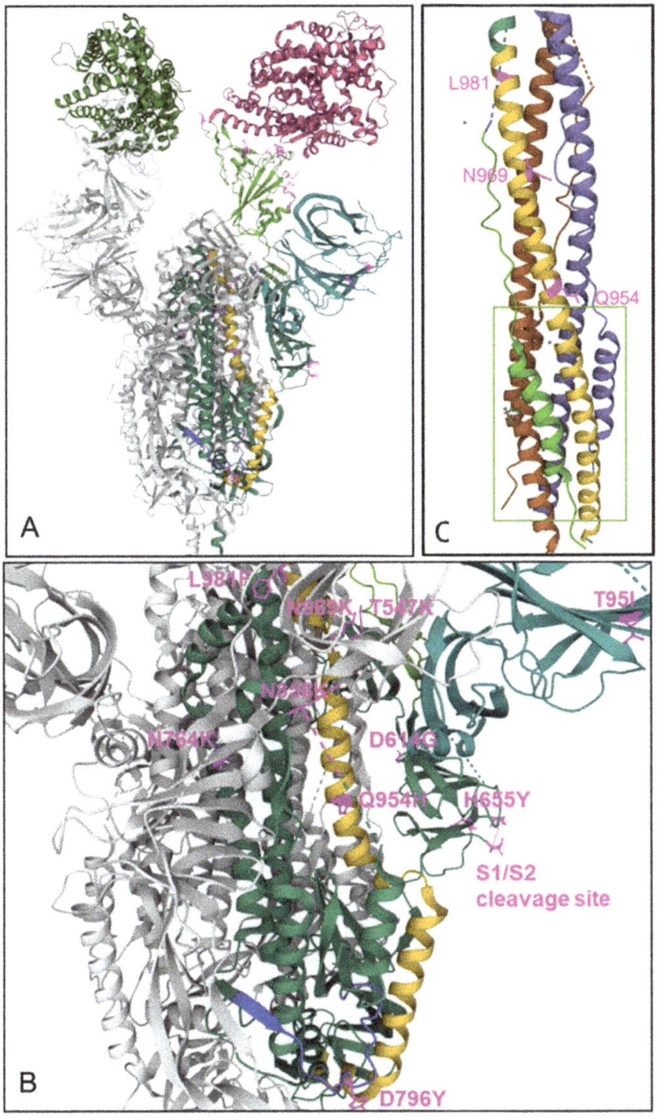

Figure 3. The Omicron spike protein trimer, binding to two ACE2 molecules (**A**), and S2 subunit (zoomed) (**B**) are shown, derived from the PDB **7T9K** [67]. Protomer 1—dark green, exept RBD—light green, binding to one ACE2 molecule (dark red, **upper right corner, A**), NTD—dark turquoise (**ride side**), HP1—yellow, FP—blue, Omicro mutations—pink, Protomers 2 & 3—grey, with one of them binding to another ACE2 (middle green, **upper left corner, A**), Two substitutions within the S1/S2 cleavage site (i.e., N679K, P681H) are not displayed in this model, (**C**) PDB 6LXT Structure of post-fusion core of wild-type 2019-nCoV S2 subunit [19]. The 6HB bundle highlighted by the green frame.

3.4. Mutations near to the S1/S2 Furin-Like Cleavage Site

The $_{861}$PRRAR$_{865}$ polybasic furin-like S1/S2 cleavage site plays a central role in the highly effective plasma membrane route of viral entry, being a necessary precedent cleavage step for the following S2′ cleavage by the cellular serine protease TMPRSS2 after binding of S1 to ACE2, provided that the cells express both ACE2 and TMRPSS2. S1/S2 cleavage at the polybasic furin-like cleavage site occurs primarily during virion release from the producer cells, but secreted furin may also enable S1/S2 cleavage of the spike protein outside the cells [90,91]. In contrast, the endosomal entry route used in cells not expressing ACE2 and/or TMPRSS2 does not require spike cleavage at the furin cleavage site but was described to be approximately 100–1000× fold less effective [13,90,92] due to restricting factors in the endosomes and enhanced recognition by the host innate immune receptors resulting in activation of antiviral cellular pathways. The highly efficient plasma membrane entry route with furin and TMPRSS2 cleavage correlates with enhanced cell fusion, leading to the formation of syncytia between multiple virus producer cells which is expected to significantly enhance viral production and pathogenicity of the virus [90–95]. Using a reverse genetic system, a SARS-CoV-2 mutant that lacked the furin cleavage site (ΔPRRA) in the S protein was generated. The deletion of PRRA reduced S protein cleavage but augmented viral replication in Vero E6 cells, which are deficient in TMPRRS2. Ectopic expression of TMPRSS2 in Vero E6 cells removed the fitness advantage for ΔPRRA SARS-CoV-2. By contrast, the ΔPRRA mutant was attenuated in a human respiratory cell line and had reduced viral pathogenesis in both hamsters and K18-hACE2 transgenic mice (which express human ACE2) [96].

The ACE2/TMPRSS2 pathway is the preferable pathway for SARS-CoV-2 to enter lung cells, such as alveolar AT1 and AT2 pneumocytes, whereas upper airway cells (expressing significantly lower amounts of ACE2 and/or TMPRSS2) seem to employ preferably or exclusively the endosomal entry route [3]. Structurally, the loop containing the S1/S2 cleavage site is largely flexible and extends outwards, exposing the cleavage site for furin in both the Delta and Omicron models. Omicron has three amino acid substitutions near the S1/S2 cleavage site: H655Y, N679K, and P681H. The substitution N679K is distinct to Omicron with no effects on the S1/S2 cleavage described so far. The H655Y substitution is also found in the Gamma variant, and substitutions at the P681 position are found in various previous VoC, i.e., P681R in the case of Delta and P681H in the Alpha variant, similar to Omicron. Substitution of the P681 (in the original Wuhan variant) by positively charged amino acids such as Arg in the case of Delta has been shown to significantly increase spike protein cleavage, enhanced syncytia formation leading to higher viral transmission, and higher pathogenicity compared to the D614G Wuhan-1 spike [94,97,98].

Furthermore, P681R was also shown to stimulate NF-κB and AP-1 signaling in human monocytic THP1 cells and to induce significantly higher levels of pro-inflammatory cytokines [98]. Different to P681R, as present in Delta, Omicron has the P681H substitution also found in the Alpha variant. Studies on the effect of the P681H substitution in the Alpha variant showed a moderate tendency for an increase in its cleavability by furin-like proteases, but that did not translate into increased virus entry or membrane fusion, which were roughly equal to the Wuhan wild-type [98]. On the other hand, improved viral fitness was suggested for both P681H and P681R SARS-CoV-2 Gamma variants [99]. Accordingly, the P681H in Omicron could be expected to have a similar effect as in the Alpha variant, with no major effect on transmissibility, and at least no negative impact.

Surprisingly, the Omicron variant was found to have significantly decreased S1/S2 cleavage, significantly lower infectivity in TMPRRS2-rich Calu-3 cells, and equal or higher infectivity to TMPRRS2-deficient H1299 cells, compared to wild-type or Delta variants, as well as almost completely absent cell fusion, compared to pronounced cell fusion and syncytia formation after the wild-type or Delta infection. Omicron spike pseudotyped virus (PV) entry into lower airway organoids and Calu-3 lung cells was impaired. In lung cells expressing TMPRSS2, the Omicron virus showed significantly lower replication in comparison to Delta. Cell–cell fusion mediated by spike glycoprotein is known to require S1/S2 cleavage and the presence of TMPRSS2. Fusogenicity of the Omicron BA.1 spike was severely impaired despite TMPRSS2 expression, leading to marked reduction in syncytium formation compared to Delta spike. These data indicate that suboptimal Omicron S1/S2 cleavage reduces efficient infection of lower airway cells expressing TMPRSS2, but not in TMPRSS2 negative cells, such as those found in the upper airways [3]. These results were rather unexpected from the molecular modelling perspective. Whereas the lower S1/S2 cleavage in Omicron compared to Delta (P681R) can be explained by the different substitutions for Omicron (similar as Alpha) P681H, the significantly lower S1/S2 cleavage compared to wild-type cannot be explained by this substitution. Moreover, if a difference was expected, then a slightly increased cleavage would have to be expected. On the other hand, these data correlate with recent preclinical and clinical data for Omicron [7,8].

4. Lower Pathogenicity of Omicron Compared to Previous VoCs

There are accumulating data that Omicron, despite a significantly higher transmissibility and infectivity, shows lower numbers of severe clinical courses compared to previous VoCs, in particular compared to the Delta variant [8]. Whereas a lower number of severe clinical outcomes in Africa could also be a result of the younger average age in the African population or other continent specific factors [100], the early reports about lower clinical severity of the SARS-CoV-2 Omicron variant in South Africa are meanwhile supported by concordant reports from other geographical regions of the Omicron pandemic (e.g., UK and USA), showing high transmissibility but significantly lower pathogenicity [101,102].

A retrospective cohort study of electronic health record (EHR) data of 577,938 first-time SARS-CoV-2 infected patients from a multicenter, nationwide database in the US during 1 September 2021–24 December 2021, including 14,054 who had their first infection during the 15 December 2021–24 December 2021 period, when the Omicron variant emerged ("**Emergent Omicron cohort**") and 563,884 who had their first infection during the 1 September 2021–15 December 2021 period when the Delta variant was predominant ("**Delta cohort**") was conducted. The 3-day risks of four outcomes (ED visit, hospitalization, ICU admission, and mechanical ventilation) were compared. The 3-day risks in the Emergent Omicron cohort outcomes were consistently less than half those in the Delta cohort for all parameters tested: ED visit: 4.55% vs. 15.22% (risk ratio or RR: 0.30, 95% CI: 0.28–0.33); hospitalization: 1.75% vs. 3.95% (RR: 0.44, 95% CI: 0.38–0.52]); ICU admission: 0.26% vs. 0.78% (RR: 0.33, 95% CI:0.23–0.48); mechanical ventilation: 0.07% vs. 0.43% (RR: 0.16, 95% CI: 0.08–0.32). In children under 5 years old, the overall risks of ED visits and hospitalization in the Emergent Omicron cohort were 3.89% and 0.96%, respectively, significantly lower than 21.01% and 2.65% in the matched Delta cohort (RR for ED visit: 0.19, 95% CI: 0.14–0.25; RR for hospitalization: 0.36, 95% CI: 0.19–0.68). Similar trends were observed for other pediatric age groups (5–11, 12–17 years), adults (18–64 years), and older adults ($\geq$65 years). In summary, the data indicate that first time SARS-CoV-2 infections occurring at a time when the Omicron variant was rapidly spreading were associated with significantly less severe outcomes than first-time infections when the Delta variant predominated [102].

This clinical picture of attenuated severity is also supported by animal data showing high, TMPRSS2-independent, replication in the upper airway organs, but lower pathogenicity in animal studies. The ability of multiple B.1.1.529 Omicron isolates to cause infection and disease in immunocompetent and human ACE2 (hACE2)-expressing mice and hamsters was studied. Despite modeling and binding data suggesting that the B.1.1.529 spike can bind more avidly to murine ACE2, the authors observed attenuation of infection in three different mouse models, i.e., 129, C57BL/6, and BALB/c mice, as compared with previous SARS-CoV-2 variants, with limited weight loss and lower viral burden in the upper and lower respiratory tracts [7]. Although K18-hACE2 transgenic mice sustained an infection in the lungs, these animals did not lose weight. In wild-type and hACE2 transgenic hamsters, lung infection, clinical disease, and pathology with Omicron also were milder compared to historical isolates or other VoCs. Overall, these studies using several different Omicron isolates demonstrated attenuated lung disease in rodents, which parallels preliminary human clinical data [7–9]. A lower virus replication in lungs, together with a faster replication of the upper respiratory system, such as nasopharyngeal and bronchi [9], could explain to a large extent Omicron's greater ability for transmission between people while apparently causing less severe COVID-19 disease. However, the molecular mechanisms responsible for this reciprocal change in tropism to upper vs. lower respiratory system in the Omicron variant are so far not defined. The answer to this question may be hidden in the involved signaling pathways and the characteristics and the distribution of the triggering receptors, as discussed below.

4.1. NF-κB Pathway Activation by SARS-CoV-2

There are increasing data for the central role of the NF-κB signaling pathways for the SARS-CoV-2 infection-induced proinflammatory cytokine/chemokine response, and severity and lethality of COVID-19. This exaggerated NF-κB-triggered proinflammatory response in acute COVID-19 is shared also with other acute respiratory viral infections caused by the highly pathogenic influenza A virus of H1N1 (e.g., Spanish flu) and H5N1 (avian flu origin), SARS-CoV, and MERS-CoV [40–46]. As early as one day post-infection with a SARS-CoV-2 infection, pluripotent stem cell-derived human lung alveolar type 2 cells have been shown to start a rapid epithelial-intrinsic inflammatory response with transcriptomic change in infected cells, characterized by a shift to an inflammatory phenotype with upregulation of NF-κB signaling and loss of the mature alveolar program [39]. Furthermore, characterization of bronchoalveolar lavage fluid immune cells from patients with COVID-19, were compared to healthy donors by using single-cell RNA sequencing and demonstrated proinflammatory monocyte-derived macrophages with an M1 profile with enhanced expression of NF-κB and STAT1/1, accompanied by high cytokine and chemokine expression [103].

4.2. SARS-CoV-2 Spike Protein Induces NF-κB

Several studies have studied which part(s) of SARS-CoV-2 are responsible for the massive NF-κB pathway activation. *Khan* et al., showed that the spike (S) protein potently induces inflammatory cytokines and chemokines, including IL-6, IL-1β, TNFα, CXCL1, CXCL2, and CCL2, but not IFNs in human and mouse macrophages. No inflammatory response was observed in response to membrane (M), envelope (E), or nucleocapsid (N) proteins. When stimulated with extracellular spike protein, A549 human lung epithelial cells also produced inflammatory cytokines and chemokines. The spike protein was shown to trigger inflammation via activation of the NF-κB pathway in a MyD88-dependent manner. Both S1 and S2 triggered NF-κB activation, with S2 showing higher potency on an equimolar basis [104].

In a second study upregulation of TLR4, IL1R, NF-κB signaling pathway molecules in COVID-19 patients were found, associated with the altered immune responses to viral components, host damage-associated molecular pattern (DAMP) signals, and cytokine signaling activation, resembling those seen with bacterial sepsis. When testing for different components of SARS-CoV-2, the nucleocapsid (NC) and the S2 subunit of spike proteins were found to activate TLR4 and NF-κB pathways with an expression of multiple proinflammatory cytokines and chemokines [105].

In a third study, the spike protein was demonstrated to promote an angiotensin II type 1 receptor (AT1)-mediated signaling cascade, and to activate NF-κB and AP-1/c-Fos via MAPK activation, and IL-6 release [106]. A fourth study demonstrated that the SARS-CoV-2 spike protein S1 subunit induces high levels of NF-κB activation, production of proinflammatory cytokines, and epithelial damage in human bronchial epithelial cells. NF-κB activation required S1 interaction with the human ACE2 receptor and early activation of endoplasmic reticulum (ER) stress and associated un-folded protein response and MAP kinase signaling pathways [107].

In another study, human peripheral blood mononuclear cells (PBMCs) showed significant release of TNFα, IL-6, IL-1β, and IL-8 following stimulation with spike S1 protein. Activation of the NF-κB pathway was demonstrated by phosphorylation of NF-κB p65, IκBα degradation, and increased DNA binding of NF-κB p65 after stimulation with spike

S1 protein. NF-κB activation and cytokine release were blocked by treatment with dexamethasone or the specific NF-κB inhibitor BAY11-7082 [108].

Furthermore, SARS-CoV-2 S protein was suggested to bind to LPS. Spike protein, when combined with low levels of LPS, boosted NF-κB activation in monocytic THP-1 cells and cytokine responses in human blood and PBMC, respectively. The study demonstrated that the S protein modulated the aggregation state of LPS, providing a potential molecular link between excessive inflammation during infection with SARS-CoV-2 and comorbidities involving increased levels of bacterial endotoxins [109].

In a mouse model, the S1 subunit of the spike protein was demonstrated to elicit strong pulmonary and systemic inflammatory responses in transgenic K18-hACE2 mice after intratracheal installation, accompanied by loss in body weight, increased white blood cell count, and protein concentration in bronchoalveolar lavage fluid (BALF), and upregulation of multiple inflammatory cytokines by activation of NF-κB and from the signal transducer and activator of transcription 3 (STAT3) [110].

Similar to SARS-CoV-2, the clinical picture of severe acute respiratory syndrome (SARS) is characterized by an overexuberant immune response with lung lymphomononuclear cell infiltration that may account for tissue damage more than the direct effect of viral replication. In addition, SARS-CoV purified recombinant S protein was shown to stimulate murine macrophages to produce proinflammatory cytokines (IL-6 and TNFα) and the chemokine IL-8, which were dependent on NF-κB activation [111]. Overall, these data demonstrate that the spike protein of both SARS-CoV-1 and SARS-CoV-2 S induces powerful NF-κB activation, showing strong similarity to data recorded for the SARS-CoV S protein.

4.3. NF-κB Is Essential for SARS-CoV-2 Replication

In addition to the multiple lines of evidence showing the critical role of the NF-κB signaling pathway in cytokine/chemokine release and hyper-immune activation, there is an additional set of data indicating that NF-κB is essential for viral replication of SARS-CoV-2 in the host cell. Epigenetic and single-cell transcriptomic analyses showed an early NF-κB transcriptional signature comprised of chemokines (e.g., CXCL8, CXCL10, CXCL11, and CCL20) and proinflammatory cytokines (e.g., IL-1A and IL-6) and upregulated NFKB1A, phospohorylation of IκBa, and NF-κB p65 in a relative absence of an ISG response. There was significantly enhanced enrichment of the NF-κB–related DNA-binding motifs and corresponding increase in genomic accessibility for REL, NKFB1, and RELA, but not for IRF3 and IRF7. Disruption of NF-κB signaling through the silencing of the NF-κB transcription factor p65 or p50 resulted in loss of virus replication that was rescued upon reconstitution. Furthermore, A549-ACE2 cells pre-treated with BAY11-7082 (an inhibitor of IκBα phosphorylation), MG115 (a proteasome inhibitor preventing proteolytic degradation of IκBα), prior to infection with SARS-CoV-2, showed significant inhibition of viral replication following BAY11-7082 treatment and an almost complete loss of viral protein and RNA expression in response to MG115. In addition, a significant reduction of secreted proinflammatory cytokines and chemokines in response to SARS-CoV-2 infection in A549-ACE2 cells after BAY11-7082 was found [112].

There is an analogy with IAV where NF-κB pathway was also found to support IAV infection by enhancing caspase-mediated nuclear export of viral ribonucleoproteins [113,114].

These data suggest that SARS-CoV-2 triggers both hyper-immune activation and viral replication via activation of the NF-κB signaling pathway.

4.4. NF-κB, Cytokines, and Hypoxia Enhance Furin Expression

As well as its central involvement in immune hyper-activation and SARS-CoV-2 replication, the NF-κB signaling pathway may also be involved in the modulation of the SARS-CoV-2 host cell type tropism by modulation of the furin-mediated cleavage of the spike protein, with respect to the availability of sufficient protease in the virus producer cells. Within the SARS-CoV-2 replication cycle, the cleavage at the furin site between S1 and

S2 most likely occurs during virion assembly, or just before release. This timing correlates with the virus' passage through the Golgi or lysosomes. Notably, furin has been found primarily in the trans-Golgi-network (TGN)—a late Golgi structure that is responsible for sorting secretory pathway proteins to their final destinations. From the TGN, furin follows trafficking through several TGN/endosomal compartments to the cell surface. The proteolytic activity of furin shows a broad pH optimum, with high enzymatic activity between (pH 5–8)—a pH range covering both TGN and lysosomes [115,116].

By cutting the bond between the S1 and S2 subunits, the furin cut triggers conformational changes in the virion spike protein so that at binding to the next host cell it is accessible to the second cut by TMPRSS2, which exposes the hydrophobic area that introduces into the host cell membrane. If spike protein is not clipped by furin, it bypasses the TMPRRS2 cleavage, and the virus can enter only via the slower and less-efficient endosomal pathway, which results in lower transmissibility to TMPRSS2-dependent cells, such as lung cells, in contrast to TMPRRS2 non-dependent cells, such as cells of the upper airway tissues [117]. Therefore, this process of furin cleavage at the S1/S2 site actually depends on two factors, i.e., the presence of a suitable cleavage site and the availability of the protease. Although furin is rather ubiquitously expressed across most tissues, it is usually expressed at very low levels [90,118], with the exception of few cell types in the brain, salivary gland, pancreas, kidney, and placenta [119].

Whereas furin is expressed usually at very low basic levels, it can be induced in response to hypoxia and cytokine stimulation. Furin is induced by IL-12 in T cells [120,121], with the IL-12 expression depending on NF-κB pathway signaling [122,123]. In contrast to the very low expression level of furin in most normal cells, elevated levels are found in many cancer cells, where furin seems to be closely related to tumor formation and migration [124]. p38 activation in cervical cancer cells was shown to induce NF-κB-dependent expressions of furin. Furin expression and cell motility was impeded by blockades to MKK3/6, p38α/β, or NF-κB signaling [125]. Another study correlated the osteopontin-p38 MAPKinase–NF-κB-furin expression with diabetes mellitus progression and increased risk of diabetes-linked premature mortality and a more severe clinical picture in diabetic patients after SARS-CoV-2 infection [126].

In this context, the role of Hypoxia-induced Factor-1 alpha (HIF-1α) and its connection to NF-κB pathway and furin expression may be important. Hypoxia was shown to stimulate furin expression, with direct HIF-1α action on the furin promoter as a canonical hypoxia-responsive element site with enhancer capability [127]. Regarding the initial signal triggering, this NF-κB-HIF-1α-Furin expression axis, there are various studies demonstrating the cross-talk between Toll-like receptor/NF-κB pathway activation and HIF-1α. The HIF-1α promotor was shown to contain an active NF-κB binding site, upstream of the transcription start site [128]. Extensive cross-talk between hypoxia and inflammation signaling have been described, showing that activation of TLR3 and TLR4 stimulated the expression of HIF-1α through NF-κB [129]. Peroxiredoxin (Prx1), a TLR4 agonist, was shown to stimulate increased NF-κB interaction with the HIF-1α promoter, leading to enhanced promoter activity and increase in HIF-1α mRNA levels, and augmented HIF-1 activity. In turn, Prx1-induced HIF-1α also promoted NF-κB activity, suggesting the presence of a positive feedback loop [130].

Furthermore, agonists of TLR4 (e.g., LPS) and TLR2 (e.g., lipoteichoic acid) have been demonstrated to induce in a NF-κB dependent way the expression of HIF-1α in human monocyte-derived dendritic cells under normoxic conditions [131]. Furthermore, activation of TLR4 by LPS was demonstrated to raise the levels of HIF-1α in macrophages. HIF-1α was shown to be a critical determinant of sepsis promoting the production of inflammatory cytokines, including TNF-α, IL-1, IL-4, IL-6, and IL-12, which reach harmful levels during early sepsis [132]. This is in line with data showing the critical role of TLR4 and NF-κB activation in HIF-1α activation during trauma/hemorrhagic shock-induced acute lung injury after lymph infusion in wild-type mice, in comparison to mice that harbor a TLR4 mutation and/or NF-κB inhibitors [133]. TLR4 was demonstrated to also promote HIF-

1α activity by triggering reactive oxygen species in cervical cancer cells by mechanisms involving activation of lipid rafts/NADPH oxidase signaling [134].

Considering NF-κB or HIF-1α-induced furin expression, these studies suggest that NF-κB activation following TLR signaling induced during SARS-CoV-2 infection may be essential for at least three mechanisms: (i) increasing viral replication, (ii) immune hyper activation, and (iii) modulation of cell tropism via alteration of proteases expression such as furin, necessary for the TMPRRS-dependent cellular entry pathway of SARS-CoV-2.

4.5. Acute Viral Infections Such as Highly Pathogenic IAV and SARS-CoV-2 Depend on Furin and Stimulate Furin Expression

The correlation between furin cleavage and pathogenicity is not only seen for the *coronaviridae* family but is found much broader among different virus types, including dsDNA viruses (*Herpesviridae* e.g., Human cytomegalovirus), *Papillomaviridae*), (+) ssRNA viruses (including *Coronaviridae, Flaviviridae* (e.g., Yelow fever virus, Dengue virus type 2), (−)ssRNA viruses (including *Filoviridae* (e.g., Ebola virus, Marburg virus), *Orthomyxoviridae* (e.g., avian IAV H5, H7, H9), *Paramyxoviridae* (e.g., Mumps virus, Measles virus, Respiratory syncytial virus), ss-RNA-RT viruses *Retroviridae* (e.g., HIV, Feline foamy virus), and dsDNA-RT viruses *Hepadnoviridae* (e.g., Hepatitis B virus). Cleavage of the furin-like site is dependent on the actual amino acid sequence of the polybasic furin cleavage motif, with the prototypical motif for furin (i.e., PCSK3, PACE = paired basic amino acid cleaving enzyme)) being R-X-K/R-R↓, and for other proprotein convertases PCSK1-7 being K/R-X_n-K/R↓ (n = 0-,2-,4-, or 6-amino acid spacer). Accordingly, different PCSKs can be involved in cleavage, and are dependent on the actual amino acid sequence around the cleavage site and availability of PCSKs. A well-known example is the haemagglutinin (HA) of different Influenza A viruses (IAV). Generally, HA of mammalian and low pathogenic avian IAV cannot be cleaved by furin (or other PCSKs), as they usually only harbour a mono- or dibasic-cleavage site. Instead, they depend on trypsin-like proteases such as human airway trypsin-like protease (HAT). Expression of such trypsin-like proteases is largely restricted to the respiratory and gastrointestinal tract. In contrast, HA of highly pathogenic avian H5 and H7 influenza A viruses can be cleaved by furin or PCSK5, which are present in many cell types. This is because they acquired a polybasic cleavage site upon insertion of additional lysine and/or arginine residues [135]. Notably, a subset of H9N2 low pathogenic avian influenza A virus strains also harbours R-S-K-R↓ or R-S-R-R↓ sites that are not only cleaved by trypsin-like proteases, but also by PCSKs. However, their cleavage is only efficient in the presence of very high amounts of furin or upon mutation of a glycosylation site in HA. Thus, the ability to exploit furin for efficient HA cleavage is not only determined by the presence of a furin consensus target site, but also by the amounts of furin [136]. Interestingly, beside the dependency of highly pathogenic avian IAV replication on furin, infection with IAV was shown to elevate furin expression in the lungs. Wild-type C57BL/6 mice were infected with the mouse-adapted IAV strain A/PuertoRico/8/1934 H1N1 (PR8). Using quantitative RT-PCR, a significant increase in the expression of furin in whole lung tissue of PR8-infected C57BL/6 mice on Day 6 after infection was found, indicating that the IAV-induced inflammation induces furin expression [137].

Notably, SARS-CoV-2 Gene Expression Omnibus records of up-regulated genes suggest that coronavirus infection triggers an increased expression of both ACE2 and FURIN genes 4 days after infection. These findings were corroborated by the increased furin expression documented in the PBMC of patients with severe acute respiratory syndrome. Furthermore, gene expression data identified numerous significantly enriched records of common human disorders manifesting up-regulation of either ACE or FURIN genes correlating with the risk for clinically severe and lethal coronavirus infection [138].

The expression, methylation, mutation rate, and functional enrichment of furin together with the survival rate and COVID-19 outcomes were analyzed in normal and cancer tissues. The furin expression in tumors was significantly increased in several cancer types, such as esophageal carcinoma (ESCA) and testicular germ cell tumors (TGCT). Furin mutations mostly increased expression of ACE2 and TMPRSS2 in various cancers, indicating furin mutations might facilitate COVID-19 cell entry in cancer patients. In addition, high expression of furin was significantly inversely correlated with long overall survival in various cancer types and correlated with increased susceptibility to SARS-CoV-2 and higher severity of COVID-19 symptoms in cancer patients [138].

The relationship between circulating furin levels, disease severity, and inflammation was studied in 52 SARS-CoV-2 patients vs. 36 healthy control participants. The mean furin and IL-6 levels were significantly higher in the peripheral blood of SARS-CoV-2 compared to the controls ($p < 0.001$). There was a close positive relationship between serum furin and IL-6, and furin and disease severity ($r = 0.793$, $p < 0001$ and $r = 0{,}533$, $p < 0.001$, respectively) in patients with SARS-CoV-2. These results suggest that furin may contribute to the exacerbation of SARS-CoV-2 infection and increased inflammation and could be used as a predictor of disease severity in COVID-19 patients [139].

Furthermore, higher furin expression was also found in diseases known to predispose a person to severe COVID-19 symptoms, such as severe asthma and in people such as COPD smokers and COPD ex-smokers. ACE2 levels were significantly increased in sputum of severe asthma compared to mild-moderate asthma. Sputum furin levels were significantly related the presence of severe asthma and were strongly associated with neutrophilic inflammation and inflammasome activation, indicating the potential for a greater morbidity and mortality outcome from SARS-CoV-2 infection in neutrophilic severe asthma [140]. Furthermore, ACE2, furin, and TMPRSS2 expression was significantly increased in small airway epithelium (SAE) and type 2 pneumocytes in smokers with COPD (COPD-CS), and ex-smokers with COPD (COPD-ES), compared to the control group that never smoked. (NC) ($p < 0.001$). Importantly, significant changes were observed for tissue co-expression of furin and TMPRSS2 with ACE2 in SAE, type 2 pneumocytes, and alveolar macrophages (AMs). These markers also negatively correlated with lung function parameters. The increased expression of ACE2, TMPRSS2, and furin in COPD patients are detrimental to lung function and indicate that these patients are more susceptible to severe COVID-19 infection. Increased type 2 pneumocytes suggest that these patients are also vulnerable to developing post-COVID-19 interstitial pulmonary fibrosis or fibrosis in general [141].

5. SARS-CoV-2 Activates Innate PRRs

The initial signals for these pathways triggered by SARS-CoV-2 can be expected in the danger signaling of the innate immune system. The host innate immune system can recognize pathogen-associated molecular patterns (PAMPs) via pattern recognition receptors (PRRs) during infection to induce inflammatory responses to eliminate pathogens. The PRR families include Toll-like receptors (TLRs), nucleotide-binding oligomerization domain (NOD)-like receptors, retinoic acid-inducible gene-I (RIG-I)-like receptors, C-type lectin receptors, and the absent in melanoma 2 (AIM2)-like receptors. Typical PAMPs are cell wall components of pathogens, such as bacterial lipopolysaccharide (LPS) and lipoproteins, glycans, and conserved proteins such as flagellin or pathogenic nucleic acids, including viral RNA and DNA. PAMPs comprise moieties which are generally conserved among a broader range of pathogenic species but are distinct from host components. Several PRRs have been reported to be involved in sensing β-coronavirus infection, including melanoma differentiation-associated protein 5 (MDA5) [142], TLR7 [143,144], and NLR family pyrin domain containing 3 NLRP3 [145].

In this context, the MyD88 adaptor protein is known to couple multiple upstream sensors (e.g., TLR) with downstream inflammatory signaling pathways such as NF-κB or IFN-induced response factors following β-coronavirus infection [146]. To determine whether MyD88 or another TLR adapter TRIF (TIR-domain-containing adapter-inducing interferon-β) play a role in SARS-CoV-2-induced inflammatory responses and pathogenesis, a publicly available dataset [147] was analyzed for MyD88 and TRIF expression in patients with differing severities of COVID-19 and showed a positive correlation between MyD88 expression and severity of COVID-19, suggesting that MyD88 is associated with COVID-19 pathogenesis in humans. By contrast, TRIF was significantly elevated only in patients with critical COVID-19. MyD88 is a key adapter shared by all TLRs, with the exception of TLR3, which signals exclusively through TRIF, with all other TLRs utilizing MyD88 to trigger inflammatory cytokine production [148]. Parallel to the expression of MyD88, the expression of TLR1, TLR2, TLR4, TLR5, TLR8 and TLR9 was significantly elevated in patients with severe and critical COVID-19. In contrast, expression of TLR3 did not show any correlation with the disease development of COVID-19, and the expression of TLR7 was increased only in patients with moderate COVID-19. Together, these data suggest an association of MyD88 and a panel of TLRs (i.e., TLR1, TLR2, TLR4, TLR5, TLR8, and TLR9) with disease progression in patients with COVID-19 [149].

In this context, it is interesting to take into consideration the expression profile and expression levels of the different TLR on the relevant target cells. From the Human Protein Atlas [119], the expression profiles for relevant receptors on different target cells normalized to nTPM (i.e., Transcripts per million protein coding genes) provide a good overview of the general expression pattern of different molecules on different host cells. In Figure 4, the expression of different TLRs together with other relevant molecules involved in SARS-CoV-2 pathogenesis are depicted. TMPRSS2 is expressed at significant levels by alveolar cells type 1 and 2, but only at minute quantities on macrophages and endothelial cells. Furin and TLR3 are expressed more broadly by many cell types present in the lungs, including alveolar cells, macrophages, and endothelial cells, at low to moderate levels. In contrast, TLR1, 2, 4, 6, and 7 and Cathepsin L (which is essential in the endosomal uptake of SARS-CoV-2) are expressed predominantly on macrophages. Regarding expressions other than macrophages, TLR4 and MD-2 (i.e., the second component of the TLR4-MD-2 complex) are also expressed on endothelial cells, and TLR2 is also expressed on alveolar type 2 cells. Alveolar type 1 cells generally show very low TLR expression. The highest expression levels are found for Cathepsin L, followed by TLR4/MD-2 and TLR2 on macrophages and endothelial cells, whereas ~10fold lower expression levels are seen for TLR1, 6, and 7, which may correlate with the different pathophysiological relevance in COVID-19, as discussed later.

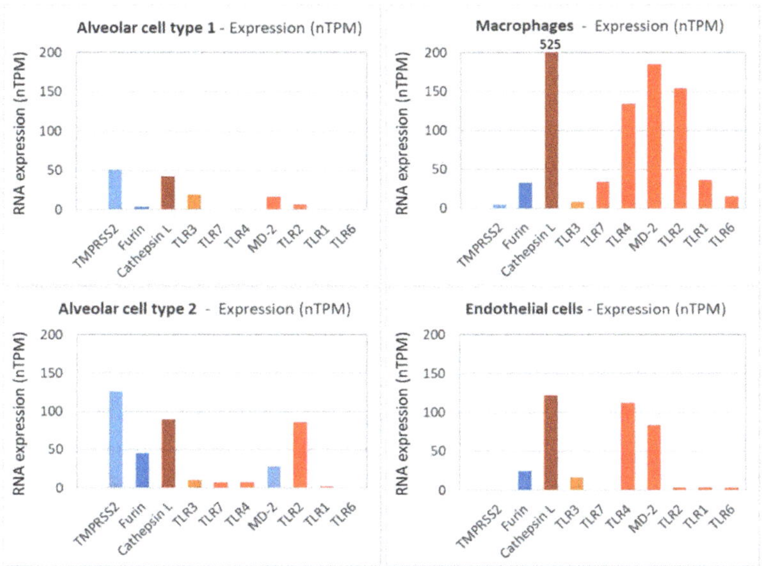

Figure 4. The expression of TMPSS2, cathepsin L, furin, TLR1, 2, 3, 4, 6, 7, and MD-2 normalized to nTPM (i.e., Transcripts per million protein coding genes) for representative cell types from the lungs are shown. Data are derived from the Human Protein Atlas [119].

5.1. SARS-CoV-2 Envelop E and Spike Protein Activate TLR2 and NF-κB

Zheng et al., have demonstrated for TLR2 and MyD88 expression their correlation with COVID-19 disease severity. TLR2 was shown to recognize the SARS-CoV-2 envelope E protein as its ligand and resulted in the TLR2-dependent cytokine (TNFα, GM-CSF, G-CSF, IL-6) and chemokine (CXCL10, MCP-1) release and lung damage in K18-hACE2 transgenic mice after infection with SARS-CoV-2. Notably, blocking TLR2 signaling in vivo provided protection against the pathogenesis of SARS-CoV-2 infection [149]. Regarding molecular characterization, the E protein was found to interact physically with the TLR2 receptor in a specific and dose-dependent manner. This interaction was able to engage the TLR2 signalling pathway as demonstrated by its capacity to activate the NF-κB transcription factor and to stimulate the production of the CXCL8 inflammatory chemokine in a TLR2-dependent manner. Inhibition of NF-κB led to significant inhibition of CXCL8 production, whereas the blockade of P38 and ERK1/2 MAP kinases resulted only in a partial CXCL8 inhibition [150].

On the other hand, Khan et al., investigated the direct inflammatory functions of major structural proteins of SARS-CoV-2 and showed that the spike protein potently induces inflammatory cytokines and chemokines, including IL-6, IL-1β, TNFα, CXCL1, CXCL2, and CCL2, but not IFNs in human and mouse macrophages. When stimulated with extracellular spike protein, A549 human lung epithelial cells also produced inflammatory cytokines and chemokines. Interestingly, epithelial cells expressing spike protein intracellularly were non-inflammatory but elicited an inflammatory response in co-cultured macrophages. Biochemical studies revealed that the spike protein triggers inflammation via activation of the NF-κB pathway in a MyD88-dependent manner. Furthermore, activation of the NF-κB pathway was abrogated in TLR2-deficient macrophages. Consistently, administration of the spike protein induced IL-6, TNFα, and IL-1β in wild-type but not in TLR2-deficient mice. In this study, both S1 and S2 subunits were demonstrated to show high NF-κB activation, with S2 showing higher potency on an equimolar basis [104].

5.2. SARS-CoV-2 Spike Protein Activates TLR4 and NF-κB

The involvement of another TLR, i.e., TLR4 in the pathogenies of COVID-19 has been shown [151]. TLR4 recognizes multiple pathogen-associated molecular patterns (PAMPs) from bacteria, viruses, and other pathogens. In addition, it recognizes certain damage-associated molecular patterns (DAMPs), such as high mobility group box 1 (HMGB1) and heat shock proteins (HSPs) released from dying or lytic cells during host tissue injury or viral infection [152,153]. TLR4 is mainly expressed on immune cells such as macrophages and dendritic cells where it plays a role in the regulation of acute inflammation, but also on some tissue-resident cell populations, for cell defense in case of infection and/or to regulate their fibrotic phenotype in cases of tissue damage [153,154]. The archetypal PAMP agonist for TLR4 is the gram-negative bacterial lipopolysaccharide [155]. Activation of TLR4 by pathogenic components leads to the production of proinflammatory cytokines via the canonical pathway and/or the production of type I interferons and anti-inflammatory cytokines via the alternative pathway. Unlike other TLRs, TLR4 is present at both the cell surface (main site), where it recognizes viral proteins before they enter the cell, and also in endosomes [156].

TLR4 is important in initiating inflammatory responses, and its overstimulation can be detrimental, leading to hyper-inflammation. Dysregulation of TLR4 signalling has been shown to play a role in the initiation and/or progression of various diseases, such as ischaemia-reperfusion injury, atherosclerosis, hypertension, cancer, and neuropsychiatric and neurodegenerative disorders [157–160]. Moreover, TLR4 is also important in the induction of the host immune response against infectious diseases such as bacterial, fungal and viral infections, and malaria [161].

Recently, there have been several studies pointing to the role of TLR4 in the pathogenesis of COVID-19 [151,162–165]. Interestingly, in silico studies have indicated that TLR4 has the strongest protein–protein interaction with the spike glycoprotein of SARS-CoV-2 compared to other TLRs [162]. A recently published study demonstrated that the induction of IL1β by SARS-CoV-2 was completely blocked by the TLR4-specific inhibitor Resatorvid. A surface plasmon resonance (SPR) assay showed that SARS-CoV-2 spike trimer directly bound to TLR4 with an affinity of ~300 nM, comparable to many virus-receptor interactions. THP-1 cells were treated with either the spike protein trimer, the N-terminal domain (NTD), or the receptor-binding domain (RBD) of spike protein, respectively. Only the trimeric protein induced IL1β and IL6, which could largely be blocked by the TLR4 inhibitor Resatorvid. Moreover, spike protein was also able to induce IL1β production in the murine macrophage cell line in a TLR4- and MyD88-dependent manner. Consistently, spike protein induced production of IL1β in the primary bone marrow-derived macrophages and peritoneal macrophages from wild-type, but not from TLR4-deficient mice. The NF-κB inhibitor (JSH-23) was able to suppress IL1β induced by spike protein. Collectively, the SARS-CoV-2 spike protein is capable of interacting with and activating TLR4. Furthermore, macrophages from ACE2-deficient or human ACE2-transgenic mice were treated with spike protein. Interestingly, deficiency of ACE2 or overexpression of human ACE2 did not affect the induction of IL1β. Treatment with an ACE2 inhibitor (MLN-4760) or soluble ACE2 was not able to inhibit the induction of IL1β by LPS or spike protein. Moreover, TMPRSS2-specific inhibitor (Bromhexine hydrochloride) did not alter the induction of IL1β by spike protein. Thus, activation of TLR4 by spike protein was not regulated by ACE2 and TMPRSS2 or virus entry. Notably, the induction of IL1β by trimeric spike proteins from SARS-CoV-2 or SARS-CoV was comparable to LPS treatment [163].

In response to exposure to the SARS-CoV-2 spike protein S1, subunit murine peritoneal exudate macrophages produced pro-inflammatory mediators, including TNF-α, IL-6, IL-1β, and nitric oxide. Exposure to S1 also activated NF-κB and c-Jun N-terminal kinase (JNK) signaling pathways. Pro-inflammatory cytokine induction by S1 was suppressed by selective inhibitors of NF-κB (BAY 11-7082) and JNK pathways (SP600125). Treatment of murine peritoneal exudate macrophages and human THP-1 cell-derived macrophages with a TLR4 antagonist attenuated pro-inflammatory cytokine induction and the activation of intracellular signaling by S1 and lipopolysaccharide. Similar results were obtained in experiments using TLR4 siRNA-transfected murine RAW264.7 macrophages. These results suggest that the SARS-CoV-2 spike protein S1 subunit activates TLR4 signaling to induce pro-inflammatory responses in murine and human macrophages [164].

Furthermore, the SARS-CoV-2 spike S1 domain was shown to act as a TLR4 agonist in rat and human cells and to induce a pro-inflammatory M1 macrophage phenotype in human THP-1 monocyte-derived macrophages. Adult rat cardiac tissue resident macrophage-derived fibrocytes (rcTMFs) were treated with either bacterial LPS or recombinant SARS-CoV-2 spike S1 glycoprotein. THP-1 monocytes were differentiated into M1 or M2 macrophages with LPS/IFNγ, S1/IFNγ, or IL-4. TLR4 activation by spike S1 or LPS resulted in the upregulation of ACE2 in rcTMFs. Likewise, spike S1 caused TLR4-mediated induction of the inflammatory and wound-healing marker COX-2 and concomitant down-regulation of the fibrosis markers CTGF and Col3a1, similar to LPS. The specific TLR4 inhibitor CLI-095 (Resatorvid®), blocked the effects of spike S1 and LPS, confirming the spike S1 subunit as a TLR4 agonist. Confocal immunofluorescence microscopy confirmed 1:1 stoichiometric spike S1 co-localization with TLR4 in rat and human cells. Furthermore, proximity ligation assays confirmed spike S1 and TLR4 binding in human and rat cells. Spike S1/IFN-γ treatment of THP-1-derived macrophages induced pro-inflammatory M_1 polarization, as shown by an increase in IL-1β and IL-6 mRNA [165].

A model has been proposed in which the SARS-CoV-2 spike glycoprotein binds TLR4 and activates TLR4 signalling, resulting in increased cell surface expression of ACE2 facilitating virus entry. Furthermore, SARS-CoV-2-induced myocarditis and multiple-organ injury may be due to TLR4 activation, aberrant TLR4 signalling, and hyperinflammation in COVID-19 patients. Therefore, TLR4 may be assumed to contribute significantly to the pathogenesis of SARS-CoV-2. TLR4 appears to be a promising therapeutic target in

COVID-19, supported by the fact that TLR4 antagonists have been previously used in sepsis and in other antiviral contexts [165].

5.3. TLR Activation during Different Highly Pathogenic Viral Infections

There is a growing list of viruses that induce an inflammatory response during acute infection through TLR4 activation. Known TLR4-activating viral proteins include the RSV fusion protein (F), the EBOV glycoprotein, the vesicular stomatitis virus glycoprotein (VSV G), and the dengue virus (DENV) nonstructural protein 1 (NS1). Notably, all infections by these viruses are also characterized by excessive inflammatory responses, which are characterized by elevated levels of a broad array of pro-inflammatory cytokines and chemokines and are associated with serious morbidity and mortality. Examples include acute lung injury caused by infections with respiratory syncytial virus (RSV), highly pathogenic IAV, or SARS-CoV. Excessive inflammatory responses induced by viral infections are not restricted to the lung but can be systemic, as found for Ebola virus (EBOV) disease and severe dengue fever [166–171] The Ebola virus glycoprotein was demonstrated to activate the innate immune response in vivo via TLR4 activation, accompanied by multiple cytokine and chemokine expression, which could be inhibited by TLR4 antagonists [172] and was accompanied by pronounced NF-κB activation [173]. There are several commonalities between these viral TLR4 activators. These proteins are all membrane-associated. VSV G, RSV F, and EBOV glycoprotein as well as the spike proteins of SARS-CoV and SARS-CoV-2 are classical viral glycoproteins that are exposed on the surface of viral particles and mediate fusion with host cell membranes through a hydrophobic fusion peptide. The fusion domain is only exposed after conformational changes that occur at the plasma membrane (RSV F, SARS-CoV, SARS-CoV-2) or in the endosome (VSV G, EBOV glycoprotein) [174]. DENV NS1 exists in multiple forms, including a secreted, membrane-bound form [175,176]. The hydrophobic fusion peptide in the RSV fusion protein has been suggested to bind into the deep hydrophobic pocket of MD-2, similar to LPS, to mediate TLR4 activation [171]. TLR4 is stimulated by membrane-bound EBOV glycoprotein and a secreted, cleaved form (shed glycoprotein), both of which retain the hydrophobic fusion domain, but not by a different secreted version of EBOV glycoprotein, i.e., soluble glycoprotein, which lacks the fusion peptide [177,178]. In addition, although DENV NS1 lacks a fusion peptide, it contains exposed hydrophobic domains that mediate membrane interaction and could play a role in TLR4 activation [176].

TLR4 antagonists which suppress LPS-induced TLR4 signaling through competitive interaction with MD-2, such as LPS from the bacterium Rhodobacter sphaeroides (LPS-RS) and Eritoran, also suppress RSV F-, EBOV glycoprotein-, and DENV NS1-mediated TLR4 activation [171,172,175,179–181], suggesting a similar mechanism of action. It remains to be determined how each of these glycoproteins interact with the TLR4 receptor complex and in what way the hydrophobic regions are made accessible for interaction with MD-2 and TLR4 leading to dimerization of the TLR4-MD-2 complex (see below). VSV G, RSV F, EBOV glycoprotein, and DENV NS1 are all glycosylated. So far, glycosylation of EBOV glycoprotein seems to be required for TLR4 activation, but it is not known whether this is also the case for the other viral glycoproteins [182].

5.4. Activation of TLR4 by LPS

TLR4, which is mainly expressed on cells of the innate immune system, including monocytes, macrophages, and dendritic cells, has long been recognized as the PRR that senses lipopolysaccharide (LPS), a component of the outer membrane of gram-negative bacteria, which can be regarded as the archetypal PAMP agonist for TLR4 [154]. Accordingly, activation of TLR4 by LPS has been studied in great detail. During the initial step, the LPS binding protein (LBP) extracts LPS from bacterial membranes and transfers it to the TLR4 co-receptor cluster of differentiation 14 (CD14). CD14 breaks down LPS aggregates and transfers monomeric LPS into a hydrophobic pocket on myeloid differentiation factor 2 (MD-2, Lymphocyte antigen 96, Ly96), which is part of the MD-2/TLR4 complex. The high-affinity binding of LPS leads to dimerization and activation of the TLR4-MD-2 complex. Dimerization of the TLR4-MD-2 complex results in the recruitment of the intracellular adaptor protein MyD88. The MyD88 aggregation signal is transmitted to IL-1 receptor kinase (IRAK) through an interaction between the death domain of MyD88 and IRAK. Phosphorylation of the signaling kinases eventually activates the transcription factors, NF-κB and activator protein 1 (AP-1) via a signaling cascade, ultimately resulting in the expression and secretion of pro-inflammatory mediators. Apart from the MyD88-dependent pathway, TLR4 dimerization can also activate the TRIF (TIR-domain containing adaptor inducing interferon-β) pathway, which activates interferon response factors to produce and secrete type-I interferons [182].

The archetypal TLR4 agonist LPS is a macromolecular glycolipid composed of the hydrophobic Lipid A attached to a long and branched carbohydrate chain. The Lipid A portion, which is responsible for most of the immunologic activity of LPS, is composed of a phosphorylated diglucosamine backbone with four to seven acyl chains attached to it. The carbohydrate region of LPS comprises the core and the O-specific chain composed of multiple carbohydrate repeating units. Removal of the entire carbohydrate chain by acid hydrolysis has only a minimal effect on the inflammatory activity of LPS, demonstrating only a minor role in recognition by host immune receptors [183].

The extracellular domain of TLR4, TLR1, TLR2, and TLR6 belongs to the Leucine-Rich Repeat proteins (LRR proteins) and is responsible for ligand binding and receptor dimerization. The structure of TLR4 is defined by the conformation of the LRR motifs. Its N-terminal and central domains provide charge complementarity for binding of its surface to its co-receptor MD-2, forming a stable heterodimer. MD-2 is smaller than TLR4 and is the main LPS binding module of the TLR4/MD-2 receptor complex. MD-2 has a β-cup fold structure, composed of two antiparallel β sheets. The two sheets are separated from each other so that the hydrophobic interior is accessible for interaction with ligands. This large internal pocket is ideally shaped for binding flat hydrophobic ligands such as LPS. MD-2 binds to TLR4 primarily via hydrogen bonds and charge interactions, and a few hydrophobic residues in the binding interface.

The crystal structure of the TLR4/MD-2 complex bound to E. coli LPS has been determined and is available at RCSB PDB 3VQ2 [184]. LPS binding induces the formation of the 'M' shaped 2:2:2 TLR4/MD-2 and LPS complex (Figure 5). The acyl chains of Lipid A are inserted into the MD-2 pocket and the two phosphate groups of Lipid A form charge and hydrogen bond interactions with charged and polar amino acid residues of the two TLR4 molecules. LPS binding causes dimerization of the TLR4/MD-2 complexes because Lipid A creates an additional binding interface between TLR4(II) and MD-2(I) coming from the two preexisting TLR4/MD-2 (see Figure 5) complexes, respectively. The dimerization interface of MD-2 interacts with a convex surface provided by a small hydrophobic patch in the C-terminal domain of the second TLR4 (Figure 5, TLR4 (II), brown). This dimerization is supported by the interaction between Lipid A inserted in the hydrophobic pocket of MD-2(I) with TLR4(II) (Figure 5). TLR4 agonists are presumable all interacting at this dimerization interface actually enabling dimerization.

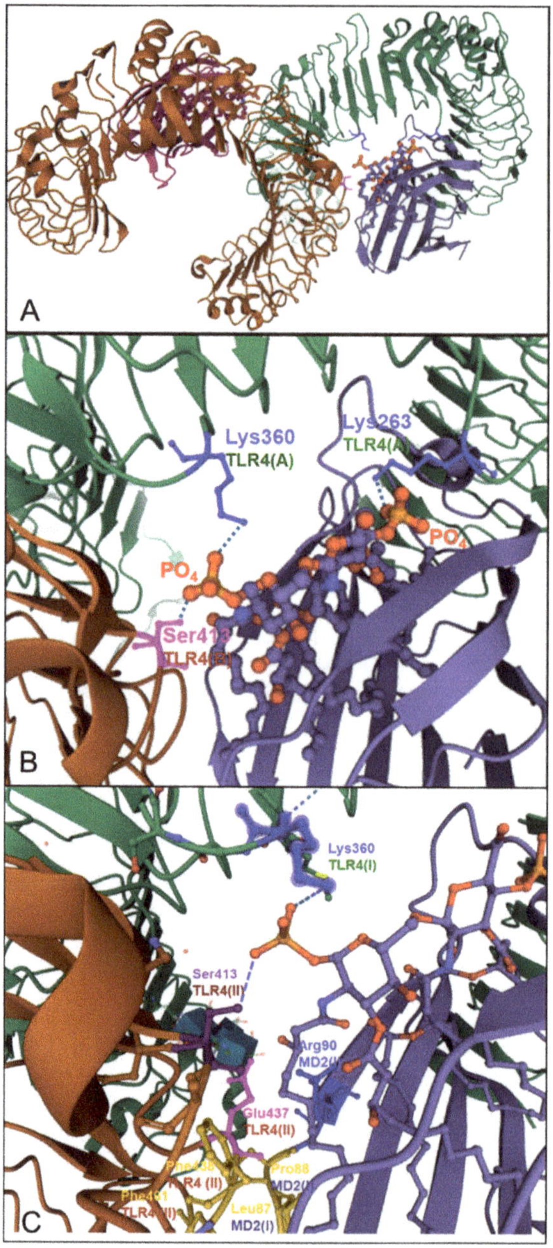

Figure 5. Dimerization of TLR4/MD-2 complexes by LPS is shown based on RCSB PDB 3VQ2 Crystal structure of mouse TLR4/MD-2/LPS complex (DOI: 10.2210/pdb3VQ2/pdb [184]. (**A**) At higher magnification: Insertion of LPS (blue/red ball stick model) into the MD-2 hydrophobic pocket (**blue**) and charged and polar interaction of the negatively charged phosphate groups of Lipid A with Lys360 and Lys263 of TLR4(I) (**green**), (**B**) and with Ser413 of the second TLR4(II) (**brown**) (**B**,**C**) and hydrophobic amino acid residues (**yellow**) in MD-2(I) (blue) and TLR4(II) (**brown**), (**C**) in the dimerization interface are depicted.

The structure–activity relationships of LPS have been studied using natural and chemically modified LPS. The crystal structure of TLRD/MD-2–LPS provides an explanation as to why LPS with six lipid chains is optimal for activation of TLR4 signaling. In the crystal structure, five of the six lipid chains of *E. coli* LPS are completely buried inside the pocket, but the remaining chain is partially exposed to the MD-2 surface and forms the hydrophobic interaction interface together with hydrophobic surface residues of TLR4(II) [183]. The two phosphate groups attached to the glucosamine of LPS seem to be essential for the formation of the stable TLRD/MD-2 complex by making charge- and hydrogen- bond interactions simultaneously with the two TLRs in the complex. Removal of the phosphate groups dramatically reduces or even completely abolishes the inflammatory activity of LPS. Studies on the interdependence of molecular charge and conformation of natural and chemically modified LPS or Lipid A and IL-6 production after stimulation of whole blood or PBMCs have shown that the number, nature, and location of negative charges strongly modulate the molecular conformation of endotoxin and biologic activity. Whereas monophosphorylated Lipid A exerts approximately 60-fold lower activity to induce IL-6 in blood cells, the phosphate-free Lipid A (e.g., dephosphorylated LPS of E. coli or phosphate-free Lipid A) almost completely loses their activity [185]. Furthermore, the number of the acyl chains in Lipid A was shown to greatly impact the stimulation activity of *E.coli*-derived Lipid A. Whereas hexaacyl Lipid A exhibited full agonistic activity, pentaacyl or tetraacyl Lipid A from *E.coli* lost their agonistic activity but kept full antagonistic activity [186]. Consequently, Lipid A derivatives with four lipid chains have antagonistic activity for the TLRD/MD-2 complex because all four lipid chains are completely submerged inside the hydrophobic MD-2 pocket, but cannot provide a hydrophobic dimerization surface that can be used for interaction with TLR4 [183].

In addition to TLR4, the structures of several other human TLRs in conjunction with their physiological or synthetic ligands have been described. TLR2 is unique among human TLRs because it can form heterodimers with other TLRs, such as TLR1 and TLR6. The principal ligands of the TLR1–TLR2 complex are triacyl lipopeptides, whereas the interaction with diacyl lipopeptides is substantially weaker. By contrast, the TLR2–TLR6 complex is able to bind to diacyl lipopeptides. These lipoproteins and lipopeptides are functionally and structurally diverse bacterial proteins anchored to the membrane by two or three covalently attached lipid chains [183]. Notably, typical TLR2 activators, such as Lipoteichoic acid also have multiple negatively charged phosphate groups.

6. Discussion of an Integrated Mechanistic Model

Many of the substitutions and deletions in the NTD and in the RBD of the spike protein can be well correlated with the observed increased infectivity and transmissibility of the Omicron variant, and with the high evasion potential from therapeutic monoclonal antibodies and vaccine-induced polyclonal neutralizing antibodies. Although three mutations near the S1/S2 furin cleavage site were expected to favor cleavage, a substantially lower cleavage efficacy has been found for the Omicron variant compared to the Delta variant and also compared to the Wuhan wild-type virus, correlating with significantly lower TMPRSS2-dependent replication in the lungs, but not in upper airway tissue, and lower syncytium formation [3,9], and a switch in primary cellular uptake pathway from membrane-based to endosomal cathepsin L dependent uptake [187], and lower pathogenicity in animal studies [7] and clinics [8].

A lower virus replication in lungs together with a faster replication in the upper respiratory system, such as nasopharyngeal and bronchi, can to a large extent explain Omi-cron's greater ability for transmission between people while apparently causing less frequent acute respiratory distress syndrome (ARDS) of the lungs and systemic symptoms of COVID-19. However, the molecular mechanisms responsible for the reciprocal changes in cellular tropism of Omicron regarding upper and lower respiratory system [3,9,187] are not sufficiently defined at this moment. The following hypothesis tries to connect the various findings into an integrated model to explain the molecular and cellular pathways for the changed cellular tropism, and lower pathogenicity of the Omicron variant in comparison to all previous VoCs.

The following paragraph summarizes the most relevant findings described in the previous sections:

(1) Different components of SARS-CoV-2, in particular of the spike protein, have been demonstrated to activate TLRs, in particular TLR4 and TLR2.
(2) Dimerization represents the general principle underlying the activation of TLRs, with activating PAMPs serving as molecular linkers promoting dimerization.
(3) For dimerization, there is a minimal number of hydrophobic chains necessary which have to fit into hydrophobic pockets in order to provide sufficient hydrophobic interactions, as demonstrated for TLR4-MD-2, TLR2-TLR1, and TLR2-TLR6 complexes, respectively.
(4) There is a common feature of the TLR activating viral glycoproteins (also including the SARS-CoV-2 spike protein) with all of them being membrane-bound proteins which contain hydrophobic domains necessary for fusion with the host cell membrane and having the potential to interact with the hydrophobic pockets of TLR complexes.
(5) Negatively charged groups have been shown to be essential for dimerization, as illustrated for TLR4-MD-2/LPS complexes.
(6) Interaction and dimerization of respective TLR complexes triggers the inherent downstream signaling pathways, mainly the NF-κB pathway.

Based on these findings, we have developed the following hypothesis:

- Some hydrophobic domains of the SARS-CoV-2 spike protein can interact with the hydrophobic pockets of TLR-complexes leading to dimerization and activation. In particular, the hydrophobic six-helix bundle fusion core structure (6HB) in the post-fusion state of the SARS-CoV-2 spike protein can be hypothesized to fit into the hydrophobic pockets of MD-2-TLR4. Other hydrophobic domains of the spike protein, such as the three hydrophobic stretches in the S2 subunit of the trimer in prefusion state, may be speculated to fit for binding to TLR2-TLR1/6 complexes.
- Distribution of charged amino acids can greatly affect binding to and dimerization of TLR complexes. The changed charge distribution on the Omicron spike protein with high accumulations of positively charged amino acid residues in the RBD and in the S2 subunit, together with loss of several negatively charged amino acids by substitutions in the Omicron spike protein, may prevent high affinity binding to the TLR complexes and/or insufficient dimerization of TLR complexes, leading to lower downstream signaling and lower pro-inflammatory activation, lower NF-kB activation, and related lower furin expression. Indeed, a lower NF-κB activation by the Omicron variant vs. a whole panel of previous variants including the D614G, Delta, Lambda, and Mu variant has been shown recently [187].

How Will this Impact the Virus Replication, Cellular Tropism, and Pathogenicity?

The SARS-CoV-2 binds via RBD of the spike protein to its high affinity receptor on the host cells. Binding to ACE2 triggers a series of conformational changes in the S2 subunit, i.e., the proteolytic cleavage between S1 and S2 by TMRRSS2, exposing hydrophobic parts of S2 to fuse the viral membrane with the membrane of the host cell, and enabling the penetration of viral RNA into the host cell. Alternatively, the virus can be taken up via clathrin-coated pits into endosomes, where proteolytic cleavage is taken over by cathepsin L. Within the host cell, the viral RNA is translated into non-structural proteins (NSPs) with massive translation viral RNA into viral non-structural and structural proteins. This process occurs mainly in bubble-like structures (i.e., double-membrane vesicles, DMVs) after remodeling of the cell's endoplasmatic reticulum (ER). The newly produced viral components assemble into complete virus particles which leave the cell via transition through the Golgi apparatus (Figure 6). Most likely at this prefinal step, the host cell protease furin mainly presents in the TGN and cleaves the spike protein at the polybasic PRRAR site, preparing the spike protein for uptake into the next host cell. The massive remodeling of the host cell ER and viral RNA replication and translation, together with the fusogenic activity of the SARS-CoV-2 spike protein leading to excessive syncytia formation, will lead to a damage of the infected cells and to DAMPs formation, which are expected to trigger the cellular PPP alert mechanism.

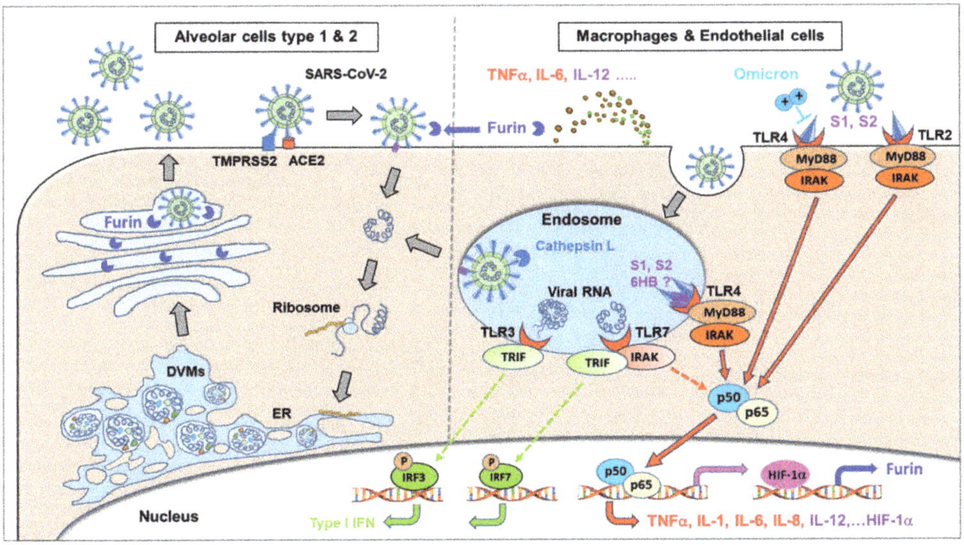

Figure 6. The SARS-CoV-2 binds via RBD of the spike protein to ACE2. Binding triggers proteolytic cleavage between S1and S2 by TMPRRS2, resulting in formation of 6HB and fusion with the host cell membrane allowing penetration of viral RNA into the host cell. Viral RNA transcription and translation of viral non-structural and structural proteins occurs in double-membrane vesicles (DMVs) after remodeling of the endoplasmatic reticulum (ER). The newly produced viral components assemble into complete virus particles which leave the cells via the Golgi apparatus where the spike protein undergoes proteolytic cleavage by furin. This process preferably occurs in virus producer cells with high TMPRSS2 expression, e.g., alveolar cells (left side). Alternatively, the virus can be taken up via clathrin-coated pits into endosomes, where proteolytic cleavage is taken over by cathepsin L. The endosomal uptake is predominant in TMPRSS2-negative cells, but not in cathepsin L-rich cells, such as innate immune cells and endothelial cells (right side). Several components of the SARS-CoV-2 virus act as PAMPs activating various Toll-like receptors (TLRs), resulting in massive activation of the NF-kB (p50/p65) pathway. The relative cellular distribution/expression differs greatly for the different TLRs with high levels of TLR4 and TLR2 on macrophages and endothelial cells, whereas only low levels are expressed on alveolar lung cells. The activation of TLRs will trigger activation of NF-κB pathway, which triggers HIF-1α activation and expression of cytokine, including TNFα, IL-1, IL-6, and IL-12. HIF-1α and IL-12 have been shown to increase furin expression. As a hypothesis for TLR activation, interaction of hydrophobic domains (and a distinct charged amino acid pattern) of the SARS-CoV-2 spike protein may be necessary for dimerization and activation of MD2-TLR4 or TLR2-TLR1/6, triggering the TLR-typical downstream signaling pathways, as has been shown so far for the whole spike protein, S1 and S2. As a hypothesis (?), the hydrophobic six-helix bundle fusion core structure (6HB) in the post-fusion state of the SARS-CoV-2 spike protein may fit into the hydrophobic pockets of MD-2-TLR4 or TLR2-TLR1/6 complexes, respectively. The significant change in the charge distribution in the Omicron spike protein, with multiple additional positively charged amino acid substitutions may prevent high affinity binding to the TLR complexes and/or insufficient dimerization of TLR complexes, leading to lower NF-kB signaling with lower expression of cytokines, HIF-1α, resulting in lower furin expression and insufficient S1/S2 cleavage, despite the presence of the polybasic furin cleavage motif.

Furthermore, different parts of the SARS-CoV-2 virus themselves act as PAMPs, triggering activation of a variety of pattern recognition receptors, in particular TLRs, leading to excessive activation of their intrinsic signal mechanisms, in particular the NF-κB pathway. In this context, the differentiated impact of various TLRs has to be taken into consideration. Whereas single-stranded viral RNA of SARS-CoV-2 or their double-stranded replication intermediates are expected to activate TLR7/TLR8 and TLR3, respectively, there are so far no indications that this should differ qualitatively from similar processes after infection by other coronaviruses, including the four seasonal low pathogenic viruses. Indeed, analysis of available data and their correlation with COVID-19 severity have shown no correlation for TLR3, whereas TLR7 correlated only with moderate COVID-19 severity, in contrast to other TLRs where activation increased with COVID-19 severity [147]. This may be due to the lack of MyD88 activation by TLR3, and a rather balanced induction of TLR7 downstream signaling leading to both MyD88 proinflammatory and TRIF- induced IRF activation, but may also correlate with the generally relatively low expression levels of TLR3 and TLR7/8 (less than 20 or 35 nTPM, respectively) compared to TLR2 or TLR4 (more than 160 and 120 nTPM, respectively, see Figure 4).

In contrast to low pathogenic coronaviruses, SARS-CoV-2 induces often an exuberated hyper inflammatory signature with cytokine/chemokine storm, massive coagulation disturbances, and systemic pro-inflammatory status in the endothelium, correlating with massive M1/Th1 cytokine release, with pathological and clinical feature shared with other highly pathogenic acute RNA virus infections, such as SARS-CoV, MERS-CoV, H5N1, and (Spanish type) H1N1 [45]. The underlying mechanism of these exaggerated pro-inflammatory reactions may rely on an additional, unbalanced and excessive NF-κB pathway activation due to powerful additional upstream signal triggers, which seem to be common for a variety of highly pathogenic acute RNA virus infections. In this context, binding and activation of TLR2 and TLR4 may play a major role. Various components of the SARS-CoV-2 virus have been demonstrated to activate TLR2 and TLR4, with the majority of reports showing massive activation by the spike proteins (and few reports for the envelope protein), always associated by excessive NF-κB activation. The excessive NF-κB activation can result from three major routes, i.e., (1) from "physiological" TLR7 and TLR3 activation by single-stranded RNA and/or double-stranded intermediates in the infected cells, (2) from excessive ER remodeling in the infected virus producer cells (i.e., primarily ACE2/TPRSS2 positive cells), and (3) from additional TLR4 and/or TLR2 activation in infected or even in non-productively infected innate cells (e.g., macrophages, endothelial cells). In this context, the relative cellular and tissue distribution/expression of the different TLRs will be a deciding factor with regard to their impact on COVID-19 severity. According to the Human Protein Atlas, TLR4 and TLR2 are expressed at high levels on different types of macrophages (TLR2 and TLR4) and endothelial cells (TLR4), and TLR2 in alveolar type 2 cells, whereas only very low levels are expressed on alveolar type 1 cells. This correlates with reports showing that cathepsin L expression, but not TMPRRS2 expression, is particularly prominent in macrophages and endothelial cells, with a demonstrated correlation between circulating levels of cathepsin L and disease course and severity in patients with COVID-19 [188].

The activation of the various types of TLRs, dominated by the highly expressed TLR4 and TLR2 on macrophages and endothelial cells, will trigger via their intrinsic signaling pathways excessive gene expression for a broad range of pro-inflammatory cytokines and chemokines, adhesion molecules, and acute phase proteins. Furthermore, the highly activated NF-κB pathway, directly or via HIF-1α activation or via cytokine release (such as IL-12), is expected to stimulate furin expression in the cells. Whereas furin can be expressed in most tissues, the basic expression levels seem to be very low. Viral infections, cancer, hypoxia, HIF-1α, and cytokines (e.g., IL-12) have been found to significantly increase furin expression. Activation of the NF-κB pathway triggering HIF-1α expression and cytokine release may play a central role in stimulating furin expression. Since cleavage at the furin-like site is essential for highly efficient TMPRSS2-dependent membrane uptake of SARS-CoV-2, the positive feedback from TLR-NF-κB-HIF-1α/IL-12-furin activation may play a significant role in providing enough of the enzyme necessary for site specific S1/S2 cleavage.

In this context the amino acid changes in the Omicron variant must be analyzed. For various mutations in the non-structural proteins of Omicron, their involvement in reducing an excessive innate response is not expected. In contrast, the rather suppressing effects of non-structural SARS-CoV-2 proteins, including nsp1, nsp3, nsp6, nsp8, ORF3b, and ORF8 resulting in suppression of IFN response have been described [16,147,189–193].

Regarding other viral structural proteins, there is only one amino acid substitution in the envelope protein T9I. In contrast, there are more than 30 mutations in the spike protein in the Omicron variant. Whereas the increased infectivity, transmissibility, and escape from immune reactions can largely be explained by substitutions or deletions in the RBD and NTD, the lower S1/S2 cleavage at the furin-like site, correlating with lower (or almost absent syncitia formation), leading to a significantly changed cellular tropism and attenuated pathogenicity, cannot be explained from the amino acid substitutions in the S1/S2 region. In this context, the interaction and activation of the TLR2-TLR1/6 and TLR4-MD-2 receptors may be the missing link. Hydrophobic and polar-charged interactions of the archetypic TLR4 agonist, LPS, with the TLR4-MD2 complex may serve as a prototype for interaction with other ligands, including the SARS-CoV-2 spike protein. Interaction of the hydrophobic domains of the SARS-CoV-2 spike protein with hydrophobic pockets, e.g., in the MD2-TLR4 or TLR2-TLR1/6, will depend on hydrophobic structures and appropriately localized and charged amino acids necessary for efficient interaction with TLR-complexes and their dimerization. Typical agonists for TLR4-MD-2 and TLR2, such as Lipid A or Lipoteichoic acid, respectively, have typical hydrophobic chains linked to exposed negatively charged phosphate groups.

We speculate that the hydrophobic six-helix bundle fusion core structure (6HB) in the post-fusion state of the SARS-CoV-2 spike protein with the central coiled-coil fusion core formed by the three HR1 domains surrounded by the three HR2 provides hydrophobic bundle-like structures, and resembling to some extent the hydrophobic structure of fatty acids in Lipid A, may be able to interact with the TLR4-MD-2 complex. Alternatively, hydrophobic stretches on the trimer in the pre-fusion state may fit into the hydrophobic pockets of MD-2-TLR4 or TLR2-TLR1/6 complexes, respectively, leading to dimerization and triggering the TLR-typical downstream signaling pathways. In this context, the significant change in the charge distribution in the Omicron spike protein, with multiple additional positively charged amino acid substitutions accompanied by loss of several negatively charged amino acids, and in the molecular vicinity of hydrophobic 3 or 6 bundled coils, may prevent high affinity binding to the TLR complexes and/or insufficient dimerization of TLR complexes.

Although probably only some of the TLR, i.e., TLR4 and/or TLR2, pathways are primarily affected by the charged electrostatic charge pattern of Omicron, this shortfall likely concerns the most powerful pathways, because of the high relative expression of TLR2 and TLR4 on innate inflammatory cells and the exceptionally high pro-inflammatory capacity of the affected innate immune cells, in comparison to the primarily SARS-CoV-2 infected bronchial or alveolar cells. This reduced innate immune cell activation, leading to lower NF-κB and HIF-1α activation, counterbalances the advantage from the preserved (or even slightly enhanced) polybasic furin cleavage site in the spike protein of Omicron because of the limited proprotein convertase (furin) availability, which may explain to a large extent the changed cellular tropism and the lower pathogenicity of the Omicron variant.

7. Conclusions

The recently appearing Omicron variant shows surprising reciprocal changes in cellular tropism of the regarding upper and lower respiratory system, which cannot be explained simply by the changed binding to the ACE2 receptor or immune escape. Here we present a new hypothesis proposing that the changed distribution of charged amino acids in the spike protein of the Omicron variant compared to all other VoCs may disturb the recognition by innate Pattern-recognition receptors (PRRs), in particular of certain Toll-like receptors (TLR), resulting in lower activation of the NF-κB pathway and related signaling pathways, and also resulting in lower furin expression, lower viral replication in the lungs, and lower systemic immune hyperactivation.

Author Contributions: Conceptualization, methodology, data curation, writing—original draft preparation, review and editing, visualization, R.K. Review and Editing, Critical Discussion, Literature and Funding, O.P. All authors have read and agreed to the published version of the manuscript.

Funding: This research received no external funding.

Institutional Review Board Statement: Not applicable.

Informed Consent Statement: Not applicable.

Acknowledgments: We acknowledge the support by the University of Tübingen.

Conflicts of Interest: The authors declare no conflict of interest.

References

1. Viana, R.; Moyo, S.; Amoako, D.G.; Tegally, H.; Scheepers, C.; Althaus, C.L.; Anyaneji, U.J.; Bester, P.A.; Boni, M.F.; Chand, M.; et al. Rapid epidemic expansion of the SARS-CoV-2 Omicron variant in southern Africa. *Nature* **2022**, *603*, 679–686. [CrossRef]
2. Dejnirattisai, W.; Huo, J.; Zhou, D.; Zahradnik, J.; Supasa, P.; Liu, C.; Duyvesteyn, H.M.E.; Ginn, H.M.; Mentzer, A.J.; Tuekprakhon, A.; et al. SARS-CoV-2 Omicron-B.1.1.529 leads to widespread escape from neutralizing antibody responses. *Cell* **2022**, *185*, 467–484.e15. [CrossRef]
3. Meng, B.; Abdullahi, A.; Ferreira, I.; Goonawardane, N.; Saito, A.; Kimura, I.; Yamasoba, D.; Gerber, P.P.; Fatihi, S.; Rathore, S.; et al. Altered TMPRSS2 usage by SARS-CoV-2 Omicron impacts infectivity and fusogenicity. *Nature* **2022**, *603*, 706–714. [CrossRef]
4. Kannan, S.R.; Spratt, A.N.; Sharma, K.; Chand, H.S.; Byrareddy, S.N.; Singh, K. Omicron SARS-CoV-2 variant: Unique features and their impact on pre-existing antibodies. *J. Autoimmun.* **2022**, *126*, 102779. [CrossRef]
5. Ye, G.; Liu, B.; Li, F. Cryo-EM structure of a SARS-CoV-2 omicron spike protein ectodomain. *Nat. Commun.* **2022**, *13*, 1214. [CrossRef]
6. Shah, M.; Woo, H.G. Omicron: A Heavily Mutated SARS-CoV-2 Variant Exhibits Stronger Binding to ACE2 and Potently Escapes Approved COVID-19 Therapeutic Antibodies. *Front. Immunol.* **2021**, *12*, 830527. [CrossRef]
7. Halfmann, P.J.; Iida, S.; Iwatsuki-Horimoto, K.; Maemura, T.; Kiso, M.; Scheaffer, S.M.; Darling, T.L.; Joshi, A.; Loeber, S.; Singh, G.; et al. SARS-CoV-2 Omicron virus causes attenuated disease in mice and hamsters. *Nature* **2022**, *603*, 687–692. [CrossRef]
8. Wolter, N.; Jassat, W.; Walaza, S.; Welch, R.; Moultrie, H.; Groome, M.; Amoako, D.G.; Everatt, J.; Bhiman, J.N.; Scheepers, C.; et al. Early assessment of the clinical severity of the SARS-CoV-2 omicron variant in South Africa: A data linkage study. *Lancet* **2022**, *399*, 437–446. [CrossRef]
9. Willett, B.J.; Grove, J.; MacLean, O.A.; Wilkie, C.; Logan, N.; Lorenzo, G.D.; Furnon, W.; Scott, S.; Manali, M.; Szemiel, A.; et al. The hyper-transmissible SARS-CoV-2 Omicron variant exhibits significant antigenic change, vaccine escape and a switch in cell entry mechanism. *medRxiv* **2022**. [CrossRef]
10. Abdelrahman, Z.; Li, M.; Wang, X. Comparative Review of SARS-CoV-2, SARS-CoV, MERS-CoV, and Influenza A Respiratory Viruses. *Front. Immunol.* **2020**, *11*, 552909. [CrossRef]
11. Eslami, N.; Aghbash, P.S.; Shamekh, A.; Entezari-Maleki, T.; Nahand, J.S.; Sales, A.J.; Baghi, H.B. SARS-CoV-2: Receptor and Co-receptor Tropism Probability. *Curr Microbiol.* **2022**, *79*, 133. [CrossRef]
12. Cantuti-Castelvetri, L.; Ojha, R.; Pedro, L.D.; Djannatian, M.; Franz, J.; Kuivanen, S.; van der Meer, F.; Kallio, K.; Kaya, T.; Anastasina, M.; et al. Neuropilin-1 facilitates SARS-CoV-2 cell entry and infectivity. *Science* **2020**, *370*, 856–860. [CrossRef]
13. Matsuyama, S.; Ujike, M.; Morikawa, S.; Tashiro, M.; Taguchi, F. Protease-mediated enhancement of severe acute respiratory syndrome coronavirus infection. *Proc. Natl. Acad. Sci. USA* **2005**, *102*, 12543–12547. [CrossRef]
14. Simmons, G.; Gosalia, D.N.; Rennekamp, A.J.; Reeves, J.D.; Diamond, S.L.; Bates, P. Inhibitors of cathepsin L prevent severe acute respiratory syndrome coronavirus entry. *Proc. Natl. Acad. Sci. USA* **2005**, *102*, 11876–11881. [CrossRef]
15. Belouzard, S.; Millet, J.K.; Licitra, B.N.; Whittaker, G.R. Mechanisms of coronavirus cell entry mediated by the viral spike protein. *Viruses* **2012**, *4*, 1011–1033. [CrossRef]
16. Shah, M.; Woo, H.G. Molecular Perspectives of SARS-CoV-2: Pathology, Immune Evasion, and Therapeutic Interventions. *Mol. Cell* **2021**, *44*, 408–421. [CrossRef]
17. Watanabe, Y.; Allen, J.D.; Wrapp, D.; McLellan, J.S.; Crispin, M. Site-specific glycan analysis of the SARS-CoV-2 spike. *Science* **2020**, *369*, 330–333. [CrossRef]
18. Shajahan, A.; Pepi, L.E.; Rouhani, D.S.; Heiss, C.; Azadi, P. Glycosylation of SARS-CoV-2: Structural and functional insights. *Anal. Bioanal. Chem.* **2021**, *413*, 7179–7193. [CrossRef]

19. Xia, S.; Zhu, Y.; Liu, M.; Lan, Q.; Xu, W.; Wu, Y.; Ying, T.; Liu, S.; Shi, Z.; Jiang, S.; et al. Fusion mechanism of 2019-nCoV and fusion inhibitors targeting HR1 domain in spike protein. *Cell. Mol. Immunol.* **2020**, *17*, 765–767. [CrossRef]
20. Sah, P.; Fitzpatrick, M.C.; Zimmer, C.F.; Abdollahi, E.; Juden-Kelly, L.; Moghadas, S.M.; Singer, B.H.; Galvani, A.P. Asymptomatic SARS-CoV-2 infection: A systematic review and meta-analysis. *Proc. Natl. Acad. Sci. USA* **2021**, *118*, e2109229118. [CrossRef]
21. Hoffmann, M.; Kleine-Weber, H.; Schroeder, S.; Kruger, N.; Herrler, T.; Erichsen, S.; Schiergens, T.S.; Herrler, G.; Wu, N.H.; Nitsche, A.; et al. SARS-CoV-2 Cell Entry Depends on ACE2 and TMPRSS2 and Is Blocked by a Clinically Proven Protease Inhibitor. *Cell* **2020**, *181*, 271–280.e8. [CrossRef]
22. Varga, Z.; Flammer, A.J.; Steiger, P.; Haberecker, M.; Andermatt, R.; Zinkernagel, A.S.; Mehra, M.R.; Schuepbach, R.A.; Ruschitzka, F.; Moch, H. Endothelial cell infection and endotheliitis in COVID-19. *Lancet* **2020**, *395*, 1417–1418. [CrossRef]
23. Ackermann, M.; Verleden, S.E.; Kuehnel, M.; Haverich, A.; Welte, T.; Laenger, F.; Vanstapel, A.; Werlein, C.; Stark, H.; Tzankov, A.; et al. Pulmonary Vascular Endothelialitis, Thrombosis, and Angiogenesis in Covid-19. *N. Engl. J. Med.* **2020**, *383*, 120–128. [CrossRef]
24. Sokolowska, M.; Lukasik, Z.M.; Agache, I.; Akdis, C.A.; Akdis, D.; Akdis, M.; Barcik, W.; Brough, H.A.; Eiwegger, T.; Eljaszewicz, A.; et al. Immunology of COVID-19: Mechanisms, clinical outcome, diagnostics, and perspectives-A report of the European Academy of Allergy and Clinical Immunology (EAACI). *Allergy* **2020**, *75*, 2445–2476. [CrossRef]
25. Huang, C.; Wang, Y.; Li, X.; Ren, L.; Zhao, J.; Hu, Y.; Zhang, L.; Fan, G.; Xu, J.; Gu, X.; et al. Clinical features of patients infected with 2019 novel coronavirus in Wuhan, China. *Lancet* **2020**, *395*, 497–506. [CrossRef]
26. Xu, Z.; Shi, L.; Wang, Y.; Zhang, J.; Huang, L.; Zhang, C.; Liu, S.; Zhao, P.; Liu, H.; Zhu, L.; et al. Pathological findings of COVID-19 associated with acute respiratory distress syndrome. *Lancet Respir. Med.* **2020**, *8*, 420–422. [CrossRef]
27. Wang, D.; Hu, B.; Hu, C.; Zhu, F.; Liu, X.; Zhang, J.; Wang, B.; Xiang, H.; Cheng, Z.; Xiong, Y.; et al. Clinical Characteristics of 138 Hospitalized Patients With 2019 Novel Coronavirus-Infected Pneumonia in Wuhan, China. *J. Am. Med. Assoc.* **2020**, *323*, 1061–1069. [CrossRef]
28. Zheng, Z.; Peng, F.; Xu, B.; Zhao, J.; Liu, H.; Peng, J.; Li, Q.; Jiang, C.; Zhou, Y.; Liu, S.; et al. Risk factors of critical & mortal COVID-19 cases: A systematic literature review and meta-analysis. *J. Infect.* **2020**, *81*, e16–e25.
29. Biswas, M.; Rahaman, S.; Biswas, T.K.; Haque, Z.; Ibrahim, B. Association of Sex, Age, and Comorbidities with Mortality in COVID-19 Patients: A Systematic Review and Meta-Analysis. *Intervirology* **2021**, *64*, 36–47. [CrossRef]
30. Kang, S.J.; Jung, S.I. Age-Related Morbidity and Mortality among Patients with COVID-19. *J. Infect. Chemother.* **2020**, *52*, 154–164. [CrossRef]
31. O'Driscoll, M.; Ribeiro Dos Santos, G.; Wang, L.; Cummings, D.A.T.; Azman, A.S.; Paireau, J.; Fontanet, A.; Cauchemez, S.; Salje, H. Age-specific mortality and immunity patterns of SARS-CoV-2. *Nature* **2021**, *590*, 140–145. [CrossRef]
32. Carsana, L.; Sonzogni, A.; Nasr, A.; Rossi, R.S.; Pellegrinelli, A.; Zerbi, P.; Rech, R.; Colombo, R.; Antinori, S.; Corbellino, M.; et al. Pulmonary post-mortem findings in a series of COVID-19 cases from northern Italy: A two-centre descriptive study. *Lancet Infect. Dis.* **2020**, *20*, 1135–1140. [CrossRef]
33. Acharya, D.; Liu, G.; Gack, M.U. Dysregulation of type I interferon responses in COVID-19. *Nat. Rev. Immunol.* **2020**, *20*, 397–398. [CrossRef]
34. Tay, M.Z.; Poh, C.M.; Renia, L.; MacAry, P.A.; Ng, L.F.P. The trinity of COVID-19: Immunity, inflammation and intervention. *Nat. Rev. Immunol.* **2020**, *20*, 363–374. [CrossRef]
35. Schett, G.; Sticherling, M.; Neurath, M.F. COVID-19: Risk for cytokine targeting in chronic inflammatory diseases? *Nat. Rev. Immunol.* **2020**, *20*, 271–272. [CrossRef]
36. Moore, J.B.; June, C.H. Cytokine release syndrome in severe COVID-19. *Science* **2020**, *368*, 473–474. [CrossRef]
37. Huang, Y.; Tu, M.; Wang, S.; Chen, S.; Zhou, W.; Chen, D.; Zhou, L.; Wang, M.; Zhao, Y.; Zeng, W.; et al. Clinical characteristics of laboratory confirmed positive cases of SARS-CoV-2 infection in Wuhan, China: A retrospective single center analysis. *Travel Med. Infect. Dis.* **2020**, *36*, 101606. [CrossRef]
38. Chua, R.L.; Lukassen, S.; Trump, S.; Hennig, B.P.; Wendisch, D.; Pott, F.; Debnath, O.; Thurmann, L.; Kurth, F.; Volker, M.T.; et al. COVID-19 severity correlates with airway epithelium-immune cell interactions identified by single-cell analysis. *Nat. Biotechnol.* **2020**, *38*, 970–979. [CrossRef]
39. Huang, J.; Hume, A.J.; Abo, K.M.; Werder, R.B.; Villacorta-Martin, C.; Alysandratos, K.D.; Beermann, M.L.; Simone-Roach, C.; Lindstrom-Vautrin, J.; Olejnik, J.; et al. SARS-CoV-2 Infection of Pluripotent Stem Cell-Derived Human Lung Alveolar Type 2 Cells Elicits a Rapid Epithelial-Intrinsic Inflammatory Response. *Cell Stem Cell* **2020**, *27*, 962–973.e7. [CrossRef]
40. Hojyo, S.; Uchida, M.; Tanaka, K.; Hasebe, R.; Tanaka, Y.; Murakami, M.; Hirano, T. How COVID-19 induces cytokine storm with high mortality. *Inflamm. Regen.* **2020**, *40*, 37. [CrossRef]
41. Hirano, T.; Murakami, M. COVID-19: A New Virus, but a Familiar Receptor and Cytokine Release Syndrome. *Immunity* **2020**, *52*, 731–733. [CrossRef]
42. Hong, K.S.; Ahn, J.H.; Jang, J.G.; Lee, J.H.; Kim, H.N.; Kim, D.; Lee, W. GSK-LSD1, an LSD1 inhibitor, quashes SARS-CoV-2-triggered cytokine release syndrome in-vitro. *Signal Transduct. Target. Ther.* **2020**, *5*, 267. [CrossRef]
43. McAleavy, M.; Zhang, Q.; Ehmann, P.J.; Xu, J.; Wipperman, M.F.; Ajithdoss, D.; Pan, L.; Wakai, M.; Simonson, R.; Gadi, A.; et al. The Activin/FLRG Pathway Associates with Poor COVID-19 Outcomes in Hospitalized Patients. *Mol. Cell. Biol.* **2022**, *42*, e0046721. [CrossRef]

44. Hariharan, A.; Hakeem, A.R.; Radhakrishnan, S.; Reddy, M.S.; Rela, M. The Role and Therapeutic Potential of NF-kappa-B Pathway in Severe COVID-19 Patients. *Inflammopharmacology* **2021**, *29*, 91–100. [CrossRef]
45. Kircheis, R.; Haasbach, E.; Lueftenegger, D.; Heyken, W.T.; Ocker, M.; Planz, O. NF-kappaB Pathway as a Potential Target for Treatment of Critical Stage COVID-19 Patients. *Front. Immunol.* **2020**, *11*, 598444. [CrossRef]
46. Neufeldt, C.J.; Cerikan, B.; Cortese, M.; Frankish, J.; Lee, J.Y.; Plociennikowska, A.; Heigwer, F.; Prasad, V.; Joecks, S.; Burkart, S.S.; et al. SARS-CoV-2 infection induces a pro-inflammatory cytokine response through cGAS-STING and NF-kappaB. *Commun. Biol.* **2022**, *5*, 45. [CrossRef]
47. Lo, M.W.; Kemper, C.; Woodruff, T.M. COVID-19: Complement, Coagulation, and Collateral Damage. *J. Immunol.* **2020**, *205*, 1488–1495. [CrossRef]
48. Holter, J.C.; Pischke, S.E.; de Boer, E.; Lind, A.; Jenum, S.; Holten, A.R.; Tonby, K.; Barratt-Due, A.; Sokolova, M.; Schjalm, C.; et al. Systemic complement activation is associated with respiratory failure in COVID-19 hospitalized patients. *Proc. Natl. Acad. Sci. USA* **2020**, *117*, 25018–25025. [CrossRef]
49. Polycarpou, A.; Howard, M.; Farrar, C.A.; Greenlaw, R.; Fanelli, G.; Wallis, R.; Klavinskis, L.S.; Sacks, S. Rationale for targeting complement in COVID-19. *EMBO Mol. Med.* **2020**, *12*, e12642. [CrossRef]
50. Yu, J.; Yuan, X.; Chen, H.; Chaturvedi, S.; Braunstein, E.M.; Brodsky, R.A. Direct activation of the alternative complement pathway by SARS-CoV-2 spike proteins is blocked by factor D inhibition. *Blood* **2020**, *136*, 2080–2089. [CrossRef]
51. Fletcher-Sandersjoo, A.; Bellander, B.M. Is COVID-19 associated thrombosis caused by overactivation of the complement cascade? A literature review. *Thromb. Res.* **2020**, *194*, 36–41. [CrossRef]
52. Mussbacher, M.; Salzmann, M.; Brostjan, C.; Hoesel, B.; Schoergenhofer, C.; Datler, H.; Hohensinner, P.; Basilio, J.; Petzelbauer, P.; Assinger, A.; et al. Cell Type-Specific Roles of NF-kappaB Linking Inflammation and Thrombosis. *Front. Immunol.* **2019**, *10*, 85. [CrossRef]
53. McCarthy, C.G.; Wilczynski, S.; Wenceslau, C.F.; Webb, R.C. A new storm on the horizon in COVID-19: Bradykinin-induced vascular complications. *Vascul. Pharmacol.* **2021**, *137*, 106826. [CrossRef]
54. Leppkes, M.; Knopf, J.; Naschberger, E.; Lindemann, A.; Singh, J.; Herrmann, I.; Sturzl, M.; Staats, L.; Mahajan, A.; Schauer, C.; et al. Vascular occlusion by neutrophil extracellular traps in COVID-19. *Ebiomedicine* **2020**, *58*, 102925. [CrossRef]
55. De Buhr, N.; von Kockritz-Blickwede, M. The Balance of Neutrophil Extracellular Trap Formation and Nuclease Degradation: An Unknown Role of Bacterial Coinfections in COVID-19 Patients? *mBio* **2021**, *12*, e03304-20. [CrossRef]
56. Kvietys, P.R.; Fakhoury, H.M.A.; Kadan, S.; Yaqinuddin, A.; Al-Mutairy, E.; Al-Kattan, K. COVID-19: Lung-Centric Immunothrombosis. *Front. Cell. Infect. Microbiol.* **2021**, *11*, 679878. [CrossRef]
57. Gando, S.; Wada, T. Pathomechanisms Underlying Hypoxemia in Two COVID-19-Associated Acute Respiratory Distress Syndrome Phenotypes: Insights From Thrombosis and Hemostasis. *Shock* **2022**, *57*, 1–6. [CrossRef]
58. Thachil, J.; Tang, N.; Gando, S.; Falanga, A.; Cattaneo, M.; Levi, M.; Clark, C.; Iba, T. ISTH interim guidance on recognition and management of coagulopathy in COVID-19. *J. Thromb. Haemost.* **2020**, *18*, 1023–1026. [CrossRef]
59. Gu, S.X.; Tyagi, T.; Jain, K.; Gu, V.W.; Lee, S.H.; Hwa, J.M.; Kwan, J.M.; Krause, D.S.; Lee, A.I.; Halene, S.; et al. Thrombocytopathy and endotheliopathy: Crucial contributors to COVID-19 thromboinflammation. *Nat. Rev. Cardiol.* **2021**, *18*, 194–209. [CrossRef]
60. Zuo, Y.; Estes, S.K.; Ali, R.A.; Gandhi, A.A.; Yalavarthi, S.; Shi, H.; Sule, G.; Gockman, K.; Madison, J.A.; Zuo, M.; et al. Prothrombotic autoantibodies in serum from patients hospitalized with COVID-19. *Sci. Transl. Med.* **2020**, *12*, eabd3876. [CrossRef]
61. Klok, F.A.; Kruip, M.; van der Meer, N.J.M.; Arbous, M.S.; Gommers, D.; Kant, K.M.; Kaptein, F.H.J.; van Paassen, J.; Stals, M.A.M.; Huisman, M.V.; et al. Confirmation of the high cumulative incidence of thrombotic complications in critically ill ICU patients with COVID-19: An updated analysis. *Thromb. Res.* **2020**, *191*, 148–150. [CrossRef]
62. Skendros, P.; Mitsios, A.; Chrysanthopoulou, A.; Mastellos, D.C.; Metallidis, S.; Rafailidis, P.; Ntinopoulou, M.; Sertaridou, E.; Tsironidou, V.; Tsigalou, C.; et al. Complement and tissue factor-enriched neutrophil extracellular traps are key drivers in COVID-19 immunothrombosis. *J. Clin. Investig.* **2020**, *130*, 6151–6157. [CrossRef]
63. George, J.N.; Nester, C.M. Syndromes of thrombotic microangiopathy. *N. Engl. J. Med.* **2014**, *371*, 1847–1848. [CrossRef]
64. Brocklebank, V.; Wood, K.M.; Kavanagh, D. Thrombotic Microangiopathy and the Kidney. *Clin. J. Am. Soc. Nephrol.* **2018**, *13*, 300–317. [CrossRef]
65. Menter, T.; Haslbauer, J.D.; Nienhold, R.; Savic, S.; Hopfer, H.; Deigendesch, N.; Frank, S.; Turek, D.; Willi, N.; Pargger, H.; et al. Postmortem examination of COVID-19 patients reveals diffuse alveolar damage with severe capillary congestion and variegated findings in lungs and other organs suggesting vascular dysfunction. *Histopathology* **2020**, *77*, 198–209. [CrossRef]
66. Gobeil, S.M.; Henderson, R.; Stalls, V.; Janowska, K.; Huang, X.; May, A.; Speakman, M.; Beaudoin, E.; Manne, K.; Li, D.; et al. Structural diversity of the SARS-CoV-2 Omicron spike. *bioRxiv*, 2022; *in press*.
67. Mannar, D.; Saville, J.W.; Zhu, X.; Srivastava, S.S.; Berezuk, A.M.; Tuttle, K.S.; Marquez, A.C.; Sekirov, I.; Subramaniam, S. SARS-CoV-2 Omicron variant: Antibody evasion and cryo-EM structure of spike protein-ACE2 complex. *Science* **2022**, *375*, 760–764. [CrossRef]
68. Yin, W.; Xu, Y.; Xu, P.; Cao, X.; Wu, C.; Gu, C.; He, X.; Wang, X.; Huang, S.; Yuan, Q.; et al. Structures of the Omicron spike trimer with ACE2 and an anti-Omicron antibody. *Science* **2022**, *375*, 1048–1053. [CrossRef]

69. Barton, M.I.; MacGowan, S.A.; Kutuzov, M.A.; Dushek, O.; Barton, G.J.; van der Merwe, P.A. Effects of common mutations in the SARS-CoV-2 Spike RBD and its ligand, the human ACE2 receptor on binding affinity and kinetics. *eLife* **2021**, *10*, e70658. [CrossRef]
70. Luan, B.; Wang, H.; Huynh, T. Enhanced binding of the N501Y-mutated SARS-CoV-2 spike protein to the human ACE2 receptor: Insights from molecular dynamics simulations. *FEBS Lett.* **2021**, *595*, 1454–1461. [CrossRef]
71. Mansbach, R.A.; Chakraborty, S.; Nguyen, K.; Montefiori, D.C.; Korber, B.; Gnanakaran, S. The SARS-CoV-2 Spike variant D614G favors an open conformational state. *Sci. Adv.* **2021**, *7*, eabf3671. [CrossRef]
72. Zhang, L.; Jackson, C.B.; Mou, H.; Ojha, A.; Peng, H.; Quinlan, B.D.; Rangarajan, E.S.; Pan, A.; Vanderheiden, A.; Suthar, M.S.; et al. SARS-CoV-2 spike-protein D614G mutation increases virion spike density and infectivity. *Nat. Commun.* **2020**, *11*, 6013. [CrossRef]
73. Tarke, A.; Sidney, J.; Kidd, C.K.; Dan, J.M.; Ramirez, S.I.; Yu, E.D.; Mateus, J.; da Silva Antunes, R.; Moore, E.; Rubiro, P.; et al. Comprehensive analysis of T cell immunodominance and immunoprevalence of SARS-CoV-2 epitopes in COVID-19 cases. *Cell Rep. Med.* **2021**, *2*, 100204. [CrossRef]
74. McCallum, M.; Czudnochowski, N.; Rosen, L.E.; Zepeda, S.K.; Bowen, J.E.; Walls, A.C.; Hauser, K.; Joshi, A.; Stewart, C.; Dillen, J.R.; et al. Structural basis of SARS-CoV-2 Omicron immune evasion and receptor engagement. *Science* **2022**, *375*, 864–868. [CrossRef]
75. Wang, P.; Nair, M.S.; Liu, L.; Iketani, S.; Luo, Y.; Guo, Y.; Wang, M.; Yu, J.; Zhang, B.; Kwong, P.D.; et al. Antibody resistance of SARS-CoV-2 variants B.1.351 and B.1.1.7. *Nature* **2021**, *593*, 130–135. [CrossRef]
76. McCarthy, K.R.; Rennick, L.J.; Nambulli, S.; Robinson-McCarthy, L.R.; Bain, W.G.; Haidar, G.; Duprex, W.P. Recurrent deletions in the SARS-CoV-2 spike glycoprotein drive antibody escape. *Science* **2021**, *371*, 1139–1142. [CrossRef]
77. Rani, P.R.; Imran, M.; Lakshmi, J.V.; Jolly, B.; Jain, A.; Surekha, A.; Senthivel, V.; Chandrasekhar, P.; Divakar, M.K.; Srinivasulu, D.; et al. Symptomatic reinfection of SARS-CoV-2 with spike protein variant N440K associated with immune escape. *J. Med. Virol.* **2021**, *93*, 4163–4165. [CrossRef]
78. Weisblum, Y.; Schmidt, F.; Zhang, F.; DaSilva, J.; Poston, D.; Lorenzi, J.C.; Muecksch, F.; Rutkowska, M.; Hoffmann, H.H.; Michailidis, E.; et al. Escape from neutralizing antibodies by SARS-CoV-2 spike protein variants. *eLife* **2020**, *9*, e61312. [CrossRef]
79. Greaney, A.J.; Starr, T.N.; Gilchuk, P.; Zost, S.J.; Binshtein, E.; Loes, A.N.; Hilton, S.K.; Huddleston, J.; Eguia, R.; Crawford, K.H.D.; et al. Complete Mapping of Mutations to the SARS-CoV-2 Spike Receptor-Binding Domain that Escape Antibody Recognition. *Cell Host Microbe* **2021**, *29*, 44–57.e9. [CrossRef]
80. Liu, Z.; VanBlargan, L.A.; Bloyet, L.M.; Rothlauf, P.W.; Chen, R.E.; Stumpf, S.; Zhao, H.; Errico, J.M.; Theel, E.S.; Liebeskind, M.J.; et al. Identification of SARS-CoV-2 spike mutations that attenuate monoclonal and serum antibody neutralization. *Cell Host Microbe* **2021**, *29*, 477–488.e4. [CrossRef]
81. Arora, P.; Rocha, C.; Kempf, A.; Nehlmeier, I.; Graichen, L.; Winkler, M.S.; Lier, M.; Schulz, S.; Jack, H.M.; Cossmann, A.; et al. The spike protein of SARS-CoV-2 variant A.30 is heavily mutated and evades vaccine-induced antibodies with high efficiency. *Cell. Mol. Immunol.* **2021**, *18*, 2673–2675. [CrossRef]
82. Starr, T.N.; Greaney, A.J.; Dingens, A.S.; Bloom, J.D. Complete map of SARS-CoV-2 RBD mutations that escape the monoclonal antibody LY-CoV555 and its cocktail with LY-CoV016. *Cell Rep. Med.* **2021**, *2*, 100255. [CrossRef]
83. Baum, A.; Fulton, B.O.; Wloga, E.; Copin, R.; Pascal, K.E.; Russo, V.; Giordano, S.; Lanza, K.; Negron, N.; Ni, M.; et al. Antibody cocktail to SARS-CoV-2 spike protein prevents rapid mutational escape seen with individual antibodies. *Science* **2020**, *369*, 1014–1018. [CrossRef]
84. Shekhar, N.; Sarma, P.; Prajapat, M.; Avti, P.; Kaur, H.; Raja, A.; Singh, H.; Bhattacharya, A.; Sharma, S.; Kumar, S.; et al. In Silico Structure-Based Repositioning of Approved Drugs for Spike Glycoprotein S2 Domain Fusion Peptide of SARS-CoV-2: Rationale from Molecular Dynamics and Binding Free Energy Calculations. *mSystems* **2020**, *5*, e00382-20. [CrossRef]
85. Schutz, D.; Ruiz-Blanco, Y.B.; Munch, J.; Kirchhoff, F.; Sanchez-Garcia, E.; Muller, J.A. Peptide and peptide-based inhibitors of SARS-CoV-2 entry. *Adv. Drug Deliv. Rev.* **2020**, *167*, 47–65. [CrossRef]
86. Shang, J.; Wan, Y.; Luo, C.; Ye, G.; Geng, Q.; Auerbach, A.; Li, F. Cell entry mechanisms of SARS-CoV-2. *Proc. Natl. Acad. Sci. USA* **2020**, *117*, 11727–11734. [CrossRef]
87. Kleine-Weber, H.; Elzayat, M.T.; Hoffmann, M.; Pohlmann, S. Functional analysis of potential cleavage sites in the MERS-coronavirus spike protein. *Sci. Rep.* **2018**, *8*, 16597. [CrossRef]
88. Ujike, M.; Nishikawa, H.; Otaka, A.; Yamamoto, N.; Yamamoto, N.; Matsuoka, M.; Kodama, E.; Fujii, N.; Taguchi, F. Heptad repeat-derived peptides block protease-mediated direct entry from the cell surface of severe acute respiratory syndrome coronavirus but not entry via the endosomal pathway. *J. Virol.* **2008**, *82*, 588–592. [CrossRef]
89. Ni, D.; Lau, K.; Turelli, P.; Raclot, C.; Beckert, B.; Nazarov, S.; Pojer, F.; Myasnikov, A.; Stahlberg, H.; Trono, D. Structural analysis of the Spike of the Omicron SARS-COV-2 variant by cryo-EM and implications for immune evasion. *bioRxiv* **2021**. [CrossRef]
90. Whittaker, G.R. SARS-CoV-2 spike and its adaptable furin cleavage site. *Lancet Microbe* **2021**, *2*, e488–e489. [CrossRef]
91. Hoffmann, M.; Kleine-Weber, H.; Pohlmann, S. A Multibasic Cleavage Site in the Spike Protein of SARS-CoV-2 Is Essential for Infection of Human Lung Cells. *Mol. Cell* **2020**, *78*, 779–784.e5. [CrossRef]
92. Sasaki, M.; Uemura, K.; Sato, A.; Toba, S.; Sanaki, T.; Maenaka, K.; Hall, W.W.; Orba, Y.; Sawa, H. SARS-CoV-2 variants with mutations at the S1/S2 cleavage site are generated in vitro during propagation in TMPRSS2-deficient cells. *PLoS Pathog.* **2021**, *17*, e1009233. [CrossRef]

93. Bestle, D.; Heindl, M.R.; Limburg, H.; Van Lam van, T.; Pilgram, O.; Moulton, H.; Stein, D.A.; Hardes, K.; Eickmann, M.; Dolnik, O.; et al. TMPRSS2 and furin are both essential for proteolytic activation of SARS-CoV-2 in human airway cells. *Life Sci. Alliance* **2020**, *3*, e202000786. [CrossRef]
94. Peacock, T.P.; Goldhill, D.H.; Zhou, J.; Baillon, L.; Frise, R.; Swann, O.C.; Kugathasan, R.; Penn, R.; Brown, J.C.; Sanchez-David, R.Y.; et al. The furin cleavage site in the SARS-CoV-2 spike protein is required for transmission in ferrets. *Nat. Microbiol.* **2021**, *6*, 899–909. [CrossRef]
95. Sasaki, M.; Toba, S.; Itakura, Y.; Chambaro, H.M.; Kishimoto, M.; Tabata, K.; Intaruck, K.; Uemura, K.; Sanaki, T.; Sato, A.; et al. SARS-CoV-2 Bearing a Mutation at the S1/S2 Cleavage Site Exhibits Attenuated Virulence and Confers Protective Immunity. *mBio* **2021**, *12*, e0141521. [CrossRef]
96. Johnson, B.A.; Xie, X.; Bailey, A.L.; Kalveram, B.; Lokugamage, K.G.; Muruato, A.; Zou, J.; Zhang, X.; Juelich, T.; Smith, J.K.; et al. Loss of furin cleavage site attenuates SARS-CoV-2 pathogenesis. *Nature* **2021**, *591*, 293–299. [CrossRef]
97. Saito, A.; Irie, T.; Suzuki, R.; Maemura, T.; Nasser, H.; Uriu, K.; Kosugi, Y.; Shirakawa, K.; Sadamasu, K.; Kimura, I.; et al. Enhanced fusogenicity and pathogenicity of SARS-CoV-2 Delta P681R mutation. *Nature* **2022**, *602*, 300–306. [CrossRef]
98. Lubinski, B.; Fernandes, M.H.V.; Frazier, L.; Tang, T.; Daniel, S.; Diel, D.G.; Jaimes, J.A.; Whittaker, G.R. Functional evaluation of the P681H mutation on the proteolytic activation of the SARS-CoV-2 variant B.1.1.7 (Alpha) spike. *iScience* **2022**, *25*, 103589. [CrossRef]
99. Yang, X.-J. SARS-COV-2 γ variant acquires spike P681H or P681R for improved viral fitness. *bioRxiv* **2021**. [CrossRef]
100. Kircheis, R.; Schuster, M.; Planz, O. COVID-19: Mechanistic Model of the African Paradox Supports the Central Role of the NF-kappaB Pathway. *Viruses* **2021**, *13*, 1887. [CrossRef]
101. Iacobucci, G. Covid-19: Runny nose, headache, and fatigue are commonest symptoms of omicron, early data show. *Br. Med. J.* **2021**, *375*, n3103. [CrossRef]
102. Wang, L.; Berger, N.A.; Kaelber, D.C.; Davis, P.B.; Volkow, N.D.; Xu, R. Comparison of outcomes from COVID infection in pediatric and adult patients before and after the emergence of Omicron. *medRxiv* **2022**. [CrossRef]
103. Liao, M.; Liu, Y.; Yuan, J.; Wen, Y.; Xu, G.; Zhao, J.; Cheng, L.; Li, J.; Wang, X.; Wang, F.; et al. Single-cell landscape of bronchoalveolar immune cells in patients with COVID-19. *Nat. Med.* **2020**, *26*, 842–844. [CrossRef]
104. Khan, S.; Shafiei, M.S.; Longoria, C.; Schoggins, J.W.; Savani, R.C.; Zaki, H. SARS-CoV-2 spike protein induces inflammation via TLR2-dependent activation of the NF-kappaB pathway. *eLife* **2021**, *10*, e68563. [CrossRef]
105. Sohn, K.M.; Lee, S.G.; Kim, H.J.; Cheon, S.; Jeong, H.; Lee, J.; Kim, I.S.; Silwal, P.; Kim, Y.J.; Paik, S.; et al. COVID-19 Patients Upregulate Toll-like Receptor 4-mediated Inflammatory Signaling That Mimics Bacterial Sepsis. *J. Korean Med. Sci.* **2020**, *35*, e343. [CrossRef]
106. Patra, T.; Meyer, K.; Geerling, L.; Isbell, T.S.; Hoft, D.F.; Brien, J.; Pinto, A.K.; Ray, R.B.; Ray, R. SARS-CoV-2 spike protein promotes IL-6 trans-signaling by activation of angiotensin II receptor signaling in epithelial cells. *PLoS Pathog.* **2020**, *16*, e1009128. [CrossRef]
107. Hsu, A.C.Y.; Wang, G.; Reid, A.T.; Veerati, P.C.; Pathinayake, P.S.; Daly, K.; Mayall, J.R.; Hansbro, P.M.; Horvat, J.C.; Wang, F.; et al. SARS-CoV-2 Spike protein promotes hyper-inflammatory response that can be ameliorated by Spike-antagonistic peptide and FDA-approved ER stress and MAP kinase inhibitors *in vitro*. *bioRxiv* **2020**. [CrossRef]
108. Olajide, O.A.; Iwuanyanwu, V.U.; Lepiarz-Raba, I.; Al-Hindawi, A.A. Induction of Exaggerated Cytokine Production in Human Peripheral Blood Mononuclear Cells by a Recombinant SARS-CoV-2 Spike Glycoprotein S1 and Its Inhibition by Dexamethasone. *Inflammation* **2021**, *44*, 1865–1877. [CrossRef]
109. Petruk, G.; Puthia, M.; Petrlova, J.; Samsudin, F.; Stromdahl, A.C.; Cerps, S.; Uller, L.; Kjellstrom, S.; Bond, P.J.; Schmidtchen, A.A. SARS-CoV-2 spike protein binds to bacterial lipopolysaccharide and boosts proinflammatory activity. *J. Mol. Cell Biol.* **2020**, *12*, 916–932. [CrossRef]
110. Colunga Biancatelli, R.M.L.; Solopov, P.A.; Sharlow, E.R.; Lazo, J.S.; Marik, P.E.; Catravas, J.D. The SARS-CoV-2 spike protein subunit S1 induces COVID-19-like acute lung injury in Kappa18-hACE2 transgenic mice and barrier dysfunction in human endothelial cells. *Am. J. Physiol. Lung. Cell Mol. Physiol.* **2021**, *321*, L477–L484. [CrossRef]
111. Wang, W.; Ye, L.; Ye, L.; Li, B.; Gao, B.; Zeng, Y.; Kong, L.; Fang, X.; Zheng, H.; Wu, Z.; et al. Up-regulation of IL-6 and TNF-alpha induced by SARS-coronavirus spike protein in murine macrophages via NF-kappaB pathway. *Virus Res.* **2007**, *128*, 1–8. [CrossRef]
112. Nilsson-Payant, B.E.; Uhl, S.; Grimont, A.; Doane, A.S.; Cohen, P.; Patel, R.S.; Higgins, C.A.; Acklin, J.A.; Bram, Y.; Chandar, V.; et al. The NF-kappaB Transcriptional Footprint Is Essential for SARS-CoV-2 Replication. *J. Virol.* **2021**, *95*, e0125721. [CrossRef]
113. Ehrhardt, C.; Ruckle, A.; Hrincius, E.R.; Haasbach, E.; Anhlan, D.; Ahmann, K.; Banning, C.; Reiling, S.J.; Kuhn, J.; Strobl, S.; et al. The NF-kappaB inhibitor SC75741 efficiently blocks influenza virus propagation and confers a high barrier for development of viral resistance. *Cell. Microbiol.* **2013**, *15*, 1198–1211. [CrossRef]
114. Haasbach, E.; Reiling, S.J.; Ehrhardt, C.; Droebner, K.; Ruckle, A.; Hrincius, E.R.; Leban, J.; Strobl, S.; Vitt, D.; Ludwig, S.; et al. The NF-kappaB inhibitor SC75741 protects mice against highly pathogenic avian influenza A virus. *Antivir. Res.* **2013**, *99*, 336–344. [CrossRef]
115. Thomas, G. Furin at the cutting edge: From protein traffic to embryogenesis and disease. *Nat. Rev. Mol. Cell Biol.* **2002**, *3*, 753–766. [CrossRef]
116. Shapiro, J.; Sciaky, N.; Lee, J.; Bosshart, H.; Angeletti, R.H.; Bonifacino, J.S. Localization of endogenous furin in cultured cell lines. *J. Histochem. Cytochem.* **1997**, *45*, 3–12. [CrossRef]

117. Scudellari, M. How the coronavirus infects cells—And why Delta is so dangerous. *Nature* **2021**, *595*, 640–644. [CrossRef]
118. Bourne, G.L.; Grainger, D.J. Development and characterisation of an assay for furin activity. *J. Immunol. Methods* **2011**, *364*, 101–108. [CrossRef]
119. Uhlen, M.; Fagerberg, L.; Hallstrom, B.M.; Lindskog, C.; Oksvold, P.; Mardinoglu, A.; Sivertsson, A.; Kampf, C.; Sjostedt, E.; Asplund, A.; et al. Proteomics. Tissue-based map of the human proteome. *Science* **2015**, *347*, 1260419. [CrossRef]
120. Pesu, M.; Muul, L.; Kanno, Y.; O'Shea, J.J. Proprotein convertase furin is preferentially expressed in T helper 1 cells and regulates interferon gamma. *Blood* **2006**, *108*, 983–985. [CrossRef]
121. Oksanen, A.; Aittomaki, S.; Jankovic, D.; Ortutay, Z.; Pulkkinen, K.; Hamalainen, S.; Rokka, A.; Corthals, G.L.; Watford, W.T.; Junttila, I.; et al. Proprotein convertase FURIN constrains Th2 differentiation and is critical for host resistance against Toxoplasma gondii. *J. Immunol.* **2014**, *193*, 5470–5479. [CrossRef]
122. Murphy, T.L.; Cleveland, M.G.; Kulesza, P.; Magram, J.; Murphy, K.M. Regulation of interleukin 12 p40 expression through an NF-kappa B half-site. *Mol. Cell. Biol.* **1995**, *15*, 5258–5267. [CrossRef]
123. Liu, T.; Zhang, L.; Joo, D.; Sun, S.C. NF-kappaB signaling in inflammation. *Signal Transduct. Target. Ther.* **2017**, *2*, 17023.
124. Zhu, L.; Liu, H.W.; Yang, Y.; Hu, X.X.; Li, K.; Xu, S.; Li, J.B.; Ke, G.; Zhang, X.B. Near-Infrared Fluorescent Furin Probe for Revealing the Role of Furin in Cellular Carcinogenesis and Specific Cancer Imaging. *Anal. Chem.* **2019**, *91*, 9682–9689. [CrossRef]
125. Kumar, V.; Behera, R.; Lohite, K.; Karnik, S.; Kundu, G.C. p38 kinase is crucial for osteopontin-induced furin expression that supports cervical cancer progression. *Cancer Res.* **2010**, *70*, 10381–10391. [CrossRef]
126. Adu-Agyeiwaah, Y.; Grant, M.B.; Obukhov, A.G. The Potential Role of Osteopontin and Furin in Worsening Disease Outcomes in COVID-19 Patients with Pre-Existing Diabetes. *Cells* **2020**, *9*, 2528. [CrossRef]
127. McMahon, S.; Grondin, F.; McDonald, P.P.; Richard, D.E.; Dubois, C.M. Hypoxia-enhanced expression of the proprotein convertase furin is mediated by hypoxia-inducible factor-1: Impact on the bioactivation of proproteins. *J. Biol. Chem.* **2005**, *280*, 6561–6569. [CrossRef]
128. Taylor, C.T.; Cummins, E.P. The role of NF-kappaB in hypoxia-induced gene expression. *Ann. N. Y. Acad Sci.* **2009**, *1177*, 178–184. [CrossRef]
129. Han, S.; Xu, W.; Wang, Z.; Qi, X.; Wang, Y.; Ni, Y.; Shen, H.; Hu, Q.; Han, W. Crosstalk between the HIF-1 and Toll-like receptor/nuclear factor-kappaB pathways in the oral squamous cell carcinoma microenvironment. *Oncotarget* **2016**, *7*, 37773–37789. [CrossRef]
130. Riddell, J.R.; Maier, P.; Sass, S.N.; Moser, M.T.; Foster, B.A.; Gollnick, S.O. Peroxiredoxin 1 stimulates endothelial cell expression of VEGF via TLR4 dependent activation of HIF-1alpha. *PLoS ONE* **2012**, *7*, e50394. [CrossRef]
131. Spirig, R.; Djafarzadeh, S.; Regueira, T.; Shaw, S.G.; von Garnier, C.; Takala, J.; Jakob, S.M.; Rieben, R.; Lepper, P.M. Effects of TLR agonists on the hypoxia-regulated transcription factor HIF-1alpha and dendritic cell maturation under normoxic conditions. *PLoS ONE* **2010**, *5*, e0010983. [CrossRef]
132. Peyssonnaux, C.; Cejudo-Martin, P.; Doedens, A.; Zinkernagel, A.S.; Johnson, R.S.; Nizet, V. Cutting edge: Essential role of hypoxia inducible factor-1alpha in development of lipopolysaccharide-induced sepsis. *J. Immunol.* **2007**, *178*, 7516–7519. [CrossRef]
133. Jiang, H.; Hu, R.; Sun, L.; Chai, D.; Cao, Z.; Li, Q. Critical role of Toll-like receptor 4 in hypoxia-inducible factor 1alpha activation during trauma/hemorrhagic shocky induced acute lung injury after lymph infusion in mice. *Shock* **2014**, *42*, 271–278. [CrossRef]
134. Yang, X.; Chen, G.T.; Wang, Y.Q.; Xian, S.; Zhang, L.; Zhu, S.M.; Pan, F.; Cheng, Y.X. TLR4 promotes the expression of HIF-1alpha by triggering reactive oxygen species in cervical cancer cells in vitro-implications for therapeutic intervention. *Mol. Med. Rep.* **2018**, *17*, 2229–2238.
135. Braun, E.; Sauter, D. Furin-mediated protein processing in infectious diseases and cancer. *Clin. Transl. Immunol.* **2019**, *8*, e1073. [CrossRef]
136. Tse, L.V.; Hamilton, A.M.; Friling, T.; Whittaker, G.R. A novel activation mechanism of avian influenza virus H9N2 by furin. *J. Virol.* **2014**, *88*, 1673–1683. [CrossRef]
137. De Greef, J.C.; Slutter, B.; Anderson, M.E.; Hamlyn, R.; O'Campo Landa, R.; McNutt, E.J.; Hara, Y.; Pewe, L.L.; Venzke, D.; Matsumura, K.; et al. Protective role for the N-terminal domain of alpha-dystroglycan in Influenza A virus proliferation. *Proc. Natl. Acad. Sci. USA* **2019**, *116*, 11396–11401. [CrossRef]
138. Li, D.; Liu, X.; Zhang, L.; He, J.; Chen, X.; Liu, S.; Fu, J.; Fu, S.; Chen, H.; Fu, J.; et al. COVID-19 disease and malignant cancers: The impact for the furin gene expression in susceptibility to SARS-CoV-2. *Int. J. Biol. Sci.* **2021**, *17*, 3954–3967. [CrossRef]
139. Kocyigit, A.; Sogut, O.; Durmus, E.; Kanimdan, E.; Guler, E.M.; Kaplan, O.; Yenigun, V.B.; Eren, C.; Ozman, Z.; Yasar, O. Circulating furin, IL-6, and presepsin levels and disease severity in SARS-CoV-2-infected patients. *Sci. Prog.* **2021**, *104*, 368504211026119. [CrossRef]
140. Kermani, N.Z.; Song, W.J.; Badi, Y.; Versi, A.; Guo, Y.; Sun, K.; Bhavsar, P.; Howarth, P.; Dahlen, S.E.; Sterk, P.J.; et al. Sputum ACE2, TMPRSS2 and FURIN gene expression in severe neutrophilic asthma. *Respir. Res.* **2021**, *22*, 10. [CrossRef]
141. Brake, S.J.; Eapen, M.S.; McAlinden, K.D.; Markos, J.; Haug, G.; Larby, J.; Chia, C.; Hardikar, A.; Singhera, G.K.; Hackett, T.L.; et al. SARS-CoV-2 (COVID-19) Adhesion Site Protein Upregulation in Small Airways, Type 2 Pneumocytes, and Alveolar Macrophages of Smokers and COPD—Possible Implications for Interstitial Fibrosis. *Int. J. Chron. Obstruct. Pulmon. Dis.* **2022**, *17*, 101–115. [CrossRef]
142. Zalinger, Z.B.; Elliott, R.; Rose, K.M.; Weiss, S.R. MDA5 Is Critical to Host Defense during Infection with Murine Coronavirus. *J. Virol.* **2015**, *89*, 12330–12340. [CrossRef]

143. Cervantes-Barragan, L.; Zust, R.; Weber, F.; Spiegel, M.; Lang, K.S.; Akira, S.; Thiel, V.; Ludewig, B. Control of coronavirus infection through plasmacytoid dendritic-cell-derived type I interferon. *Blood* **2007**, *109*, 1131–1137. [CrossRef]
144. Channappanavar, R.; Fehr, A.R.; Zheng, J.; Wohlford-Lenane, C.; Abrahante, J.E.; Mack, M.; Sompallae, R.; McCray, P.B., Jr.; Meyerholz, D.K.; Perlman, S. IFN-I response timing relative to virus replication determines MERS coronavirus infection outcomes. *J. Clin. Investig.* **2019**, *129*, 3625–3639. [CrossRef]
145. Zheng, M.; Williams, E.P.; Malireddi, R.K.S.; Karki, R.; Banoth, B.; Burton, A.; Webby, R.; Channappanavar, R.; Jonsson, C.B.; Kanneganti, T.D. Impaired NLRP3 inflammasome activation/pyroptosis leads to robust inflammatory cell death via caspase-8/RIPK3 during coronavirus infection. *J. Biol. Chem.* **2020**, *295*, 14040–14052. [CrossRef]
146. Sheahan, T.; Morrison, T.E.; Funkhouser, W.; Uematsu, S.; Akira, S.; Baric, R.S.; Heise, M.T. MyD88 is required for protection from lethal infection with a mouse-adapted SARS-CoV. *PLoS Pathog.* **2008**, *4*, e1000240. [CrossRef]
147. Hadjadj, J.; Yatim, N.; Barnabei, L.; Corneau, A.; Boussier, J.; Smith, N.; Pere, H.; Charbit, B.; Bondet, V.; Chenevier-Gobeaux, C.; et al. Impaired type I interferon activity and inflammatory responses in severe COVID-19 patients. *Science* **2020**, *369*, 718–724. [CrossRef]
148. El-Zayat, S.R.; Sibaii, H.; Mannaa, F.A. Toll-like receptors activation, signaling, and targeting: An overview. *Bull. Natl. Res. Cent.* **2019**, *43*, 187. [CrossRef]
149. Zheng, M.; Karki, R.; Williams, E.P.; Yang, D.; Fitzpatrick, E.; Vogel, P.; Jonsson, C.B.; Kanneganti, T.D. TLR2 senses the SARS-CoV-2 envelope protein to produce inflammatory cytokines. *Nat. Immunol.* **2021**, *22*, 829–838. [CrossRef]
150. Planès, R.; Bert, J.-B.; Tairi, S.; Benmohamed, L.; Bahraoui, E. SARS-CoV-2 Envelope protein (E) binds and activates TLR2: A novel target for COVID-19 interventions. *bioRxiv* **2021**. [CrossRef]
151. Aboudounya, M.M.; Heads, R.J. COVID-19 and Toll-Like Receptor 4 (TLR4): SARS-CoV-2 May Bind and Activate TLR4 to Increase ACE2 Expression, Facilitating Entry and Causing Hyperinflammation. *Mediat. Inflamm.* **2021**, *2021*, 8874339. [CrossRef]
152. Mogensen, T.H.; Paludan, S.R. Reading the viral signature by Toll-like receptors and other pattern recognition receptors. *J. Mol. Med.* **2005**, *83*, 180–192. [CrossRef] [PubMed]
153. Turner, N.A. Inflammatory and fibrotic responses of cardiac fibroblasts to myocardial damage associated molecular patterns (DAMPs). *J. Mol. Cell. Cardiol.* **2016**, *94*, 189–200. [CrossRef] [PubMed]
154. Bhattacharyya, S.; Wang, W.; Qin, W.; Cheng, K.; Coulup, S.; Chavez, S.; Jiang, S.; Raparia, K.; De Almeida, L.M.V.; Stehlik, C.; et al. TLR4-dependent fibroblast activation drives persistent organ fibrosis in skin and lung. *JCI Insight* **2018**, *3*, e98850. [CrossRef]
155. Chow, J.C.; Young, D.W.; Golenbock, D.T.; Christ, W.J.; Gusovsky, F. Toll-like receptor-4 mediates lipopolysaccharide-induced signal transduction. *J. Biol. Chem.* **1999**, *274*, 10689–10692. [CrossRef]
156. Kuzmich, N.N.; Sivak, K.V.; Chubarev, V.N.; Porozov, Y.B.; Savateeva-Lyubimova, T.N.; Peri, F. TLR4 Signaling Pathway Modulators as Potential Therapeutics in Inflammation and Sepsis. *Vaccines* **2017**, *5*, 34. [CrossRef]
157. Molteni, M.; Gemma, S.; Rossetti, C. The Role of Toll-Like Receptor 4 in Infectious and Noninfectious Inflammation. *Mediat. Inflamm.* **2016**, *2016*, 6978936. [CrossRef] [PubMed]
158. Roshan, M.H.; Tambo, A.; Pace, N.P. The Role of TLR2, TLR4, and TLR9 in the Pathogenesis of Atherosclerosis. *Int. J. Inflam.* **2016**, *2016*, 1532832. [CrossRef] [PubMed]
159. Patra, M.C.; Shah, M.; Choi, S. Toll-like receptor-induced cytokines as immunotherapeutic targets in cancers and autoimmune diseases. *Semin. Cancer Biol.* **2020**, *64*, 61–82. [CrossRef]
160. Garcia Bueno, B.; Caso, J.R.; Madrigal, J.L.; Leza, J.C. Innate immune receptor Toll-like receptor 4 signalling in neuropsychiatric diseases. *Neurosci. Biobehav. Rev.* **2016**, *64*, 134–147. [CrossRef]
161. Mukherjee, S.; Karmakar, S.; Babu, S.P. TLR2 and TLR4 mediated host immune responses in major infectious diseases: A review. *Braz. J. Infect. Dis.* **2016**, *20*, 193–204. [CrossRef]
162. Choudhury, A.; Mukherjee, S. In silico studies on the comparative characterization of the interactions of SARS-CoV-2 spike glycoprotein with ACE-2 receptor homologs and human TLRs. *J. Med. Virol.* **2020**, *92*, 2105–2113. [CrossRef] [PubMed]
163. Zhao, Y.; Kuang, M.; Li, J.; Zhu, L.; Jia, Z.; Guo, X.; Hu, Y.; Kong, J.; Yin, H.; Wang, X.; et al. SARS-CoV-2 spike protein interacts with and activates TLR41. *Cell Res.* **2021**, *31*, 818–820. [CrossRef]
164. Shirato, K.; Kizaki, T. SARS-CoV-2 spike protein S1 subunit induces pro-inflammatory responses via toll-like receptor 4 signaling in murine and human macrophages. *Heliyon* **2021**, *7*, e06187. [CrossRef] [PubMed]
165. Aboudounya, M.M.; Holt, M.R.; Heads, R.J. SARS-CoV-2 Spike S1 glycoprotein is a TLR4 agonist, upregulates ACE2 expression and induces pro-inflammatory M_1 macrophage polarisation. *bioRxiv* **2021**. [CrossRef]
166. Russell, C.D.; Unger, S.A.; Walton, M.; Schwarze, J. The Human Immune Response to Respiratory Syncytial Virus Infection. *Clin. Microbiol. Rev.* **2017**, *30*, 481–502. [CrossRef]
167. Guo, X.J.; Thomas, P.G. New fronts emerge in the influenza cytokine storm. *Semin. Immunopathol.* **2017**, *39*, 541–550. [CrossRef]
168. Bixler, S.L.; Goff, A.J. The Role of Cytokines and Chemokines in Filovirus Infection. *Viruses* **2015**, *7*, 5489–5507. [CrossRef]
169. Srikiatkhachorn, A.; Mathew, A.; Rothman, A.L. Immune-mediated cytokine storm and its role in severe dengue. *Semin. Immunopathol.* **2017**, *39*, 563–574. [CrossRef]
170. Channappanavar, R.; Perlman, S. Pathogenic human coronavirus infections: Causes and consequences of cytokine storm and immunopathology. *Semin. Immunopathol.* **2017**, *39*, 529–539. [CrossRef]

171. Rallabhandi, P.; Phillips, R.L.; Boukhvalova, M.S.; Pletneva, L.M.; Shirey, K.A.; Gioannini, T.L.; Weiss, J.P.; Chow, J.C.; Hawkins, L.D.; Vogel, S.N.; et al. Respiratory syncytial virus fusion protein-induced toll-like receptor 4 (TLR4) signaling is inhibited by the TLR4 antagonists Rhodobacter sphaeroides lipopolysaccharide and eritoran (E5564) and requires direct interaction with MD-2. *mBio* **2012**, *3*, e00218-12. [CrossRef]
172. Lai, C.Y.; Strange, D.P.; Wong, T.A.S.; Lehrer, A.T.; Verma, S. Ebola Virus Glycoprotein Induces an Innate Immune Response In vivo via TLR4. *Front. Microbiol.* **2017**, *8*, 1571. [CrossRef] [PubMed]
173. Okumura, A.; Pitha, P.M.; Yoshimura, A.; Harty, R.N. Interaction between Ebola virus glycoprotein and host toll-like receptor 4 leads to induction of proinflammatory cytokines and SOCS1. *J. Virol.* **2010**, *84*, 27–33. [CrossRef] [PubMed]
174. Harrison, S.C. Viral membrane fusion. *Virology* **2015**, *479–480*, 498–507. [CrossRef] [PubMed]
175. Modhiran, N.; Watterson, D.; Muller, D.A.; Panetta, A.K.; Sester, D.P.; Liu, L.; Hume, D.A.; Stacey, K.J.; Young, P.R. Dengue virus NS1 protein activates cells via Toll-like receptor 4 and disrupts endothelial cell monolayer integrity. *Sci. Transl. Med.* **2015**, *7*, 304ra142. [CrossRef]
176. Akey, D.L.; Brown, W.C.; Dutta, S.; Konwerski, J.; Jose, J.; Jurkiw, T.J.; DelProposto, J.; Ogata, C.M.; Skiniotis, G.; Kuhn, R.J.; et al. Flavivirus NS1 structures reveal surfaces for associations with membranes and the immune system. *Science* **2014**, *343*, 881–885. [CrossRef]
177. Iampietro, M.; Younan, P.; Nishida, A.; Dutta, M.; Lubaki, N.M.; Santos, R.I.; Koup, R.A.; Katze, M.G.; Bukreyev, A. Ebola virus glycoprotein directly triggers T lymphocyte death despite of the lack of infection. *PLoS Pathog.* **2017**, *13*, e1006397. [CrossRef]
178. Escudero-Perez, B.; Volchkova, V.A.; Dolnik, O.; Lawrence, P.; Volchkov, V.E. Shed GP of Ebola virus triggers immune activation and increased vascular permeability. *PLoS Pathog* **2014**, *10*, e1004509. [CrossRef]
179. Olejnik, J.; Forero, A.; Deflube, L.R.; Hume, A.J.; Manhart, W.A.; Nishida, A.; Marzi, A.; Katze, M.G.; Ebihara, H.; Rasmussen, A.L.; et al. Ebolaviruses Associated with Differential Pathogenicity Induce Distinct Host Responses in Human Macrophages. *J. Virol.* **2017**, *91*, e00179-17. [CrossRef]
180. Younan, P.; Ramanathan, P.; Graber, J.; Gusovsky, F.; Bukreyev, A. The Toll-Like Receptor 4 Antagonist Eritoran Protects Mice from Lethal Filovirus Challenge. *mBio* **2017**, *8*, e00226-17. [CrossRef]
181. Modhiran, N.; Watterson, D.; Blumenthal, A.; Baxter, A.G.; Young, P.R.; Stacey, K.J. Dengue virus NS1 protein activates immune cells via TLR4 but not TLR2 or TLR6. *Immunol. Cell Biol.* **2017**, *95*, 491–495. [CrossRef]
182. Olejnik, J.; Hume, A.J.; Muhlberger, E. Toll-like receptor 4 in acute viral infection: Too much of a good thing. *PLoS Pathog.* **2018**, *14*, e1007390. [CrossRef] [PubMed]
183. Park, B.S.; Lee, J.O. Recognition of lipopolysaccharide pattern by TLR4 complexes. *Exp. Mol. Med.* **2013**, *45*, e66. [CrossRef] [PubMed]
184. Ohto, U.; Fukase, K.; Miyake, K.; Shimizu, T. Structural basis of species-specific endotoxin sensing by innate immune receptor TLR4/MD-2. *Proc. Natl. Acad. Sci. USA* **2012**, *109*, 7421–7426. [CrossRef] [PubMed]
185. Schromm, A.B.; Brandenburg, K.; Loppnow, H.; Zahringer, U.; Rietschel, E.T.; Carroll, S.F.; Koch, M.H.; Kusumoto, S.; Seydel, U. The charge of endotoxin molecules influences their conformation and IL-6-inducing capacity. *J. Immunol.* **1998**, *161*, 5464–5471. [PubMed]
186. Schromm, A.B.; Brandenburg, K.; Loppnow, H.; Moran, A.P.; Koch, M.H.; Rietschel, E.T.; Seydel, U. Biological activities of lipopolysaccharides are determined by the shape of their lipid A portion. *Eur. J. Biochem.* **2000**, *267*, 2008–2013. [CrossRef]
187. Du, X.; Tang, H.; Gao, L.; Wu, Z.; Meng, F.; Yan, R.; Qiao, S.; An, J.; Wang, C.; Qin, F.X. Omicron adopts a different strategy from Delta and other variants to adapt to host. *Signal Transduct. Target. Ther.* **2022**, *7*, 45. [CrossRef]
188. Zhao, M.M.; Yang, W.L.; Yang, F.Y.; Zhang, L.; Huang, W.J.; Hou, W.; Fan, C.F.; Jin, R.H.; Feng, Y.M.; Wang, Y.C.; et al. Cathepsin L plays a key role in SARS-CoV-2 infection in humans and humanized mice and is a promising target for new drug development. *Signal Transduct. Target. Ther.* **2021**, *6*, 134. [CrossRef]
189. Blanco-Melo, D.; Nilsson-Payant, B.E.; Liu, W.C.; Uhl, S.; Hoagland, D.; Moller, R.; Jordan, T.X.; Oishi, K.; Panis, M.; Sachs, D.; et al. Imbalanced Host Response to SARS-CoV-2 Drives Development of COVID-19. *Cell* **2020**, *181*, 1036–1045.e9. [CrossRef]
190. Konno, Y.; Kimura, I.; Uriu, K.; Fukushi, M.; Irie, T.; Koyanagi, Y.; Sauter, D.; Gifford, R.J.; Consortium, U.-C.; Nakagawa, S.; et al. SARS-CoV-2 ORF3b Is a Potent Interferon Antagonist Whose Activity Is Increased by a Naturally Occurring Elongation Variant. *Cell Rep.* **2020**, *32*, 108185. [CrossRef]
191. Lei, X.; Dong, X.; Ma, R.; Wang, W.; Xiao, X.; Tian, Z.; Wang, C.; Wang, Y.; Li, L.; Ren, L.; et al. Activation and evasion of type I interferon responses by SARS-CoV-2. *Nat. Commun.* **2020**, *11*, 3810. [CrossRef]
192. Park, M.D. Immune evasion via SARS-CoV-2 ORF8 protein? *Nat. Rev. Immunol.* **2020**, *20*, 408. [CrossRef] [PubMed]
193. Yang, Z.; Zhang, X.; Wang, F.; Wang, P.; Kuang, E.; Li, X. Suppression of MDA5-mediated antiviral immune responses by NSP8 of SARS-CoV-2. *bioRxiv* **2020**. [CrossRef]

International Journal of
Molecular Sciences

MDPI

Review

Toll-like Receptor Mediation in SARS-CoV-2: A Therapeutic Approach

Abdul Manan [1], Rameez Hassan Pirzada [1], Muhammad Haseeb [1,2] and Sangdun Choi [1,2,*]

1 Department of Molecular Science and Technology, Ajou University, Suwon 16499, Korea
2 S&K Therapeutics, Ajou University Campus Plaza 418, 199 Worldcup-ro, Yeongtong-gu, Suwon 16502, Korea
* Correspondence: sangdunchoi@ajou.ac.kr

Abstract: The innate immune system facilitates defense mechanisms against pathogen invasion and cell damage. Toll-like receptors (TLRs) assist in the activation of the innate immune system by binding to pathogenic ligands. This leads to the generation of intracellular signaling cascades including the biosynthesis of molecular mediators. TLRs on cell membranes are adept at recognizing viral components. Viruses can modulate the innate immune response with the help of proteins and RNAs that downregulate or upregulate the expression of various TLRs. In the case of COVID-19, molecular modulators such as type 1 interferons interfere with signaling pathways in the host cells, leading to an inflammatory response. Coronaviruses are responsible for an enhanced immune signature of inflammatory chemokines and cytokines. TLRs have been employed as therapeutic agents in viral infections as numerous antiviral Food and Drug Administration-approved drugs are TLR agonists. This review highlights the therapeutic approaches associated with SARS-CoV-2 and the TLRs involved in COVID-19 infection.

Keywords: TLR; immune system; inflammation; antiviral; SARS-CoV-2

Citation: Manan, A.; Pirzada, R.H.; Haseeb, M.; Choi, S. Toll-like Receptor Mediation in SARS-CoV-2: A Therapeutic Approach. *Int. J. Mol. Sci.* **2022**, *23*, 10716. https://doi.org/10.3390/ijms231810716

Academic Editors: Ralf Kircheis and Oliver Planz

Received: 20 August 2022
Accepted: 10 September 2022
Published: 14 September 2022

1. Introduction

Toll-like receptors (TLRs) are central mediators of the innate and adaptive immune responses. The immune system exhibits a defense mechanism for the host against pathogenic materials (exogenous and/or endogenous) at the cellular level [1]. Pattern recognition receptors (PRRs) including DNA sensors, RIG-1-like receptors, and TLRs are part of the innate immune system that protects against microbial infection. PRRs recognize conserved pathogen-associated molecular patterns (PAMPs) from microbes and endogenous danger-associated molecular patterns (DAMPs) produced by necrotic cells [2]. PAMPs are derived from viral, bacterial, parasitic, and fungal pathogens. The chemical nature of PAMPs recognized by TLRs varies greatly among organisms. In phylogenetics, TLRs are considered the most ancient class of PRRs. A large number of TLRs have been reported across a wide range of vertebrate and invertebrate species. The signaling pathways and adaptor proteins related to TLRs are evolutionary conserved, from Porifera to mammals. Moreover, similar domain patterns can be observed in most TLR homologs [3,4].

Viruses are responsible for initiating innate immunity through TLRs. Viruses, via a combination of small and unique proteins, not only escape the innate immune system but also destabilize the paybacks of the virus [5]. Similar to other pathogens, viruses are sensed by TLRs. Some viruses encode unique proteins that target TLR signaling. The hepatitis C virus encodes proteins that inhibit TLR-mediated signaling such as NS5A and protease NS3/4A [6,7], which inhibits MyD88 and cleaves TIR-domain-containing adapter-inducing interferon-β (TRIF), respectively. Moreover, the two vaccinia virus proteins have been reported as inhibitors of the TLR system; for example, A52R was observed to inhibit TLR-mediated NF-κB activation by targeting IRAK2 [8], whereas A46R exhibited a connection with TLR signaling downregulation by employing Toll-interleukin-1 receptor (TIR) domain-containing adaptors [9]. Intracellular TLRs not only sense viral and bacterial nucleic acids, but also identify self-nucleic acids in cellular abnormalities such as autoimmunity [10].

A novel single-stranded RNA (ssRNA)-containing virus causes coronavirus disease (COVID-19), also referred to as severe acute respiratory syndrome coronavirus 2 (SARS-CoV-2), which became a pandemic after the first case was identified in Wuhan, China in December 2019. With the spread of COVID-19, the pandemic poses a global challenge [11,12]. From a clinical point of view, the virus has various manifestations ranging from patients becoming critically ill with acute respiratory distress syndrome to asymptomatic infection. In the intensive care unit, multiorgan support therapy has been essential in almost every case of COVID-19 (Figure 1). The critical disease stage is typically observed at 7–10 days of clinical infection [11,13]. Hyperinflammatory outcomes (cytokine storm) are mainly associated with clinical impediments and mortality [14]. A possible treatment methodology in the form of vaccines is being employed for the prevention of SARS-CoV-2 infection, but there is no operative therapeutic treatment option available. Consequently, exploring new drug targets is necessary. One of the most important molecular targets is TLRs. The interaction of the SARS-CoV-2 spike glycoprotein with TLR and the enhanced expression of genes associated with TLR signaling could indicate the possible involvement of these tiny molecular machines and their inflammatory cascades [14].

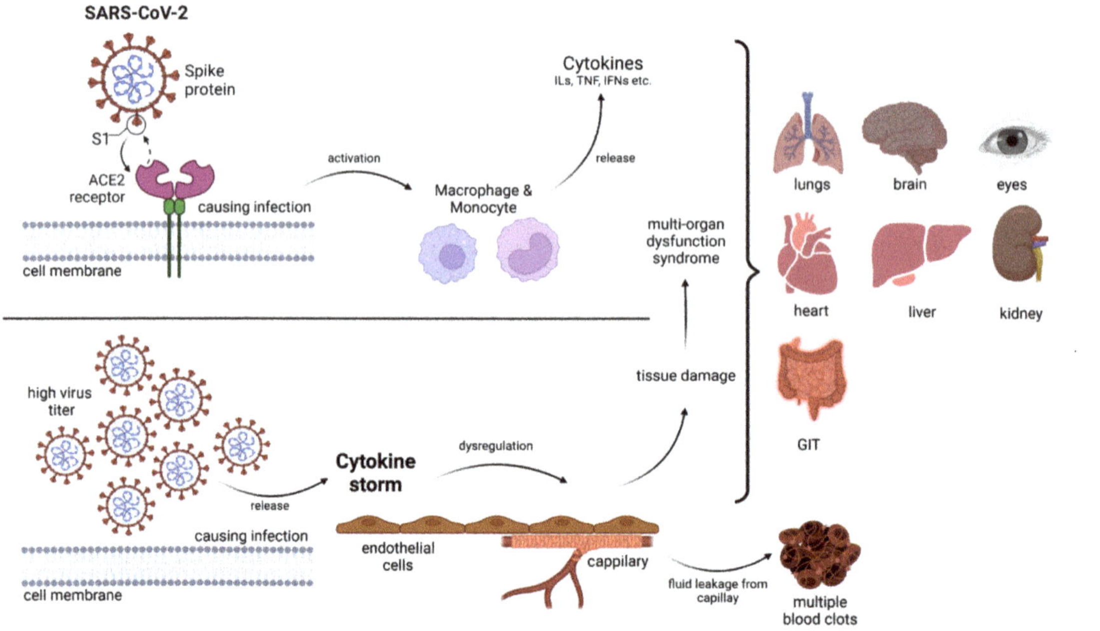

Figure 1. An overview of the SARS-CoV-2 infection pathway. During viral infection, immune cells are activated and release several cytokines as required for the biological system. A high virus titer is associated with a cytokine storm, and such dysregulation in the body of the patient may lead to multi-organ dysfunction syndrome. GIT—gastrointestinal tract.

Structurally, TLRs are type I transmembrane (TM) proteins with three distinct domains including an extracellular domain (ectodomain) that contains tandem copies of leucine-rich repeats, a single-pass TM as well as a cytoplasmic TIR downstream-signaling domain. TLRs experience either homodimerization or heterodimerization when encountering PAMPs and/or DAMPs, and adaptor proteins are employed; subsequently, a complex cellular event of downstream signal transduction is initiated, leading to the expression of inflammatory cytokines and interferons (IFN) that is observable at the molecular level [2]. The underlying TLR signaling cascades have been elucidated using structural, genetic, biochemical, and in silico methodologies [15].

Downstream signaling is made possible by the presence of cytosolic TIR domain-adaptor proteins such as TRIF (also known as TICAM1), TRAM (TICAM2), MyD88, and MAL [5,16]. The involvement of TLRs with TIR adaptors leads to the activation of cytosolic signaling complexes including IRAK and TRAF proteins. These entities are responsible for the activation of transcription factors such as IRF and NF-κB. This executes the synthesis of type I IFNs and proinflammatory cytokines [16]. IRF7 is essential for IFN-α synthesis, NF-κB is necessary for TNF and IL-6 induction, and IRF3 and NF-κB are required for IFN-β production [17].

By neutralizing internal and/or external threats administered by TLRs, the innate immune system makes defensive contributions to the survival of the host biological system. However, dysregulation and/or overactivation of this system leads to various disorders such as inflammation, cancer, and autoimmunity [18–20].

2. Structure of Coronavirus

SARS-CoV-2, a member of the β-coronavirus genus in the family Coronaviridae, has an envelope and positive-sense ssRNA genome of 29,891 nucleotides, encoding circular nucleocapsid proteins with 9860 amino acid residues [21]. The viral particle size ranges from 80 to 220 nm. Overall, 10 open reading frames (ORFs) have been identified in its genome to date (approximately 26–32 kb). The first ORF (almost 2/3 of the viral RNA) encodes polyproteins 1a (ORF1a) and 1b (ORF1b) [22]. Furthermore, these ORFs are cleaved by proteases into 16 nonstructural proteins (NSPs) that are responsible for genome replication and transcription [23]. Structural proteins (SPs) are encoded by the remaining ORFs [24,25]. The main SPs and NSPs of SARS-CoV-2 are summarized in Tables 1 and 2, respectively. The name coronavirus is derived from the appearance under the electron microscope, in which the presence of crown-like spikes on the envelope resembles the corona of the sun [26]. SPs form the viral envelope that holds the RNA genome, while NSPs are expressed in host-infected cells but are not incorporated into virion infectious particles. These NSPs include various transcription factors and enzymes such as RNA-dependent RNA polymerase (RdRp) and hemagglutinin esterase (HE). Moreover, the virion employs enzymes such as RNA replicases and viral proteases to replicate itself [22,27–29].

Various SPs have been identified including the glycoprotein membrane (M), spike (S), small envelope (E), and nucleoprotein (N), and other accessory proteins. M-glycoprotein is the most abundant, spanning the membrane bilayer thrice [30]. S-glycoprotein (150 kDa) is a type-I TM protein on the outer surface of the virus and is responsible for the binding of the virus to host cell receptors (ACE2). The S protein amino acid sequence of SARS-CoV-2 exhibits 86% similarity to that of SARS-CoV [31]. The S protein consists of oligosaccharides bound to serine amino acids through *o*-glycosides. The three major segments of S protein are the ectodomain, TM, and intracellular regions. The intracellular domain comprises the membrane fusion subunit S2 (trimeric stalk) as well as a short tail part known as the receptor-binding S1 domain (RBD; three S1 heads) [32,33]. Protein–protein interaction (PPI) between the human ACE2 and SARS-CoV-2 S protein facilitates viral attachment as well as the cellular entry of coronaviruses; thus, small-molecule blockage of these PPIs is a more inspiring therapeutic approach than inhibition via antibodies [34]. The S1 subunit of the S protein enables ACE2-mediated virus attachment, whereas the S2 subunit facilitates membrane fusion. Specifically, asparagine, glutamine, serine, phenylalanine, and leucine residues present in the S protein boost ACE2 binding [35].

Moreover, N protein bound to nucleic acids is an important structural component of the virus, which is responsible for viral replication and cellular response to infection in the host cellular machinery [31] (Table 1). The N protein comprises a serine-rich linker region sandwiched between the N-terminal domain (NTD) and the C-terminal domain

(CTD). These termini are crucial for viral entry and processing in host cells. The CTD regulates nucleocapsid formation and the NTD adheres to the viral genome in the form of orthorhombic crystals. Phosphorylation sites are also present in the linker region, which control its function [35]. In the case of SARS-CoV, the N protein enhances the activation of cyclooxygenase-2 (COX-2), resulting in the inflammation of pulmonary cells [36]. Moreover, the N protein interacts with the p42 proteasome subunit, which degrades the virion [37]. This also disables type-I IFN, which is responsible for suppressing the host immune responses produced by biological systems against viral infections [38]. The interaction of the N protein with heterogeneous nuclear ribonucleoproteins leads to increased viral RNA synthesis [39]. The N protein sequence of SARS-CoV-2 shows a 94.3% similarity to that of the SARS-CoV [31].

The smallest TM structural protein in coronaviruses is the E protein (Table 1), which comprises two different domains: the NTD (1–9 residues) as well as a hydrophobic domain (10–37 residues), with a chain at the terminal (38–76 residues) [40–42]. The E protein plays a crucial biological role, not only in the structural integrity of the virus, but also in host virulence [43]. The E protein sequence of SARS-CoV-2 shows a 96.1% similarity to that of SARS-CoV [31].

The M protein plays a crucial role in maintaining the shape of the viral envelope (Table 1). This function can be achieved by interacting with other viral proteins that exhibit PPIs [44]. The M protein is also known as the central organization of coronavirus proteins. The binding of E to M produces the virus envelope, and this interaction is sufficient for the synthesis and release of viruses [45,46]. The binding of M with S is an important event for the retention of the S protein in the endoplasmic reticulum–Golgi complex as well as its integration into new viruses [46,47]. Moreover, the interaction of N with M stabilizes the nucleocapsid (RNA–N protein complex) and the internal core of viruses, resulting in the completion of viral assembly [47,48]. The M protein amino acid sequence of SARS-CoV-2 exhibits a 96.4% similarity with that of SARS-CoV [31].

Table 1. The structural proteins (SPs) of coronaviruses and their physiological significance.

Sr. No.	SPs	PDB ID	Residues	Physiological Significance	Reference
1	E	7K3G	76–109	Virus assembly, morphogenesis, viral–host interaction, membrane permeability	[49]
2	M	8CTK	220–260	Virus assembly, protein interactions (M–M, M–S, M–N)	[50]
3	N	6VY0, 6YUN	422	Abundant RNA-binding protein, virion genome packaging	[51]
4	S	6VYB	1273	Main antigen component, triggers the host immune response	[52]

Table 2. The non-structural proteins (NSPs) of coronaviruses and their physiological significance.

Sr. No.	NSPs	PDB ID	Residues	Physiological Significance	Reference
1	NSP1	7K3N	180	Protein synthesis, prevents antiviral activity of host cells, degrades host mRNA	[53–55]
2	NSP2	7MSW	638	Genome replication, disruption of intracellular host signaling	[56–58]
3	NSP3 (Papain-like protease, PL_{pro})	7KAG, 6WEY, 6WUU, 7LG0	1945	Integral to viral replication, post-translational processing of the two polyproteins, suppresses host protein synthesis	[22,58,59]
4	NSP4	3GZF	500	Protects new replicated virions, replication and assembly of viral structures in host cell	[60,61]

Table 2. *Cont.*

Sr. No.	NSPs	PDB ID	Residues	Physiological Significance	Reference
5	NSP5 (3C-like protease, $3CL_{pro}$)	6LU7	306	Protein cleavage capacity (conserved feature)	[62,63]
6	NSP6	-	290	Induction of autophagosomes, inhibition of viral components to reach host lysosomes	[64–66]
7	NSP7	7JLT	83	Primase complex (NSP7-NSP8), hetero-oligomeric complex (NSP7-NSP8-RdRp), viral replication	[67–69]
8	NSP8	7JLT	198	Primase complex (NSP7-NSP8), hetero-oligomeric complex (NSP7-NSP8-RdRp), viral replication	[67–69]
9	NSP9	6WXD	113	RNA synthesis, carries viral RNA to the host cell, responsible for proliferation	[70–72]
10	NSP10	6ZPE	139	Cofactor activation for replicative enzymes, complex NSP10-NSP14, viral RNA proofreading	[73–75]
11	NSP11	-	13	Cleavage product of PP1a by $3CL_{pro}/M^{Pro}$	[21,76]
12	NSP12 (RNA polymerase, RdRp)	6YYT	932	RNA polymerase activity	[29,77–80]
13	NSP13	6JYT	601	Helicase activity	[29,81]
14	NSP14	7R2V	527	Viral RNA methylation, viral RNA proofreading, methyltransferase activity	[73,82–84]
15	NSP15	6WXC	346	Endoribonuclease activity	[81,85]
16	NSP16	6WVN	298	Viral replication, immune response evasion Viral RNA methylation, methyltransferase activity	[84,86,87]

3. Overview of TLR Signaling

Invading pathogens stimulate the release of proinflammatory mediators in response to infection (Figures 1 and 2). Signaling networks are necessary for the protection of the host against invading microorganisms. TLR signaling dysregulation plays a central role in the development and progression of infection. Inflammatory secretory molecules including chemokines, ILs, IFNs, and tumor necrosis factor-alpha (TNF-α) are part and parcel of TLR signaling, resulting in the modulation of cellular characteristics such as apoptosis, immune response, and proliferation [88–90]. Mitogen-activated protein kinases (MAPKs) and NF-κB are activated by TLRs. TLR3 and TLR4 are involved in the stimulation of IRF3. In contrast, IRF7 is triggered by TLR7–9 [91]. TLRs are stimulated by interactions with ligands to initiate an intracellular downstream signaling cascade, leading to activation of the host defense system [92].

The nature of the ligand and downstream adaptor molecules directs the TLR signaling cascade (Table 3). Two distinct pathways play critical roles in TLR signaling: MyD88-dependent and -independent pathways [93] (Figure 2). The former pathway employs all TLRs (except for TLR3), resulting in the biosynthesis of inflammatory cytokines [94]. In contrast, the latter pathway (also referred to as the TRIF-dependent pathway) involves TLR3 and TLR4, resulting in the expression of IFN-I [95]. In other words, the interaction of PAMP and PRR leads to the biosynthesis of proinflammatory cytokines as well as IFN-1, which is a cellular indication of the immune response [96]. Several negative regulators that enhance the activation of the innate immune response are involved in TLR-dependent signaling cascades. Hence, the overactivation of TLRs can lead to the interruption of immune cell homeostasis, resulting in the risk of inflammatory disorders [97]. Consequently, inhibitors (antagonists) targeting these receptors and/or cascades can serve as novel therapeutics to treat such disorders [98].

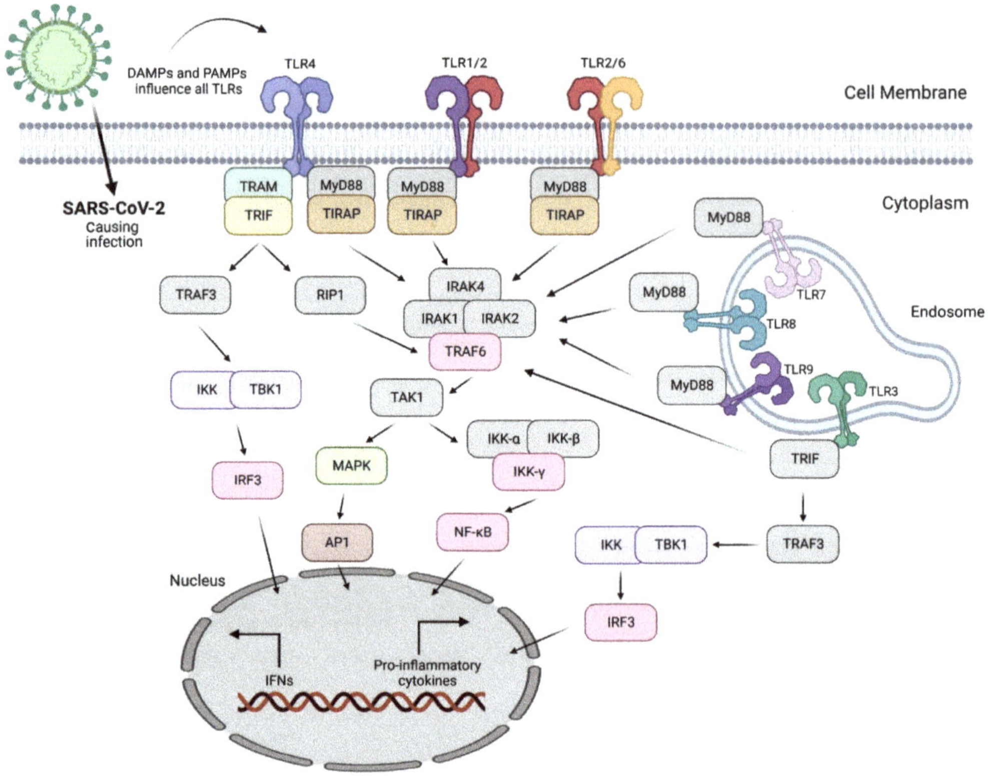

Figure 2. SARS-CoV-2 causes infection in the lungs mainly via DAMPs and PAMPs produced as a result of the action of nearly all Toll-like receptors (TLRs). Only TLRs involved in virus sensing and/or signaling are displayed here.

Table 3. Toll-like receptors (TLRs) and their physiological significance.

TLRs	Ligand Recognition	Form	Localization	Adaptor Molecules	Negative Adaptors	Response	Reference
TLR1	Triacyl lipopeptides, soluble factors	Heterodimer	Cell surface	MyD88, Mal	-	NF-κB activation and proinflammatory cytokines	[99,100]
TLR2	Hsp70, lipopeptide, HCV, Nonstructural protein 3	Heterodimer	Cell surface	MyD88, Mal	-	NF-κB activation and proinflammatory cytokines	[101,102]
TLR3	dsRNA	Homodimers	Endosomal membrane	TRIF	SARM negatively regulates TRIF	IRF activation, production of type 1 IFNs and proinflammatory cytokines	[103,104]

Table 3. *Cont.*

TLRs	Ligand Recognition	Form	Localization	Adaptor Molecules	Negative Adaptors	Response	Reference
TLR4	Lipopolysaccharide, Taxol, S protein of SARS-CoV-2	Homodimers	Cell surface	MyD88, Mal, TRIF, TRAM	SARM negatively regulates TRIF and TRAM to consequently reduce inflammation	Activation of NF-κB, pro-inflammatory cytokines, and IFN-inducible genes	[105,106]
TLR5	Flagellin	Homodimers	Cell surface	MyD88	-	Activation of NF-κB and proinflammatory cytokines	[107,108]
TLR6	Diacyl lipopeptides, lipoteichoic acid, fungal zymosan	Heterodimer	Cell surface	MyD88, Mal/TIRAP	-	Activation of NF-κB and proinflammatory cytokines	[109,110]
TLR7	SARS-CoV-2 ssRNA, imadozoquinoline	Homodimers	Endosomal membrane	MyD88	-	IRF activation, production of Type 1 IFNs and proinflammatory cytokines	[111,112]
TLR8	SARS-CoV-2 ssRNA		Endosomal membrane	MyD88	-	IRF activation, production of type 1 IFNs and proinflammatory cytokines	[113,114]
TLR9	Unmethylated CPG-containing ssDNA, hemozoin from the malaria parasite	Homodimers	Endosomal membrane	MyD88	-	IRF activation, production of type 1 IFNs and proinflammatory cytokines	[115,116]

4. Role of Antiviral Drugs Employing TLRs

When a pathogen such as a virus invades, an antiviral immune response is evident in the host cells. Various conserved molecular patterns of PAMPs have been identified. As discussed above, TLRs are the key constituents of the innate immune system, and multiple TLRs (TLR1–4, TLR6–9) identify viral ligands [17,117–119]. With respect to their functional importance, TLRs might be potentially employed to treat not only inflammatory disorders but also viral diseases. This can be explained by a deep insight into the positive and negative mediators of TLRs [97,120]. TLR agonists lack accessory molecules but can mimic natural ligands; hence, they exhibit a low molecular weight and have potential for expanded pharmacokinetics and pharmacodynamics in comparison with the parent molecule. Moreover, TLR antagonists help to deal with autoimmune and inflammatory disorders by defeating unnecessary inflammation, resulting in an antibody- or cell-mediated response that suppresses disease progression [97,121,122].

Different approaches are employed by viruses in which they weaken their recognition by masking and/or increasing the dysregulation of mediators. Viruses disturb TLR signaling through their own mechanisms. Thus, TLRs are largely involved in the molecular interaction between viruses and host cells [5]. Various PRRs are engaged in the response to viral infection, which is also the case for TLRs. A thorough understanding of this interaction has facilitated the development of various strategies to limit viral infection including antiviral immunity as well as therapeutics [5]. Moreover, viral infection activates TLRs to increase cytokine levels, resulting in an antiviral innate immune response. The interaction between viruses and TLRs at every step of the signaling pathway plays an important role in developing effective antiviral therapies as well as in identifying novel molecular targets for the advancement in antiviral drugs [123]. The regulation of invasion, replication, and immune responses is a significant factor in viral pathogenesis [117]. Viral glycoproteins and

NSPs released in the extracellular region are responsible for the stimulation of TLR2 and TLR4 due to their presence on the cellular surface [117,124,125]. In contrast, TLR3, TLR7/8, and TLR9, which are present in the endosomal compartment, contain viral double-stranded RNA (dsRNA) [126], ssRNA [114], and CpG DNA (unmethylated) [116], respectively.

TLR agonists have a positive effect on antiviral immunity and exhibit significant resistance against experimental infections [127–129]. The TLR–virus interaction involves a complex mechanism that is associated with the type of TLR as well as the type of virus. Moreover, multiple PRRs are required to initiate an immune response to various viral infections. Moreover, significant differences in TLR signaling have been reported between mice and humans. Therefore, therapeutic manipulation of TLRs requires an understanding of human cellular immunity [130]. Some examples are presented below.

TLR2 activation enhances the innate immune response to viral infections and can be used to treat viral respiratory diseases. Using the shock-and-kill strategy, immune cell recognition is enhanced and latently infected cells are eliminated [112,131]. TLRs can be used to reverse HIV-1 latency and trigger innate immune responses. In an evaluation of the effectiveness of SMU-Z1 (a novel TLR1/2 agonist), in addition to enhancing latent HIV-1 transcription (ex vivo), the NF-κB and MAPK pathways were also targeted in cells [131]. Latency-reversing agents have been employed for HIV reactivation, resulting in enhanced immune activation [112]. Dual TLR2/7 agonists were synthesized and characterized based on their latency-reversing ability, which were found to effectively reactivate the latency. TLR2 components reactivate HIV by NF-κB stimulation and the secretion of IL-22 (thereby enhancing the antiviral state and inhibiting HIV infection), whereas TLR7 components induce the secretion of TNF-α [112]. The activation of TLR2 in vivo has been assessed against rhinovirus infection [132]. Airway epithelial cells promote an extended immune response characterized by IFN-λ expression, NF-κB activation, and lymphocyte recruitment, resulting in a reduction in viral-induced inflammation and continued antiviral innate immunity [132].

TLR3 (the first identified antiviral TLR) in humans confers protective immunity against vaccinia virus (VACV) infection. In contrast, TLR3 is responsible for the detrimental effects of VACV infection in mice and TLR4 has the same effect in humans [133,134]. The recognition of dsRNA by TLR3 is further evidence of the role of TLRs in the antiviral response [119,126,135]. TLR3 signaling can be activated by a synthetic dsRNA agonist (a potent immune stimulant), resulting in protective immunity against multiple viruses including coronaviruses [136–139]. Viral-origin ssRNA sequences (rich in GU- and AU-) are detected by TLR7 and TLR8, which are functionally similar and only differ with respect to their expression patterns [113,130]. TLR7/8 expression is evident in dendritic cells, monocytes, and macrophages [140]. Additional examples are listed in Table 4.

Table 4. Reported antiviral agonists employing Toll-like receptors (TLRs).

Drugs	TLRs	Viruses	Significance	References
Pam_2CSK_4	TLR2	Parainfluenza	Reduced virus replication	[141]
INNA-051	TLR2	SARS-CoV-2	Reduces viral RNA load	[142]
PIKA	TLR3	Influenza A	Reduces virus load	[143]
Poly ICLC	TLR3	HIV	Release of IFN-α/β/γ	[144]
NA6	TLR4	Norovirus	Induction of IFN-β	[145]

Table 4. *Cont.*

Drugs	TLRs	Viruses	Significance	References
MPL	TLR4	VZV	Stimulate cytokines	[146]
Flagellin	TLR5	Influenza A	Reduces virus replication	[147]
CBLB502	TLR5	ConA	Activation of NF-κB	[148]
Pam_2CSK_4	TLR6	Parainfluenza	Reduces virus replication	[141]
INNA-051	TLR6	SARS-CoV-2	Reduces viral RNA load	[142]
GS-9620	TLR7	HIV	Reactivates latency	[112]
Vesatolimod	TLR7	HIV	Modest delay in viral rebound	[149]
R848	TLR7/8	Zika	Activation of NF-κB	[150]
GS-9688	TLR8	HBV	Activation of dendritic and natural killer cells	[151]
ODN2395	TLR9	Parainfluenza	Reduces viral replication	[141]

CBLB502—Entolimod; ConA—Concanavalin A; GS-9688—Selgantolimod; R848—Resiquimod; NA6—neoagarohexaose; VZV—Varicella-Zoster virus.

5. Possible Molecular Interactions of SARS-CoV-2 with TLRs

SARS-CoV-2 is not only associated with viral illness but also with disorders of immunopathology. DAMPs and viral components act as TLR ligands for their overactivation. TLR4 (membrane-bound) and TLR3/7/8 (endosomal) play significant roles in the production of cytokine storms. The ssRNA of SARS-CoV-2 is recognized by TLR7/8, and after replication, the viral dsRNA is recognized by TLR3, which leads to TRIF-mediated inflammatory signaling [152]. The MyD88-dependent pathway (leading to overactivation of TLRs), related to the TRIF pathway, provides a possible link between SARS-CoV-2 and TLRs [153]. The production of type I (IFN-α and IFN-β) and type III [IFN-λ (1/2/3)] IFNs by TLRs is a significant antiviral feature that can be exploited for systematized viral control as well as clearance [117,119,154]. Type I and III IFNs perform the same function (despite their structural differences) in cellular signaling, although type III IFN receptors are primarily localized to the epithelial surface (airway epithelial cells) [155]. Cytokines (type III IFN) bind to their receptors, and the signal cascade is initiated by the JAK/STAT pathway, leading to the formation of IFN-stimulated genes [156]. Activation of the JAK/STAT pathway induced by TLRs may lead to macrophage activation syndrome [157]. Virally infected cells are killed by activated dendritic cells, natural killer cells, and macrophages stimulated by IFN [158]. SARS-CoV-2 infection results in higher levels of chemokines and proinflammatory cytokines in the blood [159,160]. These biological conditions lead to host cell death and organ injury [161]. Hence, the synthesis of DAMPs amplifies inflammation by TLR binding via the MyD88-dependent pathway. Elevated TLR stimulation, signaling cascades, and NF-κB may influence the severity of COVID-19 [153]. The nutritional profile has a basic influence on immunity. Compounds with immunomodulatory, anti-inflammatory as well as antiviral characteristics can be helpful against SARS-CoV-2 infection [162]. Various studies have suggested encouraging results in the case of nutraceuticals [163,164]. Compounds including astaxanthin, curcumin, glycyrrhizin, hesperidin, lactoferrin, luteolin, quercetin as well as resveratrol may inhibit and counteract the symptoms of COVID-19 [165–172].

Accordingly, IFN has been dynamically explored as a therapeutic target for COVID-19. This is because the release of type III IFN in the lungs could be responsible for the observed immunopathology of COVID-19 [173,174]. In contrast, type I IFN in combination with antiviral drugs has exhibited the opposite results including reduced systemic inflammation and viral clearance [173–175]. The synthesis of proinflammatory cytokines is associated with MyD88-dependent pathways, whereas the activation of type I and III IFNs is linked with the TRIF-dependent pathway [176,177] (Table 3). SARS-CoV-1 dsRNA and ssRNA are not detected by TLR3 and TLR7 and show some protective dodging mechanisms; hence, the same strategy could be used by SARS-CoV-2 [178,179]. The stimulation of TLRs by SARS-CoV-2 is responsible for activation of the inflammasome and the subsequent release of IL-1β and IL-6. Moreover, enhanced inflammasome activation is linked to non-promising consequences in patients with COVID-19 [180]. TLR2 signaling is activated by SARS-CoV-2 infection (Table 5). Thus, blocking of the signaling has been proposed as a potential target for the treatment of COVID-19 [181] because the strong effect of proinflammatory cytokines leads to disease severity through the activation of TLR2 [182]. In the context of infection with β-coronaviruses, MyD88, the TLR adaptor, has been reported to be a significant factor in the release of a large number of inflammatory cytokines [183]. SARS-CoV-2 interacts with various TLRs, directly or indirectly. Multiple interacting residues have been reported in the literature considering PPI and the design of agonists/antagonists. The interacting residues are based on experimental as well as computation studies. Only TLRs involved in virus sensing and/or signaling are displayed (Table 5).

Multiple TLR (2, 4, 7, 9)-deficient macrophages were infected with the mouse hepatitis virus. TLR2 deficiency resulted in the inhibition of TNF and IL6 expression as well as inflammatory cytokine genes. In contrast, other TLR deficiencies had negligible effects on these genes [182]. In the case of SARS-CoV-2, an inhibitor of TLR2 caused a noteworthy reduction in cytokine and chemokine release. This study demonstrated the role of TLR2 in sensing viral invasion upstream of MyD88 [182]. The TRIF pathway activated by TLR3 showed a protective response against Middle East respiratory syndrome (MERS)-CoV and SARS-CoV infections [184] (Tables 3 and 5). Mice lacking TLR3, TLR4, and TRIF adaptor are exceedingly vulnerable to SARS-CoV and enhanced pulmonary infection, resulting in a risk of mortality [185]. Moreover, a role of TLR4 has been identified in the pathology of SARS-CoV-2, characterized by excessive inflammation in patients and activation of the inflammasome [186,187]. TLR4 inhibition in animal models has been shown to decrease lung injury by alleviating NF-κB pathway stimulation [188]. Viral infection and subsequent inflammation results in the production of DAMPs that act as ligands for TLR4. Heat shock proteins released from virus-infected cells act as TLR4 agonists [189]. TLR5 has been proposed as a target against SARS-CoV-2 in the development of drugs and vaccines [190].

During cytokine storms, elevated levels of IL-6 in the serum have been observed in patients with COVID-19 (Figure 1). TLR7 (activated by viral components) stimulates the MyD88-dependent pathway, resulting in the release of ILs and TNF-α, particularly IL-6 [191,192]. Structurally and phylogenetically similar receptors but different TLR7/8 agonists synthesize different cytokines [193]. ssRNA fragments in SARS-CoV-2 induced by TLR7/8 have been detected [194]. Whole-genome sequencing of SARS-CoV-2 in comparison with other coronaviruses (MERS-CoV and SARS-CoV) has revealed that TLR7 could be significantly involved in COVID-19 as the viral genome contains more ssRNA motifs that can bind to TLR7 [195]. Moreover, the TLR7 agonists imiquimod and imidazoquinolinone (with a role in TLR7activation) are under investigation as potential therapeutics against COVID-19. These drugs have been observed to decrease systemic inflammation and innate immune activation due to their antiviral effects [196,197]. RNA and DNA rich in unmethylated CpG islands can be recognized by TLR9. Both viral and mitochondrial DNA enriched in the same sequence are associated with inflammatory responses involving TLR9-mediated signaling. Moreover, the activation of p53 [198] and mammalian target of rapamycin (mTOR) is being considered as a therapeutic target against SARS-CoV-2. mTOR blockers are also associated with the MyD88 and TLR9 pathways [199].

Table 5. The interaction of Toll-like receptors (TLRs) with SARS-CoV-2 and other coronaviruses.

Coronaviruses	TLRs	Interacting Residues of TLRs	References
SARS-CoV-2	TLR2	Tyr323, Phe325, Val 348, Phe349	[182,200,201]
	TLR3	His39, His60, His108, Asn515, Asn517, His539, Asn541, Arg544, Ser571	[111,202,203]
	TLR4	Arg264, Glu266, Asp294, Tyr295, Tyr296, Thr319, Glu321, Lys341, Lys362, Gly363, Gly364, Arg382	[188,204,205]
	TLR7/8	Phe349, Tyr356, Gly379, Val381, Phe408, Asp555, Leu557, Gly584, Thr586	[114,206–208]
SARS-CoV	TLR3	His39, His60, His108, Asn515, Asn517, His539, Asn541, Arg544, Ser571	[136,185,203]
	TLR4	Arg264, Glu266, Asp294, Tyr295, Tyr296, Thr319, Glu321, Lys341, Lys362, Gly363, Gly364, Arg382	[185,205]
	TLR7/8	Phe349, Tyr356, Gly379, Val381, Phe408, Asp555, Leu557, Gly584, Thr586	[207–209]
MERS-CoV	TLR3	His39, His60, His108, Asn515, Asn517, His539, Asn541, Arg544, Ser571	[203,209]
	TLR4	Arg264, Glu266, Asp294, Tyr295, Tyr296, Thr319, Glu321, Lys341, Lys362, Gly363, Gly364, Arg382	[205,210,211]
	TLR7/8	Phe349, Tyr356, Gly379, Val381, Phe408, Asp555, Leu557, Gly584, Thr586	[207–209,212,213]

6. Promising Drug Targets in SARS-CoV-2

Possible and effective drug targets as well as therapeutic agents against SARS-CoV-2 have been suggested by various researchers worldwide [214]. For example, virulence factors, enzymes, host-specific receptors, and glycosylated-structural proteins have been identified in pathological conditions caused by the coronavirus [215]. Activators of transcription signaling pathways, proinflammatory cytokines, Janus kinase/signal transducers, and NSPs also play a crucial role in the pathology. Antiviral therapeutic strategies such as drug repurposing depend on chemical and molecular interactions between the host machinery and viral small molecules [215].

Low-molecular-weight molecules from plants (phytochemicals) have been tested for their antiviral activity. Compounds extracted from plants have been shown to exhibit antiviral activity against SARS-CoV in Vero cells. Lycorine was identified as the active ingredient of *Lindera aggregata*, and it has been suggested that the plant extract and lycorine can be a good option for the development of novel antiviral drugs [7]. In plants, secondary metabolites are produced by metabolic pathways, which are also referred to as phytochemicals [216]. These metabolites have been screened for their efficacy against microbes and viruses. Various phytochemicals have been shown to inhibit viral infection and replication [217]. Bioactive phytochemicals can improve and strengthen host immunity. For example, less vulnerability to infections and assistance in the stoppage of viral infections through host immune function have been observed with the treatment of vitamins A and C [218]. Various in vitro, in vivo, and in silico models using marine-derived natural compounds exhibiting promising anti-SARS-CoV-2 efficacy have been previously summarized [219].

Various proteins including ACE-2, RdRp, 3CLpro, PLpro, RBD, and cathepsin L could be operative therapeutic targets [67,198,220–223]. Although several molecules have been suggested as drug candidates, currently, there are no accessible operative anti-CoV mediators. Molecular interactions between ACE2 and SARS viruses are determinants of the initial infection. Hence, renin–angiotensin–aldosterone system (RAAS) inhibitors may modify ACE2 expression, resulting in reduced SARS-CoV-2 virulence. ACE2 (type I transmembrane-metallocarboxypeptidase enzyme) controls the effects of RAAS and is a key receptor for both SARS-CoV-1 and SARS-CoV-2, which facilitates entry into human lung cells through the S protein of the coronavirus [224–228]. Considering the complexity of the pathogenesis of SARS-CoV-2, clinically approved drugs that stimulate ACE2 may serve as operative anti-SARS-CoV-2 therapeutics [229]. ACE inhibitors (captopril) stimulate the ACE2/angiotensin (1–7)/receptor axis [230]. In animal models, treatment with angiotensin receptor blockers was shown to enhance ACE2 expression [231,232]. The ACE2–RBD complex is proteolytically regulated by type-2 transmembrane cellular serine protease (TMPRSS2), which leads to ACE2 cleavage and S protein activation [233]. The RBD (S protein) of SARS-CoV-2 contains more ACE2-interacting residues (Tyr473, Gln474, Cys488, Tyr489, Val524, and Cys525) than that of SARS-CoV, and is involved in loop formation. These mutations are evident in the sequence (RBD) of SARS-CoV-2 [234–236]. Moreover, two binding hotspot residues (Lys31 and Lys353) have been reported to be more sensitive to S protein binding. Lys31 and Lys353 formed salt-bridge(s) with Glu35 and Asp38, respectively, surrounded by a hydrophobic region [237]. Additionally, other studies support the development of promising ACE2 inhibitors for SARS-CoV-2 [220,238,239].

In the case of glycosylated S protein, membrane fusion inhibitors for the S2 subunit and antibodies (monoclonal) targeting the S1 subunit could be operative therapeutic mediators to treat coronavirus infection. Vaccine development has also been promoted against coronaviruses. Small-molecule inhibitors (SMIs) might be suitable for inhaled and/or oral administration, exhibit less mutation and strain sensitivity, are less immunogenic, and convenient. Novel drug-like SMIs (DRI-C23041 and DRI-C91005) have been identified. These SMIs inhibit the interaction between S protein and human ACE2 [34]. Moreover, griffithsin, a compound derived from red algae, adheres to the SARS-CoV-2 glycosylated S protein as well as to HIV [240]. Furin, a serine endoprotease, cleaves S1–S2 and may be suitable as an anti-SARS-CoV-2 agent [241]. Emodin, a *Rheum tangutica*-derived compound, not only inhibits the interaction between the ACE2 receptor and SARS-CoV-2 but also blocks SARS-CoV ORF3a [242,243]. Moreover, the host protease is employed by SARS-CoV-2 for the priming of the S protein. Camostat mesylate, an inhibitor of proteases, helps in the infection of lung cell lines [244]. Similar to the S protein, other structural proteins as well as NSPs have been highlighted as potential targets for the development of antiviral drugs. In the RBD, 14 different potent residues of the S protein have been identified that interact significantly with ACE2, resulting in the stability of the complex [245], while 15 significant residues in the RBD of the S protein have been reported in the case of the Omicron SARS-CoV-2 variant [246]. Both of these studies analyzed (in silico) natural compounds for their anti-SARS-CoV-2 bioactivity [245,246]. Additionally, terpenes (natural compounds) have been suggested as anti-SARS-CoV2 binding agents between the RBD and ACE2 receptor. Terpenes showed a strong affinity for RBD and inhibited its interaction with ACE2 [245].

Toremifene, a nonsteroidal selective estrogen receptor modulator, was found to block the viral replication of coronaviruses such as MERS-CoV and SARS-CoV [247] and the Ebola virus [248] by targeting viral membrane proteins. Hence, it has also become a potential candidate inhibitor of SARS-CoV-2 replication [248]. Moreover, a team of researchers proposed a region (residues) in the M protein for the development of novel drugs and/or peptides to block dimer formation [249]. The interaction of the M protein (heterodimer) with the S and E proteins (via PPIs) has been proposed by computational analysis, and key amino acids for the M–E complex (W55, F96, F103) and M–S complex (Y71, Y75) have been identified [250].

A high percentage of E protein is expressed inside the infected cells, which is responsible for viral assembly, maturation, budding, and proliferation [35,40,214]. The percentage similarity of the SARS-CoV-2 E protein sequence with that of other coronaviruses (96.1%) [31] demonstrates the potential for repurposing and/or development of pan-anti-corona drug candidates. Small molecules (phytochemicals) such as belachinal, vibsanol B, and macaflavanone E have been evaluated for the inhibition of E protein activity by in silico analyses [251].

The N protein exhibits essential activities such as proliferation of the virus as another important component, similar to other SPs. This provides a promising area for developing effective therapeutics to inhibit viral proliferation. For example, glycogen synthase kinase-3 (GSK-3), also known as serine/threonine protein kinase, is an important component of N protein phosphorylation. GSK-3 inhibitors inhibit N protein phosphorylation and result in damaged proliferation (infected lung epithelial cells) in SARS-CoV-2 in a cell type-dependent manner [252]. Candidate inhibitors of the N protein have been suggested by a screening method on a biochip platform using a quantum-dot (QD) RNA oligonucleotide. The novel anti-SARS potential of catechin gallate and gallocatechin has been identified. These two molecules (0.05 μg/mL) presented more than 40% inhibition activity on a QD-based RNA oligonucleotide system [253]. Computational analysis has suggested that zidovudine triphosphate is a potent inhibitor of the N protein of SARS-CoV-2 [254]. Based on an in silico approach, another repurposing study shed light on vanganciclovir, which is approved for treating patients with HIV and shows activity against N protein as well as the main protease [254].

SARS-CoV-2 depends on proteases of the Golgi apparatus to synthesize NSP1–16 in the host cell [27]. NSP3 [papain-like protease (PL_{pro})] is a multidomain protein and the largest protein in the coronavirus genome. Several regions of the NSP3 gene are involved in viral replication. NSP3 contains a SARS-unique domain that can attach to G-quadruplexes, which are guanine-rich non-canonical nucleic acid structures that are essential for SARS-CoV replication. SARS-CoV-2 NSP3 shows structural similarities [255]. By developing a protease assay and screening a custom compound library, two molecules (dihydrotanshinone I and Ro 08-2750) were identified to significantly inhibit PL_{pro} in protease. Additionally, the inhibition of viral replication was evaluated by an isopeptidase assay using cell culture [256]. Another protease, NSP5 ($3CL_{pro}$), was identified as a primary target (similar to PL_{pro}) for coronavirus drug discovery. Both of these targets are crucial and have conserved activity in the proteolytic processing of viral replicase polyproteins [257].

Coronaviruses encode two or three proteases that cleave replicase polyproteins. Many NSPs assemble into the replicase–transcriptase complex, which generates a reasonable environment for RNA synthesis and subsequent replication as well as the transcription of sub-genomic RNAs [258]. Replicase polyproteins 1a and 1ab are comprised of NSPs11 and 16, respectively [259]. M^{Pro} (the main protease), commonly known as NSP5, is employed for the cleavage of these polyproteins, exhibiting crucial events of viral assembly and maturation [259]. M^{Pro} is a dimer (306 residues) with two identical monomers, and is a significant target responsible for viral polyprotein cleavage 1ab at 11 (a major cleavage site), required for generating the NSP7–NSP8–NSP12 complex (viral replication complex) [260,261]. Residues interacting with two novel inhibitors against M^{Pro} have been identified: His41, Met49, Met165, Val186, Asp187, Arg188 as well as Gln189, exhibiting hydrophobic and H-bonding. Both inhibitors reside in the substrate-binding site and inhibit the enzymatic activity of M^{Pro} in SARS-CoV-2 [261].

Targeting highly conserved genes and/or proteins including RdRp (NSP12), M^{Pro}, and helicases is a promising antiviral drug development approach to inhibit the replication and proliferation of SARS-CoV-2 [262]. Hence, inhibitors targeting these enzymes may reduce the threat of mutation-mediated drug resistance and facilitate effective antiviral protection [262]. A conserved motif (Ser-Aps-Asp) in the RdRp domain was identified at the C-terminus. The binding and activity of RdRp were enhanced by the NSP7–NSP8 (cofactor) complex. This binding stabilizes the entire closed conformation, which is packed beside the thumb–finger interface. The binding residues between the RdRp and RNA complex and RdRp docking to develop inhibitors have been extensively studied [78–80].

The inhibition of RdRP is important as one of the key strategies for developing antiviral therapeutics. The selective inhibition of RdRp may not cause noteworthy side effects or toxicity in host cells [263]. Natural compounds and their derivatives have exhibited significant binding affinity to RdRp [264–266], with promising outcomes that require further investigation.

NSPs 7–16 are responsible for coronavirus RNA synthesis and processing, which generate two large replicase polyproteins by cleavage. SARS-CoV-2 possesses a large number of enzymes that are responsible for RNA synthesis as well as RNA processing. The genome that is expressed and replicated by enzymes is two to three times larger than that of any other RNA viruses. RdRp is an important drug target because of its vital role in generating viral RNA [77–80].

Coronaviruses possess three important virulence factors: NSP1, NSP3c, and ORF7a. These factors help in the escape of viruses from host innate immunity and may be potential drug targets [55,267]. NSP1 and NSP3c interact with the host 40S ribosomal subunit and adenosine diphosphate-ribose, respectively. This leads to the degradation of mRNA and the inhibition of type-I IFN synthesis by NSP1, while NSP3c assists viruses to counterattack the immune response of the host [55,268]. ORF7a is directly attached to bone marrow matrix antigen-2 (BST-2) and has the ability to stop its glycosylation. BST-2 plays a specific role in regulating the release of newly synthesized viruses [267,269].

7. Conclusions

SARS-CoV-2 is recognized by various TLRs. Surface (TLR2 and 4) and intracellular (TRL3, 7/8, and 9) factors have been reported to be involved in the perception of SARS-CoV-2 infection by the immune system. Multiple adaptors such as MyD88 and TRIF are recruited by TLRs to initiate downstream signaling pathways. Various protein targets from viruses and the host machinery have been suggested as potential drug targets against SARS-CoV-2. Protein targets from viruses include both structural and nonstructural proteins. Similarly, TLRs are functional protein targets during SARS-CoV-2 infection.

Author Contributions: Conceptualization, A.M. and R.H.P.; Writing—original draft preparation, A.M.; Writing—review and editing, A.M., R.H.P. and M.H.; Visualization, A.M. and M.H.; Supervision, S.C.; Funding acquisition, S.C. All authors have read and agreed to the published version of the manuscript.

Funding: This work was supported by the Korea Drug Development Fund funded by the Ministry of Science and ICT, Ministry of Trade, Industry, and Energy, and Ministry of Health and Welfare (HN21C1058). This study was also supported by the National Research Foundation of Korea (NRF-2022M3A9G1014520, 2019M3D1A1078940, and 2019R1A6A1A11051471).

Informed Consent Statement: Not applicable.

Data Availability Statement: Not applicable.

Conflicts of Interest: The authors declare no conflict of interest.

References

1. Iwasaki, A.; Medzhitov, R. Control of Adaptive Immunity by the Innate Immune System. *Nat. Immunol.* **2015**, *16*, 343–353. [CrossRef] [PubMed]
2. Patra, M.C.; Choi, S. Recent Progress in the Development of Toll-like Receptor (TLR) Antagonists. *Expert Opin. Ther. Pat.* **2016**, *26*, 719–730. [CrossRef] [PubMed]
3. Nie, L.; Cai, S.-Y.; Shao, J.-Z.; Chen, J. Toll-Like Receptors, Associated Biological Roles, and Signaling Networks in Non-Mammals. *Front. Immunol.* **2018**, *9*, 1523. [CrossRef] [PubMed]
4. Liu, G.; Zhang, H.; Zhao, C.; Zhang, H. Evolutionary History of the Toll-Like Receptor Gene Family across Vertebrates. *Genome Biol. Evol.* **2020**, *12*, 3615–3634. [CrossRef] [PubMed]
5. Carty, M.; Bowie, A.G. Recent Insights into the Role of Toll-like Receptors in Viral Infection. *Clin. Exp. Immunol.* **2010**, *161*, 397–406. [CrossRef]
6. Abe, T.; Kaname, Y.; Hamamoto, I.; Tsuda, Y.; Wen, X.; Taguwa, S.; Moriishi, K.; Takeuchi, O.; Kawai, T.; Kanto, T.; et al. Hepatitis C Virus Nonstructural Protein 5A Modulates the Toll-like Receptor-MyD88-Dependent Signaling Pathway in Macrophage Cell Lines. *J. Virol.* **2007**, *81*, 8953–8966. [CrossRef]

7. Li, S.Y.; Chen, C.; Zhang, H.Q.; Guo, H.Y.; Wang, H.; Wang, L.; Zhang, X.; Hua, S.N.; Yu, J.; Xiao, P.G.; et al. Identification of Natural Compounds with Antiviral Activities against SARS-Associated Coronavirus. *Antivir. Res.* **2005**, *67*, 18. [CrossRef]
8. Maloney, G.; Schröder, M.; Bowie, A.G. Vaccinia Virus Protein A52R Activates P38 Mitogen-Activated Protein Kinase and Potentiates Lipopolysaccharide-Induced Interleukin-10. *J. Biol. Chem.* **2005**, *280*, 30838–30844. [CrossRef]
9. Stack, J.; Haga, I.R.; Schröder, M.; Bartlett, N.W.; Maloney, G.; Reading, P.C.; Fitzgerald, K.A.; Smith, G.L.; Bowie, A.G. Vaccinia Virus Protein A46R Targets Multiple Toll-like-Interleukin-1 Receptor Adaptors and Contributes to Virulence. *J. Exp. Med.* **2005**, *201*, 1007–1018. [CrossRef]
10. Blasius, A.L.; Beutler, B. Intracellular Toll-like Receptors. *Immunity* **2010**, *32*, 305–315. [CrossRef]
11. Zhu, N.; Zhang, D.; Wang, W.; Li, X.; Yang, B.; Song, J.; Zhao, X.; Huang, B.; Shi, W.; Lu, R.; et al. A Novel Coronavirus from Patients with Pneumonia in China, 2019. *N. Engl. J. Med.* **2020**, *382*, 727–733. [CrossRef] [PubMed]
12. Onofrio, L.; Caraglia, M.; Facchini, G.; Margherita, V.; de Placido, S.; Buonerba, C. Toll-like Receptors and COVID-19: A Two-Faced Story with an Exciting Ending. *Future Sci. OA* **2020**, *6*, FSO605. [CrossRef] [PubMed]
13. Pearce, L.; Davidson, S.M.; Yellon, D.M. The Cytokine Storm of COVID-19: A Spotlight on Prevention and Protection. *Expert Opin. Ther. Targets* **2020**, *24*, 723–730. [CrossRef] [PubMed]
14. Kaushik, D.; Bhandari, R.; Kuhad, A. TLR4 as a Therapeutic Target for Respiratory and Neurological Complications of SARS-CoV-2. *Expert Opin. Ther. Targets* **2021**, *25*, 491–508. [CrossRef]
15. Kawasaki, T.; Kawai, T. Toll-like Receptor Signaling Pathways. *Front. Immunol.* **2014**, *5*, 461. [CrossRef]
16. O'Neill, L.A.J.; Bowie, A.G. The Family of Five: TIR-Domain-Containing Adaptors in Toll-like Receptor Signalling. *Nat. Rev. Immunol.* **2007**, *7*, 353–364. [CrossRef]
17. Takeuchi, O.; Akira, S. Innate Immunity to Virus Infection. *Immunol. Rev.* **2009**, *227*, 75–86. [CrossRef]
18. Jiménez-Dalmaroni, M.J.; Gerswhin, M.E.; Adamopoulos, I.E. The Critical Role of Toll-like Receptors-From Microbial Recognition to Autoimmunity: A Comprehensive Review. *Autoimmun. Rev.* **2016**, *15*, 1–8. [CrossRef]
19. Singh, N.; Baby, D.; Rajguru, J.; Patil, P.; Thakkannavar, S.; Pujari, V. Inflammation and Cancer. *Ann. Afr. Med.* **2019**, *18*, 121–126. [CrossRef]
20. Wu, Y.W.; Tang, W.; Zuo, J.P. Toll-like Receptors: Potential Targets for Lupus Treatment. *Acta Pharmacol. Sin.* **2015**, *36*, 1395–1407. [CrossRef]
21. Chan, J.F.W.; Kok, K.H.; Zhu, Z.; Chu, H.; To, K.K.W.; Yuan, S.; Yuen, K.Y. Genomic Characterization of the 2019 Novel Human-Pathogenic Coronavirus Isolated from a Patient with Atypical Pneumonia after Visiting Wuhan. *Emerg. Microbes Infect.* **2020**, *9*, 221–236. [CrossRef] [PubMed]
22. Snijder, E.J.; Decroly, E.; Ziebuhr, J. The Nonstructural Proteins Directing Coronavirus RNA Synthesis and Processing. *Adv. Virus Res.* **2016**, *96*, 59. [CrossRef] [PubMed]
23. Shi, S.T.; Lai, M.M.C. Viral and Cellular Proteins Involved in Coronavirus Replication. In *Coronavirus Replication and Reverse Genetics*; Springer: Berlin/Heidelberg, Germany, 2005; Volume 287, p. 95. [CrossRef]
24. Guo, Y.R.; Cao, Q.D.; Hong, Z.S.; Tan, Y.Y.; Chen, S.D.; Jin, H.J.; Tan, K.S.; Wang, D.Y.; Yan, Y. The Origin, Transmission and Clinical Therapies on Coronavirus Disease 2019 (COVID-19) Outbreak-an Update on the Status. *Mil. Med. Res.* **2020**, *7*, 11. [CrossRef] [PubMed]
25. Han, Y.; Du, J.; Su, H.; Zhang, J.; Zhu, G.; Zhang, S.; Wu, Z.; Jin, Q. Identification of Diverse Bat Alphacoronaviruses and Betacoronaviruses in China Provides New Insights into the Evolution and Origin of Coronavirus-Related Diseases. *Front. Microbiol.* **2019**, *10*, 1900. [CrossRef] [PubMed]
26. Park, S.E. Epidemiology, Virology, and Clinical Features of Severe Acute Respiratory Syndrome -Coronavirus-2 (SARS-CoV-2; Coronavirus Disease-19). *Clin. Exp. Pediatr.* **2020**, *63*, 119–124. [CrossRef]
27. Gasmalbari, E.; Abbadi, O.S. Non-Structural Proteins of SARS-CoV-2 as Potential Sources for Vaccine Synthesis. *Infect. Dis. Trop. Med.* **2020**, *6*, e667.
28. Gorla, U.S.; Rao, G.S.N.K. SARS-CoV-2: The Prominent Role of Non-Structural Proteins (NSPS) in COVID-19. *Indian J. Pharm. Educ. Res.* **2020**, *54*, S381–S389. [CrossRef]
29. Yoshimoto, F.K. The Proteins of Severe Acute Respiratory Syndrome Coronavirus-2 (SARS CoV-2 or n-COV19), the Cause of COVID-19. *Protein J.* **2020**, *39*, 198–216. [CrossRef]
30. Mousavizadeh, L.; Ghasemi, S. Genotype and Phenotype of COVID-19: Their Roles in Pathogenesis. *J. Microbiol. Immunol. Infect.* **2021**, *54*, 159–163. [CrossRef]
31. Mohammed, M.E.A. The Percentages of SARS-CoV-2 Protein Similarity and Identity with SARS-CoV and BatCoV RaTG13 Proteins Can Be Used as Indicators of Virus Origin. *J. Proteins Proteom.* **2021**, *12*, 81. [CrossRef] [PubMed]
32. Du, L.; He, Y.; Zhou, Y.; Liu, S.; Zheng, B.J.; Jiang, S. The Spike Protein of SARS-CoV–a Target for Vaccine and Therapeutic Development. *Nat. Rev. Microbiol.* **2009**, *7*, 226–236. [CrossRef] [PubMed]
33. Yuan, Y.; Cao, D.; Zhang, Y.; Ma, J.; Qi, J.; Wang, Q.; Lu, G.; Wu, Y.; Yan, J.; Shi, Y.; et al. Cryo-EM Structures of MERS-CoV and SARS-CoV Spike Glycoproteins Reveal the Dynamic Receptor Binding Domains. *Nat. Commun.* **2017**, *8*, 15092. [CrossRef] [PubMed]
34. Bojadzic, D.; Alcazar, O.; Chen, J.; Chuang, S.T.; Condor Capcha, J.M.; Shehadeh, L.A.; Buchwald, P. Small-Molecule Inhibitors of the Coronavirus Spike: ACE2 Protein-Protein Interaction as Blockers of Viral Attachment and Entry for SARS-CoV-2. *ACS Infect. Dis.* **2021**, *7*, 1519–1534. [CrossRef] [PubMed]

35. Satarker, S.; Nampoothiri, M. Structural Proteins in Severe Acute Respiratory Syndrome Coronavirus-2. *Arch. Med. Res.* **2020**, *51*, 482–491. [CrossRef] [PubMed]
36. Yan, X.; Hao, Q.; Mu, Y.; Timani, K.A.; Ye, L.; Zhu, Y.; Wu, J. Nucleocapsid Protein of SARS-CoV Activates the Expression of Cyclooxygenase-2 by Binding Directly to Regulatory Elements for Nuclear Factor-Kappa B and CCAAT/Enhancer Binding Protein. *Int. J. Biochem. Cell Biol.* **2006**, *38*, 1417–1428. [CrossRef]
37. Wang, Q.; Li, C.; Zhang, Q.; Wang, T.; Li, J.; Guan, W.; Yu, J.; Liang, M.; Li, D. Interactions of SARS Coronavirus Nucleocapsid Protein with the Host Cell Proteasome Subunit P42. *Virol. J.* **2010**, *7*, 99. [CrossRef]
38. Lu, X.; Pan, J.; Tao, J.; Guo, D. SARS-CoV Nucleocapsid Protein Antagonizes IFN-β Response by Targeting Initial Step of IFN-β Induction Pathway, and Its C-Terminal Region Is Critical for the Antagonism. *Virus Genes* **2011**, *42*, 37. [CrossRef]
39. Luo, H.; Chen, Q.; Chen, J.; Chen, K.; Shen, X.; Jiang, H. The Nucleocapsid Protein of SARS Coronavirus Has a High Binding Affinity to the Human Cellular Heterogeneous Nuclear Ribonucleoprotein A1. *FEBS Lett.* **2005**, *579*, 2623. [CrossRef]
40. Ruch, T.R.; Machamer, C.E. The Hydrophobic Domain of Infectious Bronchitis Virus E Protein Alters the Host Secretory Pathway and Is Important for Release of Infectious Virus. *J. Virol.* **2011**, *85*, 675–685. [CrossRef]
41. Kuo, L.; Hurst, K.R.; Masters, P.S. Exceptional Flexibility in the Sequence Requirements for Coronavirus Small Envelope Protein Function. *J. Virol.* **2007**, *81*, 2249–2262. [CrossRef] [PubMed]
42. Verdiá-Báguena, C.; Nieto-Torres, J.L.; Alcaraz, A.; Dediego, M.L.; Enjuanes, L.; Aguilella, V.M. Analysis of SARS-CoV E Protein Ion Channel Activity by Tuning the Protein and Lipid Charge. *Biochim. Biophys. Acta* **2013**, *1828*, 2026–2031. [CrossRef] [PubMed]
43. Venkatagopalan, P.; Daskalova, S.M.; Lopez, L.A.; Dolezal, K.A.; Hogue, B.G. Coronavirus Envelope (E) Protein Remains at the Site of Assembly. *Virology* **2015**, *478*, 75–85. [CrossRef] [PubMed]
44. Schoeman, D.; Fielding, B.C. Coronavirus Envelope Protein: Current Knowledge. *Virol. J.* **2019**, *16*, 69. [CrossRef] [PubMed]
45. Corse, E.; Machamer, C.E. Infectious Bronchitis Virus E Protein Is Targeted to the Golgi Complex and Directs Release of Virus-like Particles. *J. Virol.* **2000**, *74*, 4319–4326. [CrossRef] [PubMed]
46. Mortola, E.; Roy, P. Efficient Assembly and Release of SARS Coronavirus-like Particles by a Heterologous Expression System. *FEBS Lett* **2004**, *576*, 174–178. [CrossRef] [PubMed]
47. Fehr, A.R.; Perlman, S. Coronaviruses: An Overview of Their Replication and Pathogenesis. *Methods Mol. Biol.* **2015**, *1282*, 1–23. [CrossRef]
48. Narayanan, K.; Maeda, A.; Maeda, J.; Makino, S. Characterization of the Coronavirus M Protein and Nucleocapsid Interaction in Infected Cells. *J. Virol.* **2000**, *74*, 8127–8134. [CrossRef]
49. Liu, D.X.; Yuan, Q.; Liao, Y. Coronavirus Envelope Protein: A Small Membrane Protein with Multiple Functions. *Cell Mol. Life Sci* **2007**, *64*, 2043–2048. [CrossRef]
50. Arndt, A.L.; Larson, B.J.; Hogue, B.G. A Conserved Domain in the Coronavirus Membrane Protein Tail Is Important for Virus Assembly. *J. Virol.* **2010**, *84*, 11418–11428. [CrossRef]
51. Cubuk, J.; Alston, J.J.; Incicco, J.J.; Singh, S.; Stuchell-Brereton, M.D.; Ward, M.D.; Zimmerman, M.I.; Vithani, N.; Griffith, D.; Wagoner, J.A.; et al. The SARS-CoV-2 Nucleocapsid Protein Is Dynamic, Disordered, and Phase Separates with RNA. *Nat. Commun.* **2021**, *12*, 1936. [CrossRef]
52. Huang, Y.; Yang, C.; Xu, X.F.; Xu, W.; Liu, S. wen Structural and Functional Properties of SARS-CoV-2 Spike Protein: Potential Antivirus Drug Development for COVID-19. *Acta Pharmacol. Sin.* **2020**, *41*, 1141–1149. [CrossRef] [PubMed]
53. Züst, R.; Cervantes-Barragán, L.; Kuri, T.; Blakqori, G.; Weber, F.; Ludewig, B.; Thiel, V. Coronavirus Non-Structural Protein 1 Is a Major Pathogenicity Factor: Implications for the Rational Design of Coronavirus Vaccines. *PLoS Pathog.* **2007**, *3*, 1062–1072. [CrossRef] [PubMed]
54. Afsar, M.; Narayan, R.; Akhtar, M.N.; Das, D.; Rahil, H.; Nagaraj, S.K.; Eswarappa, S.M.; Tripathi, S.; Hussain, T. Drug Targeting Nsp1-Ribosomal Complex Shows Antiviral Activity against SARS-CoV-2. *Elife* **2022**, *11*, e74877. [CrossRef]
55. Kamitani, W.; Narayanan, K.; Huang, C.; Lokugamage, K.; Ikegami, T.; Ito, N.; Kubo, H.; Makino, S. Severe Acute Respiratory Syndrome Coronavirus Nsp1 Protein Suppresses Host Gene Expression by Promoting Host MRNA Degradation. *Proc. Natl. Acad. Sci. USA* **2006**, *103*, 12885–12890. [CrossRef]
56. Ma, J.; Chen, Y.; Wu, W.; Chen, Z. Structure and Function of N-Terminal Zinc Finger Domain of SARS-CoV-2 NSP2. *Virol. Sin.* **2021**, *36*, 1104–1112. [CrossRef]
57. Cornillez-Ty, C.T.; Liao, L.; Yates, J.R.; Kuhn, P.; Buchmeier, M.J. Severe Acute Respiratory Syndrome Coronavirus Nonstructural Protein 2 Interacts with a Host Protein Complex Involved in Mitochondrial Biogenesis and Intracellular Signaling. *J. Virol.* **2009**, *83*, 10314–10318. [CrossRef]
58. Angeletti, S.; Benvenuto, D.; Bianchi, M.; Giovanetti, M.; Pascarella, S.; Ciccozzi, M. COVID-2019: The Role of the Nsp2 and Nsp3 in Its Pathogenesis. *J Med Virol* **2020**, *92*, 584–588. [CrossRef]
59. Khan, M.T.; Zeb, M.T.; Ahsan, H.; Ahmed, A.; Ali, A.; Akhtar, K.; Malik, S.I.; Cui, Z.; Ali, S.; Khan, A.S.; et al. SARS-CoV-2 Nucleocapsid and Nsp3 Binding: An in Silico Study. *Arch. Microbiol.* **2021**, *203*, 59–66. [CrossRef]
60. Alsaadi, E.A.J.; Jones, I.M. Membrane Binding Proteins of Coronaviruses. *Future Virol.* **2019**, *14*, 275–286. [CrossRef]
61. Oostra, M.; te Lintelo, E.G.; Deijs, M.; Verheije, M.H.; Rottier, P.J.M.; de Haan, C.A.M. Localization and Membrane Topology of Coronavirus Nonstructural Protein 4: Involvement of the Early Secretory Pathway in Replication. *J. Virol.* **2007**, *81*, 12323–12336. [CrossRef] [PubMed]

62. Helmy, Y.A.; Fawzy, M.; Elaswad, A.; Sobieh, A.; Kenney, S.P.; Shehata, A.A. The COVID-19 Pandemic: A Comprehensive Review of Taxonomy, Genetics, Epidemiology, Diagnosis, Treatment, and Control. *J. Clin. Med.* **2020**, *9*, 1225. [CrossRef] [PubMed]
63. Stobart, C.C.; Sexton, N.R.; Munjal, H.; Lu, X.; Molland, K.L.; Tomar, S.; Mesecar, A.D.; Denison, M.R. Chimeric Exchange of Coronavirus Nsp5 Proteases (3CLpro) Identifies Common and Divergent Regulatory Determinants of Protease Activity. *J. Virol.* **2013**, *87*, 12611. [CrossRef] [PubMed]
64. Benvenuto, D.; Angeletti, S.; Giovanetti, M.; Bianchi, M.; Pascarella, S.; Cauda, R.; Ciccozzi, M.; Cassone, A. Evolutionary Analysis of SARS-CoV-2: How Mutation of Non-Structural Protein 6 (NSP6) Could Affect Viral Autophagy. *J. Infect.* **2020**, *81*, e24–e27. [CrossRef] [PubMed]
65. Cottam, E.M.; Whelband, M.C.; Wileman, T. Coronavirus NSP6 Restricts Autophagosome Expansion. *Autophagy* **2014**, *10*, 1426–1441. [CrossRef]
66. Sun, X.; Liu, Y.; Huang, Z.; Xu, W.; Hu, W.; Yi, L.; Liu, Z.; Chan, H.; Zeng, J.; Liu, X.; et al. SARS-CoV-2 Non-Structural Protein 6 Triggers NLRP3-Dependent Pyroptosis by Targeting ATP6AP1. *Cell Death Differ.* **2022**, *29*, 1240–1254. [CrossRef]
67. Biswal, M.; Diggs, S.; Xu, D.; Khudaverdyan, N.; Lu, J.; Fang, J.; Blaha, G.; Hai, R.; Song, J. Two Conserved Oligomer Interfaces of NSP7 and NSP8 Underpin the Dynamic Assembly of SARS-CoV-2 RdRP. *Nucleic Acids Res.* **2021**, *49*, 5956–5966. [CrossRef]
68. Krichel, B.; Falke, S.; Hilgenfeld, R.; Redecke, L.; Uetrecht, C. Processing of the SARS-CoV Pp1a/Ab Nsp7-10 Region. *Biochem. J.* **2020**, *477*, 1009–1019. [CrossRef]
69. Te Velthuis, A.J.W.; van den Worm, S.H.E.; Snijder, E.J. The SARS-Coronavirus Nsp7+nsp8 Complex Is a Unique Multimeric RNA Polymerase Capable of Both de Novo Initiation and Primer Extension. *Nucleic Acids Res.* **2012**, *40*, 1737–1747. [CrossRef]
70. Zeng, Z.; Deng, F.; Shi, K.; Ye, G.; Wang, G.; Fang, L.; Xiao, S.; Fu, Z.; Peng, G. Dimerization of Coronavirus Nsp9 with Diverse Modes Enhances Its Nucleic Acid Binding Affinity. *J. Virol.* **2018**, *92*, e00692-18. [CrossRef]
71. De Araújo, O.J.; Pinheiro, S.; Zamora, W.J.; Alves, C.N.; Lameira, J.; Lima, A.H. Structural, Energetic and Lipophilic Analysis of SARS-CoV-2 Non-Structural Protein 9 (NSP9). *Sci. Rep.* **2021**, *11*, 23003. [CrossRef]
72. Littler, D.R.; Gully, B.S.; Colson, R.N.; Rossjohn, J. Crystal Structure of the SARS-CoV-2 Non-Structural Protein 9, Nsp9. *iScience* **2020**, *23*, 101258. [CrossRef] [PubMed]
73. Rona, G.; Zeke, A.; Miwatani-Minter, B.; de Vries, M.; Kaur, R.; Schinlever, A.; Garcia, S.F.; Goldberg, H.V.; Wang, H.; Hinds, T.R.; et al. The NSP14/NSP10 RNA Repair Complex as a Pan-Coronavirus Therapeutic Target. *Cell Death Differ.* **2021**, *29*, 285–292. [CrossRef] [PubMed]
74. Ma, Y.; Wu, L.; Shaw, N.; Gao, Y.; Wang, J.; Sun, Y.; Lou, Z.; Yan, L.; Zhang, R.; Rao, Z. Structural Basis and Functional Analysis of the SARS Coronavirus Nsp14-Nsp10 Complex. *Proc. Natl. Acad. Sci. USA* **2015**, *112*, 9436–9441. [CrossRef]
75. Bouvet, M.; Lugari, A.; Posthuma, C.C.; Zevenhoven, J.C.; Bernard, S.; Betzi, S.; Imbert, I.; Canard, B.; Guillemot, J.C.; Lécine, P.; et al. Coronavirus Nsp10, a Critical Co-Factor for Activation of Multiple Replicative Enzymes. *J. Biol. Chem.* **2014**, *289*, 25783–25796. [CrossRef]
76. Gadhave, K.; Kumar, P.; Kumar, A.; Bhardwaj, T.; Garg, N.; Giri, R. Conformational Dynamics of 13 Amino Acids Long NSP11 of SARS-CoV-2 under Membrane Mimetics and Different Solvent Conditions. *Microb. Pathog.* **2021**, *158*, 105041. [CrossRef]
77. Elfiky, A.A. SARS-CoV-2 RNA Dependent RNA Polymerase (RdRp) Targeting: An in Silico Perspective. *J. Biomol. Struct. Dyn.* **2021**, *39*, 3204–3212. [CrossRef]
78. Khan, M.T.; Irfan, M.; Ahsan, H.; Ahmed, A.; Kaushik, A.C.; Khan, A.S.; Chinnasamy, S.; Ali, A.; Wei, D.-Q. Structures of SARS-CoV-2 RNA-Binding Proteins and Therapeutic Targets. *Intervirology* **2021**, *64*, 55–68. [CrossRef]
79. Mishra, A.; Rathore, A.S. RNA Dependent RNA Polymerase (RdRp) as a Drug Target for SARS-CoV2. *J. Biomol. Struct. Dyn.* **2022**, *40*, 6039–6051. [CrossRef]
80. Yin, W.; Mao, C.; Luan, X.; Shen, D.D.; Shen, Q.; Su, H.; Wang, X.; Zhou, F.; Zhao, W.; Gao, M.; et al. Structural Basis for Inhibition of the RNA-Dependent RNA Polymerase from SARS-CoV-2 by Remdesivir. *Science* **2020**, *368*, 1499. [CrossRef]
81. Wang, H.; Xue, S.; Yang, H.; Chen, C. Recent Progress in the Discovery of Inhibitors Targeting Coronavirus Proteases. *Virol. Sin.* **2016**, *31*, 24. [CrossRef] [PubMed]
82. Niu, X.; Kong, F.; Hou, Y.J.; Wang, Q. Crucial Mutation in the Exoribonuclease Domain of Nsp14 of PEDV Leads to High Genetic Instability during Viral Replication. *Cell Biosci.* **2021**, *11*, 106. [CrossRef]
83. Tahir, M. Coronavirus Genomic Nsp14-ExoN, Structure, Role, Mechanism, and Potential Application as a Drug Target. *J. Med. Virol.* **2021**, *93*, 4258–4264. [CrossRef]
84. Krafcikova, P.; Silhan, J.; Nencka, R.; Boura, E. Structural Analysis of the SARS-CoV-2 Methyltransferase Complex Involved in RNA Cap Creation Bound to Sinefungin. *Nat. Commun.* **2020**, *11*, 3717. [CrossRef]
85. Yoshimoto, F.K. A Biochemical Perspective of the Nonstructural Proteins (NSPs) and the Spike Protein of SARS CoV-2. *Protein J.* **2021**, *40*, 260–295. [CrossRef] [PubMed]
86. Vithani, N.; Ward, M.D.; Zimmerman, M.I.; Novak, B.; Borowsky, J.H.; Singh, S.; Bowman, G.R. SARS-CoV-2 Nsp16 Activation Mechanism and a Cryptic Pocket with Pan-Coronavirus Antiviral Potential. *Biophys. J.* **2021**, *120*, 2880–2889. [CrossRef] [PubMed]
87. Rosas-Lemus, M.; Minasov, G.; Shuvalova, L.; Inniss, N.L.; Kiryukhina, O.; Brunzelle, J.; Satchell, K.J.F. High-Resolution Structures of the SARS-CoV-2 2′-O-Methyltransferase Reveal Strategies for Structure-Based Inhibitor Design. *Sci. Signal.* **2020**, *13*, eabe1202. [CrossRef]
88. Fitzgerald, K.A.; Kagan, J.C. Toll-like Receptors and the Control of Immunity. *Cell* **2020**, *180*, 1044–1066. [CrossRef]

89. Balka, K.R.; de Nardo, D. Understanding Early TLR Signaling through the Myddosome. *J. Leukoc. Biol.* **2019**, *105*, 339–351. [CrossRef]
90. Satoh, T.; Akira, S. Toll-Like Receptor Signaling and Its Inducible Proteins. *Microbiol. Spectr.* **2016**, *4*, 4–6. [CrossRef]
91. Barton, G.M.; Kagan, J.C. A Cell Biological View of Toll-like Receptor Function: Regulation through Compartmentalization. *Nat. Rev. Immunol.* **2009**, *9*, 535–541. [CrossRef] [PubMed]
92. Wang, Y.; Song, E.; Bai, B.; Vanhoutte, P.M. Toll-like Receptors Mediating Vascular Malfunction: Lessons from Receptor Subtypes. *Pharmacol. Ther.* **2016**, *158*, 91–100. [CrossRef] [PubMed]
93. Kagan, J.C. Signaling Organelles of the Innate Immune System. *Cell* **2012**, *151*, 1168–1178. [CrossRef] [PubMed]
94. Yamamoto, M.; Sato, S.; Mori, K.; Hoshino, K.; Takeuchi, O.; Takeda, K.; Akira, S. Cutting Edge: A Novel Toll/IL-1 Receptor Domain-Containing Adapter That Preferentially Activates the IFN-Beta Promoter in the Toll-like Receptor Signaling. *J. Immunol.* **2002**, *169*, 6668–6672. [CrossRef]
95. Kawai, T.; Akira, S. The Role of Pattern-Recognition Receptors in Innate Immunity: Update on Toll-like Receptors. *Nat. Immunol.* **2010**, *11*, 373–384. [CrossRef] [PubMed]
96. Zhang, Y.; Liang, C. Innate Recognition of Microbial-Derived Signals in Immunity and Inflammation. *Sci. China Life Sci.* **2016**, *59*, 1210–1217. [CrossRef] [PubMed]
97. El-Zayat, S.R.; Sibaii, H.; Mannaa, F.A. Toll-like Receptors Activation, Signaling, and Targeting: An Overview. *Bull. Natl. Res. Cent.* **2019**, *43*, 187. [CrossRef]
98. Gao, W.; Xiong, Y.; Li, Q.; Yang, H. Inhibition of Toll-like Receptor Signaling as a Promising Therapy for Inflammatory Diseases: A Journey from Molecular to Nano Therapeutics. *Front. Physiol.* **2017**, *8*, 508. [CrossRef]
99. Lushpa, V.A.; Goncharuk, M.V.; Lin, C.; Zalevsky, A.O.; Talyzina, I.A.; Luginina, A.P.; Vakhrameev, D.D.; Shevtsov, M.B.; Goncharuk, S.A.; Arseniev, A.S.; et al. Modulation of Toll-like Receptor 1 Intracellular Domain Structure and Activity by Zn2+ Ions. *Commun. Biol.* **2021**, *4*, 1003. [CrossRef]
100. Takeuchi, O.; Sato, S.; Horiuchi, T.; Hoshino, K.; Takeda, K.; Dong, Z.; Modlin, R.L.; Akira, S. Cutting Edge: Role of Toll-Like Receptor 1 in Mediating Immune Response to Microbial Lipoproteins. *J. Immunol.* **2002**, *169*, 10–14. [CrossRef]
101. Ahmed, S.; Moawad, M.; Elhefny, R.; Chest, M.A.-E.J. *Is Toll like Receptor 4 a Common Pathway Hypothesis for Development of Lung Cancer and Idiopathic Pulmonary Fibrosis?* Elsevier: Amsterdam, The Netherlands, 2016.
102. Oliveira-Nascimento, L.; Massari, P.; Wetzler, L.M. The Role of TLR2 Ininfection and Immunity. *Front. Immunol.* **2012**, *3*, 79. [CrossRef] [PubMed]
103. Zainol, M.I.B.; Kawasaki, T.; Monwan, W.; Murase, M.; Sueyoshi, T.; Kawai, T. Innate Immune Responses through Toll-like Receptor 3 Require Human-Antigen-R-Mediated Atp6v0d2 MRNA Stabilization. *Sci. Rep.* **2019**, *9*, 20406. [CrossRef] [PubMed]
104. Chen, Y.; Lin, J.; Zhao, Y.; Ma, X.; Yi, H. Toll-like Receptor 3 (TLR3) Regulation Mechanisms and Roles in Antiviral Innate Immune Responses. *J. Zhejiang Univ. Sci. B* **2021**, *22*, 609. [CrossRef]
105. Molteni, M.; Gemma, S.; Rossetti, C. The Role of Toll-Like Receptor 4 in Infectious and Noninfectious Inflammation. *Mediators Inflamm.* **2016**, *2016*, 6978936. [CrossRef] [PubMed]
106. Zhang, Y.; Liang, X.; Bao, X.; Xiao, W.; Chen, G. Toll-like Receptor 4 (TLR4) Inhibitors: Current Research and Prospective. *Eur. J. Med. Chem.* **2022**, *235*, 114291. [CrossRef]
107. Caballero, I.; Boyd, J.; Alminanã, C.; Sánchez-López, J.A.; Basatvat, S.; Montazeri, M.; Maslehat Lay, N.; Elliott, S.; Spiller, D.G.; White, M.R.H.; et al. Understanding the Dynamics of Toll-like Receptor 5 Response to Flagellin and Its Regulation by Estradiol. *Sci. Rep.* **2017**, *7*, 40981. [CrossRef]
108. Zhang, W.; Wang, L.; Sun, X.H.; Liu, X.; Xiao, Y.; Zhang, J.; Wang, T.; Chen, H.; Zhan, Y.Q.; Yu, M.; et al. Toll-like Receptor 5-Mediated Signaling Enhances Liver Regeneration in Mice. *Mil. Med. Res.* **2021**, *8*, 16. [CrossRef]
109. Kim, J.H.; Kordahi, M.C.; Chac, D.; William DePaolo, R. Toll-like Receptor-6 Signaling Prevents Inflammation and Impacts Composition of the Microbiota during Inflammation-Induced Colorectal Cancer. *Cancer Prev. Res.* **2020**, *13*, 25–40. [CrossRef]
110. De Almeida, L.A.; Macedo, G.C.; Marinho, F.A.V.; Gomes, M.T.R.; Corsetti, P.P.; Silva, A.M.; Cassataro, J.; Giambartolomei, G.H.; Oliveira, S.C. Toll-like Receptor 6 Plays an Important Role in Host Innate Resistance to Brucella Abortus Infection in Mice. *Infect. Immun.* **2013**, *81*, 1654–1662. [CrossRef]
111. Bortolotti, D.; Gentili, V.; Rizzo, S.; Schiuma, G.; Beltrami, S.; Strazzabosco, G.; Fernandez, M.; Caccuri, F.; Caruso, A.; Rizzo, R. TLR3 and TLR7 RNA Sensor Activation during SARS-CoV-2 Infection. *Microorganisms* **2021**, *9*, 1820. [CrossRef]
112. Macedo, A.B.; Novis, C.L.; de Assis, C.M.; Sorensen, E.S.; Moszczynski, P.; Huang, S.H.; Ren, Y.; Spivak, A.M.; Jones, R.B.; Planelles, V.; et al. Dual TLR2 and TLR7 Agonists as HIV Latency-Reversing Agents. *JCI Insight* **2018**, *3*. [CrossRef] [PubMed]
113. De Marcken, M.; Dhaliwal, K.; Danielsen, A.C.; Gautron, A.S.; Dominguez-Villar, M. TLR7 and TLR8 Activate Distinct Pathways in Monocytes during RNA Virus Infection. *Sci. Signal.* **2019**, *12*, eaaw1347. [CrossRef] [PubMed]
114. Salvi, V.; Nguyen, H.O.; Sozio, F.; Schioppa, T.; Gaudenzi, C.; Laffranchi, M.; Scapini, P.; Passari, M.; Barbazza, I.; Tiberio, L.; et al. SARS-CoV-2-Associated SsRNAs Activate Inflammation and Immunity via TLR7/8. *JCI Insight* **2021**, *6*, e150542. [CrossRef] [PubMed]
115. Costa, T.J.; Potje, S.R.; Fraga-Silva, T.F.C.; da Silva-Neto, J.A.; Barros, P.R.; Rodrigues, D.; Machado, M.R.; Martins, R.B.; Santos-Eichler, R.A.; Benatti, M.N.; et al. Mitochondrial DNA and TLR9 Activation Contribute to SARS-CoV-2-Induced Endothelial Cell Damage. *Vascul. Pharmacol.* **2022**, *142*, 106946. [CrossRef] [PubMed]

116. Khan, N.S.; Lukason, D.P.; Feliu, M.; Ward, R.A.; Lord, A.K.; Reedy, J.L.; Ramirez-Ortiz, Z.G.; Tam, J.M.; Kasperkovitz, P.V.; Negoro, P.E.; et al. CD82 Controls CpG-Dependent TLR9 Signaling. *FASEB J.* **2019**, *33*, 12500–12514. [CrossRef]
117. Lester, S.N.; Li, K. Toll-Like Receptors in Antiviral Innate Immunity. *J. Mol. Biol.* **2014**, *426*, 1246. [CrossRef]
118. Gaglia, M.M. Anti-Viral and pro-Inflammatory Functions of Toll-like Receptors during Gamma-Herpesvirus Infections. *Virol. J.* **2021**, *18*, 218. [CrossRef]
119. Shah, M.; Anwar, M.A.; Kim, J.H.; Choi, S. Advances in Antiviral Therapies Targeting Toll-like Receptors. *Expert Opin. Investig. Drugs* **2016**, *25*, 437–453. [CrossRef]
120. Vijay, K. Toll-like Receptors in Immunity and Inflammatory Diseases: Past, Present, and Future. *Int. Immunopharmacol.* **2018**, *59*, 391. [CrossRef]
121. Federico, S.; Pozzetti, L.; Papa, A.; Carullo, G.; Gemma, S.; Butini, S.; Campiani, G.; Relitti, N. Modulation of the Innate Immune Response by Targeting Toll-like Receptors: A Perspective on Their Agonists and Antagonists. *J. Med. Chem.* **2020**, *63*, 13466–13513. [CrossRef]
122. O'Neill, L.A.J.; Hennessy, E.J.; Parker, A.E. Targeting Toll-like Receptors: Emerging Therapeutics? *Nat. Rev. Drug Discov.* **2010**, *9*, 293–307. [CrossRef]
123. Zheng, W.; Xu, Q.; Zhang, Y.; Xiaofei, E.; Gao, W.; Zhang, M.; Zhai, W.; Rajkumar, R.S.; Liu, Z. Toll-like Receptor-Mediated Innate Immunity against Herpesviridae Infection: A Current Perspective on Viral Infection Signaling Pathways. *Virol. J.* **2020**, *17*, 192. [CrossRef]
124. Boehme, K.W.; Compton, T. Innate Sensing of Viruses by Toll-Like Receptors. *J. Virol.* **2004**, *78*, 7867. [CrossRef] [PubMed]
125. Ge, Y.; Mansell, A.; Ussher, J.E.; Brooks, A.E.S.; Manning, K.; Wang, C.J.H.; Taylor, J.A. Rotavirus NSP4 Triggers Secretion of Proinflammatory Cytokines from Macrophages via Toll-Like Receptor 2. *J. Virol.* **2013**, *87*, 11160–11167. [CrossRef]
126. Dela Justina, V.; Giachini, F.R.; Priviero, F.; Webb, R.C. COVID-19 and Hypertension: Is There a Role for DsRNA and Activation of Toll-like Receptor 3? *Vascul. Pharmacol.* **2021**, *140*, 106861. [CrossRef] [PubMed]
127. Lau, Y.F.; Tang, L.H.; Ooi, E.E. A TLR3 Ligand That Exhibits Potent Inhibition of Influenza Virus Replication and Has Strong Adjuvant Activity Has the Potential for Dual Applications in an Influenza Pandemic. *Vaccine* **2009**, *27*, 1354–1364. [CrossRef] [PubMed]
128. Miller, R.L.; Meng, T.C.; Tomai, M.A. The Antiviral Activity of Toll-like Receptor 7 and 7/8 Agonists. *Drug News Perspect.* **2008**, *21*, 69–87. [CrossRef]
129. Rose, W.A.; McGowin, C.L.; Pyles, R.B. FSL-1, a Bacterial-Derived Toll-like Receptor 2/6 Agonist, Enhances Resistance to Experimental HSV-2 Infection. *Virol. J.* **2009**, *6*, 195. [CrossRef]
130. Bowie, A.G.; Unterholzner, L. Viral Evasion and Subversion of Pattern-Recognition Receptor Signalling. *Nat. Rev. Immunol.* **2008**, *8*, 911–922. [CrossRef]
131. Duan, S.; Xu, X.; Wang, J.; Huang, L.; Peng, J.; Yu, T.; Zhou, Y.; Cheng, K.; Liu, S. TLR1/2 Agonist Enhances Reversal of HIV-1 Latency and Promotes NK Cell-Induced Suppression of HIV-1-Infected Autologous CD4 + T Cells. *J. Virol.* **2021**, *95*, e00816-21. [CrossRef]
132. Girkin, J.; Loo, S.L.; Esneau, C.; Maltby, S.; Mercuri, F.; Chua, B.; Reid, A.T.; Veerati, P.C.; Grainge, C.L.; Wark, P.A.B.; et al. TLR2-Mediated Innate Immune Priming Boosts Lung Anti-Viral Immunity. *Eur. Respir. J.* **2021**, *58*, 2001584. [CrossRef] [PubMed]
133. Howell, M.D.; Gallo, R.L.; Boguniewicz, M.; Jones, J.F.; Wong, C.; Streib, J.E.; Leung, D.Y.M. Cytokine Milieu of Atopic Dermatitis Skin Subverts the Innate Immune Response to Vaccinia Virus. *Immunity* **2006**, *24*, 341–348. [CrossRef] [PubMed]
134. Hutchens, M.A.; Luker, K.E.; Sonstein, J.; Núñez, G.; Curtis, J.L.; Luker, G.D. Protective Effect of Toll-like Receptor 4 in Pulmonary Vaccinia Infection. *PLoS Pathog.* **2008**, *4*, e1000153. [CrossRef] [PubMed]
135. Alexopoulou, L.; Holt, A.C.; Medzhitov, R.; Flavell, R.A. Recognition of Double-Stranded RNA and Activation of NF-KappaB by Toll-like Receptor 3. *Nature* **2001**, *413*, 732–738. [CrossRef]
136. Mazaleuskaya, L.; Veltrop, R.; Ikpeze, N.; Martin-Garcia, J.; Navas-Martin, S. Protective Role of Toll-like Receptor 3-Induced Type I Interferon in Murine Coronavirus Infection of Macrophages. *Viruses* **2012**, *4*, 901–923. [CrossRef] [PubMed]
137. Zhou, Y.; Wang, X.; Liu, M.; Hu, Q.; Song, L.; Ye, L.; Zhou, D.; Ho, W. A Critical Function of Toll-like Receptor-3 in the Induction of Anti-Human Immunodeficiency Virus Activities in Macrophages. *Immunology* **2010**, *131*, 40–49. [CrossRef]
138. Isogawa, M.; Robek, M.D.; Furuichi, Y.; Chisari, F.v. Toll-like Receptor Signaling Inhibits Hepatitis B Virus Replication in Vivo. *J. Virol.* **2005**, *79*, 7269–7272. [CrossRef]
139. Guillot, L.; le Goffic, R.; Bloch, S.; Escriou, N.; Akira, S.; Chignard, M.; Si-Tahar, M. Involvement of Toll-like Receptor 3 in the Immune Response of Lung Epithelial Cells to Double-Stranded RNA and Influenza A Virus. *J. Biol. Chem.* **2005**, *280*, 5571–5580. [CrossRef]
140. Zhang, M.; Yan, Z.; Wang, J.; Yao, X. Toll-like Receptors 7 and 8 Expression Correlates with the Expression of Immune Biomarkers and Positively Predicts the Clinical Outcome of Patients with Melanoma. *Onco Targets Ther.* **2017**, *10*, 4339. [CrossRef]
141. Drake, M.G.; Evans, S.E.; Dickey, B.F.; Fryer, A.D.; Jacoby, D.B. Toll-like Receptor-2/6 and Toll-like Receptor-9 Agonists Suppress Viral Replication but Not Airway Hyperreactivity in Guinea Pigs. *Am J Respir Cell Mol. Biol.* **2013**, *48*, 790–796. [CrossRef]
142. Proud, P.C.; Tsitoura, D.; Watson, R.J.; Chua, B.Y.; Aram, M.J.; Bewley, K.R.; Cavell, B.E.; Cobb, R.; Dowall, S.; Fotheringham, S.A.; et al. Prophylactic Intranasal Administration of a TLR2/6 Agonist Reduces Upper Respiratory Tract Viral Shedding in a SARS-CoV-2 Challenge Ferret Model. *EBioMedicine* **2021**, *63*, 103153. [CrossRef] [PubMed]

143. Lau, Y.F.; Tang, L.H.; Ooi, E.E.; Subbarao, K. Activation of the Innate Immune System Provides Broad-Spectrum Protection against Influenza A Viruses with Pandemic Potential in Mice. *Virology* **2010**, *406*, 80–87. [CrossRef] [PubMed]
144. Christopher, M.; Wong, J. Use of Toll-Like Receptor 3 Agonists Against Respiratory Viral Infections. *Antiinflamm. Antiallergy Agents Med. Chem.* **2011**, *10*, 327. [CrossRef]
145. Kim, M.; Lee, J.E.; Cho, H.; Jung, H.G.; Lee, W.; Seo, H.Y.; Lee, S.H.; Ahn, D.G.; Kim, S.J.; Yu, J.W.; et al. Antiviral Efficacy of Orally Delivered Neoagarohexaose, a Nonconventional TLR4 Agonist, against Norovirus Infection in Mice. *Biomaterials* **2020**, *263*, 120391. [CrossRef] [PubMed]
146. Luchner, M.; Reinke, S.; Milicic, A. TLR Agonists as Vaccine Adjuvants Targeting Cancer and Infectious Diseases. *Pharmaceutics* **2021**, *13*, 142. [CrossRef]
147. Georgel, A.F.; Cayet, D.; Pizzorno, A.; Rosa-Calatrava, M.; Paget, C.; Sencio, V.; Dubuisson, J.; Trottein, F.; Sirard, J.C.; Carnoy, C. Toll-like Receptor 5 Agonist Flagellin Reduces Influenza A Virus Replication Independently of Type I Interferon and Interleukin 22 and Improves Antiviral Efficacy of Oseltamivir. *Antiviral Res.* **2019**, *168*, 28–35. [CrossRef]
148. Melin, N.; Sánchez-Taltavull, D.; Fahrner, R.; Keogh, A.; Dosch, M.; Büchi, I.; Zimmer, Y.; Medová, M.; Beldi, G.; Aebersold, D.M.; et al. Synergistic Effect of the TLR5 Agonist CBLB502 and Its Downstream Effector IL-22 against Liver Injury. *Cell Death Dis.* **2021**, *12*, 366. [CrossRef]
149. SenGupta, D.; Brinson, C.; DeJesus, E.; Mills, A.; Shalit, P.; Guo, S.; Cai, Y.; Wallin, J.J.; Zhang, L.; Humeniuk, R.; et al. The TLR7 Agonist Vesatolimod Induced a Modest Delay in Viral Rebound in HIV Controllers after Cessation of Antiretroviral Therapy. *Sci. Transl. Med.* **2021**, *13*, eabg3071. [CrossRef]
150. Vanwalscappel, B.; Tada, T.; Landau, N.R. Toll-like Receptor Agonist R848 Blocks Zika Virus Replication by Inducing the Antiviral Protein Viperin. *Virology* **2018**, *522*, 199–208. [CrossRef]
151. Amin, O.E.; Colbeck, E.J.; Daffis, S.; Khan, S.; Ramakrishnan, D.; Pattabiraman, D.; Chu, R.; Micolochick Steuer, H.; Lehar, S.; Peiser, L.; et al. Therapeutic Potential of TLR8 Agonist GS-9688 (Selgantolimod) in Chronic Hepatitis B: Remodeling of Antiviral and Regulatory Mediators. *Hepatology* **2021**, *74*, 55–71. [CrossRef]
152. Manik, M.; Singh, R.K. Role of Toll-like Receptors in Modulation of Cytokine Storm Signaling in SARS-CoV-2-induced COVID-19. *J. Med. Virol.* **2022**, *94*, 869. [CrossRef] [PubMed]
153. Mabrey, F.L.; Morrell, E.D.; Wurfel, M.M. TLRs in COVID-19: How They Drive Immunopathology and the Rationale for Modulation. *Innate Immun.* **2021**, *27*, 503–513. [CrossRef] [PubMed]
154. Liu, S.Y.; Sanchez, D.J.; Aliyari, R.; Lu, S.; Cheng, G. Systematic Identification of Type I and Type II Interferon-Induced Antiviral Factors. *Proc. Natl. Acad. Sci. USA* **2012**, *109*, 4239–4244. [CrossRef] [PubMed]
155. Lazear, H.M.; Schoggins, J.W.; Diamond, M.S. Shared and Distinct Functions of Type I and Type III Interferons. *Immunity* **2019**, *50*, 907–923. [CrossRef]
156. Zanoni, I.; Granucci, F.; Broggi, A. Interferon (IFN)-λ Takes the Helm: Immunomodulatory Roles of Type III IFNs. *Front. Immunol.* **2017**, *8*, 1661. [CrossRef]
157. Lotfi, M.; Rezaei, N. SARS-CoV-2: A Comprehensive Review from Pathogenicity of the Virus to Clinical Consequences. *J. Med. Virol.* **2020**, *92*, 1864. [CrossRef]
158. Teijaro, J.R. Type I Interferons in Viral Control and Immune Regulation. *Curr. Opin. Virol.* **2016**, *16*, 31–40. [CrossRef]
159. Rabaan, A.A.; Al-Ahmed, S.H.; Muhammad, J.; Khan, A.; Sule, A.A.; Tirupathi, R.; Al Mutair, A.; Alhumaid, S.; Al-Omari, A.; Dhawan, M.; et al. Role of Inflammatory Cytokines in COVID-19 Patients: A Review on Molecular Mechanisms, Immune Functions, Immunopathology and Immunomodulatory Drugs to Counter Cytokine Storm. *Vaccines* **2021**, *9*, 436. [CrossRef]
160. Darif, D.; Hammi, I.; Kihel, A.; el Idrissi Saik, I.; Guessous, F.; Akarid, K. The Pro-Inflammatory Cytokines in COVID-19 Pathogenesis: What Goes Wrong? *Microb. Pathog.* **2021**, *153*, 104799. [CrossRef]
161. Yapasert, R.; Khaw-On, P.; Banjerdpongchai, R. Coronavirus Infection-Associated Cell Death Signaling and Potential Therapeutic Targets. *Molecules* **2021**, *26*, 7459. [CrossRef]
162. Alesci, A.; Aragona, M.; Cicero, N.; Lauriano, E.R. Can Nutraceuticals Assist Treatment and Improve COVID-19 Symptoms? *Nat. Prod. Res.* **2022**, *36*, 2672–2691. [CrossRef] [PubMed]
163. Domi, E.; Hoxha, M.; Kolovani, E.; Tricarico, D.; Zappacosta, B. The Importance of Nutraceuticals in COVID-19: What's the Role of Resveratrol? *Molecules* **2022**, *27*, 2376. [CrossRef] [PubMed]
164. Parisi, G.F.; Carota, G.; Castruccio Castracani, C.; Spampinato, M.; Manti, S.; Papale, M.; di Rosa, M.; Barbagallo, I.; Leonardi, S. Nutraceuticals in the Prevention of Viral Infections, Including COVID-19, among the Pediatric Population: A Review of the Literature. *Int. J. Mol. Sci.* **2021**, *22*, 52465. [CrossRef]
165. Ramdani, L.H.; Bachari, K. Potential Therapeutic Effects of Resveratrol against SARS-CoV-2. *Acta Virol.* **2020**, *64*, 276–280. [CrossRef] [PubMed]
166. Derosa, G.; Maffioli, P.; D'Angelo, A.; di Pierro, F. A Role for Quercetin in Coronavirus Disease 2019 (COVID-19). *Phytother. Res.* **2021**, *35*, 1230–1236. [CrossRef]
167. Fan, W.; Qian, S.; Qian, P.; Li, X. Antiviral Activity of Luteolin against Japanese Encephalitis Virus. *Virus Res.* **2016**, *220*, 112–116. [CrossRef]
168. Hao, L.; Shan, Q.; Wei, J.; Ma, F.; Sun, P. Lactoferrin: Major Physiological Functions and Applications. *Curr. Protein Pept. Sci.* **2019**, *20*, 139–144. [CrossRef]

169. Haggag, Y.A.; El-Ashmawy, N.E.; Okasha, K.M. Is Hesperidin Essential for Prophylaxis and Treatment of COVID-19 Infection? *Med. Hypotheses* **2020**, *144*, 109957. [CrossRef]
170. Luo, P.; Liu, D.; Li, J. Pharmacological Perspective: Glycyrrhizin May Be an Efficacious Therapeutic Agent for COVID-19. *Int. J. Antimicrob. Agents* **2020**, *55*, 105995. [CrossRef]
171. Babaei, F.; Nassiri-Asl, M.; Hosseinzadeh, H. Curcumin (a Constituent of Turmeric): New Treatment Option against COVID-19. *Food Sci. Nutr.* **2020**, *8*, 5215–5227. [CrossRef]
172. Pasandi Pour, A.; Farahbakhsh, H. *Lawsonia Inermis* L. Leaves Aqueous Extract as a Natural Antioxidant and Antibacterial Product. *Nat. Prod. Res.* **2020**, *34*, 3399–3403. [CrossRef] [PubMed]
173. Davoudi-Monfared, E.; Rahmani, H.; Khalili, H.; Hajiabdolbaghi, M.; Salehi, M.; Abbasian, L.; Kazemzadeh, H.; Yekaninejad, M.S. A Randomized Clinical Trial of the Efficacy and Safety of Interferon β-1a in Treatment of Severe COVID-19. *Antimicrob. Agents Chemother.* **2020**, *64*, e01061-20. [CrossRef]
174. Zhou, Q.; Chen, V.; Shannon, C.P.; Wei, X.S.; Xiang, X.; Wang, X.; Wang, Z.H.; Tebbutt, S.J.; Kollmann, T.R.; Fish, E.N. Interferon-A2b Treatment for COVID-19. *Front. Immunol.* **2020**, *11*, 1061. [CrossRef] [PubMed]
175. Hung, I.F.N.; Lung, K.C.; Tso, E.Y.K.; Liu, R.; Chung, T.W.H.; Chu, M.Y.; Ng, Y.Y.; Lo, J.; Chan, J.; Tam, A.R.; et al. Triple Combination of Interferon Beta-1b, Lopinavir-Ritonavir, and Ribavirin in the Treatment of Patients Admitted to Hospital with COVID-19: An Open-Label, Randomised, Phase 2 Trial. *Lancet* **2020**, *395*, 1695–1704. [CrossRef]
176. Barbalat, R.; Lau, L.; Locksley, R.M.; Barton, G.M. Toll-like Receptor 2 on Inflammatory Monocytes Induces Type I Interferon in Response to Viral but Not Bacterial Ligands. *Nat. Immunol.* **2009**, *10*, 1200–1209. [CrossRef] [PubMed]
177. Kawai, T.; Akira, S. Toll-like Receptors and Their Crosstalk with Other Innate Receptors in Infection and Immunity. *Immunity* **2011**, *34*, 637–650. [CrossRef]
178. Menachery, V.D.; Debbink, K.; Baric, R.S. Coronavirus Non-Structural Protein 16: Evasion, Attenuation, and Possible Treatments. *Virus Res.* **2014**, *194*, 191–199. [CrossRef]
179. Bouvet, M.; Debarnot, C.; Imbert, I.; Selisko, B.; Snijder, E.J.; Canard, B.; Decroly, E. In Vitro Reconstitution of SARS-Coronavirus MRNA Cap Methylation. *PLoS Pathog.* **2010**, *6*, e1000863. [CrossRef]
180. De Rivero Vaccari, J.C.; Dietrich, W.D.; Keane, R.W.; de Rivero Vaccari, J.P. The Inflammasome in Times of COVID-19. *Front. Immunol.* **2020**, *11*, 583373. [CrossRef]
181. Sariol, A.; Perlman, S. SARS-CoV-2 Takes Its Toll. *Nat. Immunol.* **2021**, *22*, 801–802. [CrossRef]
182. Zheng, M.; Karki, R.; Williams, E.P.; Yang, D.; Fitzpatrick, E.; Vogel, P.; Jonsson, C.B.; Kanneganti, T.D. TLR2 Senses the SARS-CoV-2 Envelope Protein to Produce Inflammatory Cytokines. *Nat. Immunol.* **2021**, *22*, 829–838. [CrossRef] [PubMed]
183. Zhou, H.; Zhao, J.; Perlman, S. Autocrine Interferon Priming in Macrophages but Not Dendritic Cells Results in Enhanced Cytokine and Chemokine Production after Coronavirus Infection. *mBio* **2010**, *1*, e00219-10. [CrossRef] [PubMed]
184. Khanmohammadi, S.; Rezaei, N. Role of Toll-like Receptors in the Pathogenesis of COVID-19. *J. Med. Virol.* **2021**, *93*, 2735. [CrossRef] [PubMed]
185. Totura, A.L.; Whitmore, A.; Agnihothram, S.; Schäfer, A.; Katze, M.G.; Heise, M.T.; Baric, R.S. Toll-Like Receptor 3 Signaling via TRIF Contributes to a Protective Innate Immune Response to Severe Acute Respiratory Syndrome Coronavirus Infection. *mBio* **2015**, *6*, e00638-15. [CrossRef]
186. Cicco, S.; Cicco, G.; Racanelli, V.; Vacca, A. Neutrophil Extracellular Traps (NETs) and Damage-Associated Molecular Patterns (DAMPs): Two Potential Targets for COVID-19 Treatment. *Mediators Inflamm.* **2020**, *2020*, 7527953. [CrossRef]
187. Khadke, S.; Ahmed, N.; Ahmed, N.; Ratts, R.; Raju, S.; Gallogly, M.; de Lima, M.; Sohail, M.R. Harnessing the Immune System to Overcome Cytokine Storm and Reduce Viral Load in COVID-19: A Review of the Phases of Illness and Therapeutic Agents. *Virol. J.* **2020**, *17*, 154. [CrossRef]
188. Cao, C.; Yin, C.; Shou, S.; Wang, J.; Yu, L.; Li, X.; Chai, Y. Ulinastatin Protects Against LPS-Induced Acute Lung Injury By Attenuating TLR4/NF-KB Pathway Activation and Reducing Inflammatory Mediators. *Shock* **2018**, *50*, 595–605. [CrossRef]
189. Ohashi, K.; Burkart, V.; Flohé, S.; Kolb, H. Cutting Edge: Heat Shock Protein 60 Is a Putative Endogenous Ligand of the Toll-like Receptor-4 Complex. *J. Immunol.* **2000**, *164*, 558–561. [CrossRef]
190. Chakraborty, C.; Sharma, A.R.; Bhattacharya, M.; Sharma, G.; Lee, S.S.; Agoramoorthy, G. Consider TLR5 for New Therapeutic Development against COVID-19. *J. Med. Virol.* **2020**, *92*, 2314–2315. [CrossRef]
191. Magro, G. SARS-CoV-2 and COVID-19: Is Interleukin-6 (IL-6) the "culprit Lesion" of ARDS Onset? What Is There besides Tocilizumab? SGP130Fc. *Cytokine X* **2020**, *2*, 100029. [CrossRef]
192. Su, H.; Lei, C.T.; Zhang, C. Interleukin-6 Signaling Pathway and Its Role in Kidney Disease: An Update. *Front. Immunol.* **2017**, *8*, 405. [CrossRef] [PubMed]
193. Ghosh, T.K.; Mickelson, D.J.; Fink, J.; Solberg, J.C.; Inglefield, J.R.; Hook, D.; Gupta, S.K.; Gibson, S.; Alkan, S.S. Toll-like Receptor (TLR) 2-9 Agonists-Induced Cytokines and Chemokines: I. Comparison with T Cell Receptor-Induced Responses. *Cell Immunol.* **2006**, *243*, 48–57. [CrossRef] [PubMed]
194. Moreno-Eutimio, M.A.; López-Macías, C.; Pastelin-Palacios, R. Bioinformatic Analysis and Identification of Single-Stranded RNA Sequences Recognized by TLR7/8 in the SARS-CoV-2, SARS-CoV, and MERS-CoV Genomes. *Microbes Infect.* **2020**, *22*, 226–229. [CrossRef]

195. Van der Made, C.I.; Simons, A.; Schuurs-Hoeijmakers, J.; van den Heuvel, G.; Mantere, T.; Kersten, S.; van Deuren, R.C.; Steehouwer, M.; van Reijmersdal, S.V.; Jaeger, M.; et al. Presence of Genetic Variants Among Young Men with Severe COVID-19. *JAMA* **2020**, *324*, 663–673. [CrossRef] [PubMed]
196. Angelopoulou, A.; Alexandris, N.; Konstantinou, E.; Mesiakaris, K.; Zanidis, C.; Farsalinos, K.; Poulas, K. Imiquimod-A Toll like Receptor 7 Agonist-Is an Ideal Option for Management of COVID 19. *Environ. Res.* **2020**, *188*, 109858. [CrossRef] [PubMed]
197. Jangra, S.; de Vrieze, J.; Choi, A.; Rathnasinghe, R.; Laghlali, G.; Uvyn, A.; van Herck, S.; Nuhn, L.; Deswarte, K.; Zhong, Z.; et al. Sterilizing Immunity against SARS-CoV-2 Infection in Mice by a Single-Shot and Lipid Amphiphile Imidazoquinoline TLR7/8 Agonist-Adjuvanted Recombinant Spike Protein Vaccine. *Angew. Chem. Int. Ed. Engl.* **2021**, *60*, 9467–9473. [CrossRef]
198. Ma-Lauer, Y.; Carbajo-Lozoya, J.; Hein, M.Y.; Müller, M.A.; Deng, W.; Lei, J.; Meyer, B.; Kusov, Y.; von Brunn, B.; Bairad, D.R.; et al. P53 Down-Regulates SARS Coronavirus Replication and Is Targeted by the SARS-Unique Domain and PLpro via E3 Ubiquitin Ligase RCHY1. *Proc. Natl. Acad. Sci. USA* **2016**, *113*, E5192–E5201. [CrossRef]
199. Fekete, T.; Ágics, B.; Bencze, D.; Bene, K.; Szántó, A.; Tarr, T.; Veréb, Z.; Bácsi, A.; Pázmándi, K. Regulation of RLR-Mediated Antiviral Responses of Human Dendritic Cells by MTOR. *Front. Immunol.* **2020**, *11*, 2326. [CrossRef]
200. Khan, S.; Shafiei, M.S.; Longoria, C.; Schoggins, J.W.; Savani, R.C.; Zaki, H. SARS-CoV-2 Spike Protein Induces Inflammation via TLR2-Dependent Activation of the NF-KB Pathway. *Elife* **2021**, *10*. [CrossRef]
201. Durai, P.; Shin, H.-J.; Achek, A.; Kwon, H.-K.; Govindaraj, R.G.; Panneerselvam, S.; Yesudhas, D.; Choi, J.; No, K.T.; Choi, S. Toll-like Receptor 2 Antagonists Identified through Virtual Screening and Experimental Validation. *FEBS J.* **2017**, *284*, 2264–2283. [CrossRef]
202. Menezes, M.C.S.; Veiga, A.D.M.; Martins de Lima, T.; Kunimi Kubo Ariga, S.; Vieira Barbeiro, H.; de Lucena Moreira, C.; Pinto, A.A.S.; Brandao, R.A.; Marchini, J.F.; Alencar, J.C.; et al. Lower Peripheral Blood Toll-like Receptor 3 Expression Is Associated with an Unfavorable Outcome in Severe COVID-19 Patients. *Sci. Rep.* **2021**, *11*, 15223. [CrossRef] [PubMed]
203. Gosu, V.; Son, S.; Shin, D.; Song, K.-D. Insights into the Dynamic Nature of the DsRNA-Bound TLR3 Complex. *Sci. Rep.* **2019**, *9*, 3652. [CrossRef] [PubMed]
204. Aboudounya, M.M.; Heads, R.J. COVID-19 and Toll-Like Receptor 4 (TLR4): SARS-CoV-2 May Bind and Activate TLR4 to Increase ACE2 Expression, Facilitating Entry and Causing Hyperinflammation. *Mediators Inflamm.* **2021**, *2021*, 8874339. [CrossRef] [PubMed]
205. Krishnan, M.; Choi, J.; Jang, A.; Choi, S.; Yeon, J.; Jang, M.; Lee, Y.; Son, K.; Shin, S.Y.; Jeong, M.S.; et al. Molecular Mechanism Underlying the TLR4 Antagonistic and Antiseptic Activities of Papiliocin, an Insect Innate Immune Response Molecule. *Proc. Natl. Acad. Sci. USA* **2022**, *119*, e2115669119. [CrossRef]
206. Dyavar, S.R.; Singh, R.; Emani, R.; Pawar, G.P.; Chaudhari, V.D.; Podany, A.T.; Avedissian, S.N.; Fletcher, C.V.; Salunke, D.B. Role of Toll-like Receptor 7/8 Pathways in Regulation of Interferon Response and Inflammatory Mediators during SARS-CoV2 Infection and Potential Therapeutic Options. *Biomed. Pharmacother.* **2021**, *141*, 111794. [CrossRef]
207. Gentile, F.; Deriu, M.; Licandro, G.; Prunotto, A.; Danani, A.; Tuszynski, J. Structure Based Modeling of Small Molecules Binding to the TLR7 by Atomistic Level Simulations. *Molecules* **2015**, *20*, 8316–8340. [CrossRef]
208. Zhang, Z.; Ohto, U.; Shibata, T.; Krayukhina, E.; Taoka, M.; Yamauchi, Y.; Tanji, H.; Isobe, T.; Uchiyama, S.; Miyake, K.; et al. Structural Analysis Reveals That Toll-like Receptor 7 Is a Dual Receptor for Guanosine and Single-Stranded RNA. *Immunity* **2016**, *45*, 737–748. [CrossRef]
209. Yang, C.W.; Chen, M.F. Low Compositions of Human Toll-like Receptor 7/8-Stimulating RNA Motifs in the MERS-CoV, SARS-CoV and SARS-CoV-2 Genomes Imply a Substantial Ability to Evade Human Innate Immunity. *PeerJ* **2021**, *9*, e11008. [CrossRef]
210. Mubarak, A.; Alturaiki, W.; Hemida, M.G. Middle East Respiratory Syndrome Coronavirus (MERS-CoV): Infection, Immunological Response, and Vaccine Development. *J. Immunol. Res.* **2019**, *2019*, 6491738. [CrossRef]
211. Durai, P.; Batool, M.; Shah, M.; Choi, S. Middle East Respiratory Syndrome Coronavirus: Transmission, Virology and Therapeutic Targeting to Aid in Outbreak Control. *Exp. Mol. Med.* **2015**, *47*, e181. [CrossRef]
212. Channappanavar, R.; Fehr, A.R.; Zheng, J.; Wohlford-Lenane, C.; Abrahante, J.E.; Mack, M.; Sompallae, R.; McCray, P.B.; Meyerholz, D.K.; Perlman, S. IFN-I Response Timing Relative to Virus Replication Determines MERS Coronavirus Infection Outcomes. *J. Clin. Investig.* **2019**, *129*, 3625–3639. [CrossRef] [PubMed]
213. Cervantes-Barragan, L.; Züst, R.; Weber, F.; Spiegel, M.; Lang, K.S.; Akira, S.; Thiel, V.; Ludewig, B. Control of Coronavirus Infection through Plasmacytoid Dendritic-Cell-Derived Type I Interferon. *Blood* **2007**, *109*, 1131–1137. [CrossRef] [PubMed]
214. Gralinski, L.E.; Menachery, V.D. Return of the Coronavirus: 2019-NCoV. *Viruses* **2020**, *12*, 135. [CrossRef]
215. Wondmkun, Y.T.; Mohammed, O.A. A Review on Novel Drug Targets and Future Directions for COVID-19 Treatment. *Biologics* **2020**, *14*, 77–82. [CrossRef]
216. Hussein, R.A.; El-Anssary, A.A. Plants Secondary Metabolites: The Key Drivers of the Pharmacological Actions of Medicinal Plants. *Herb. Med.* **2018**. [CrossRef]
217. Pour, P.M.; Fakhri, S.; Asgary, S.; Farzaei, M.H.; Echeverría, J. The Signaling Pathways, and Therapeutic Targets of Antiviral Agents: Focusing on the Antiviral Approaches and Clinical Perspectives of Anthocyanins in the Management of Viral Diseases. *Front. Pharmacol.* **2019**, *10*, 1207. [CrossRef]
218. Häkkinen, S.H.; Kärenlampi, S.O.; Heinonen, I.M.; Mykkänen, H.M.; Törronen, A.R. Content of the Flavonols Quercetin, Myricetin, and Kaempferol in 25 Edible Berries. *J. Agric. Food Chem.* **1999**, *47*, 2274–2279. [CrossRef]

219. Rahman, M.; Islam, R.; Shohag, S.; Hossain, E.; Shah, M.; Shuvo, S.; Khan, H.; Chowdhury, A.R.; Bulbul, I.J.; Hossain, S.; et al. Multifaceted Role of Natural Sources for COVID-19 Pandemic as Marine Drugs. *Environ. Sci. Pollut. Res. Int.* **2022**, *29*, 46527–46550. [CrossRef]
220. Hwang, S.S.; Lim, J.; Yu, Z.; Kong, P.; Sefik, E.; Xu, H.; Harman, C.C.D.; Kim, L.K.; Lee, G.R.; Li, H.B.; et al. Cryo-EM Structure of the 2019-NCoV Spike in the Prefusion Conformation. *Science* **2020**, *367*, 1255–1260. [CrossRef]
221. Mody, V.; Ho, J.; Wills, S.; Mawri, A.; Lawson, L.; Ebert, M.C.C.J.C.; Fortin, G.M.; Rayalam, S.; Taval, S. Identification of 3-Chymotrypsin like Protease (3CLPro) Inhibitors as Potential Anti-SARS-CoV-2 Agents. *Commun. Biol.* **2021**, *4*, 93. [CrossRef]
222. Mamedov, T.; Yuksel, D.; Ilgın, M.; Gürbüzaslan, I.; Gulec, B.; Mammadova, G.; Ozdarendeli, A.; Yetiskin, H.; Kaplan, B.; Islam Pavel, S.T.; et al. Production and Characterization of Nucleocapsid and Rbd Cocktail Antigens of SARS–CoV-2 in Nicotiana Benthamiana Plant as a Vaccine Candidate against COVID-19. *Vaccines* **2021**, *9*, 1337. [CrossRef] [PubMed]
223. Zhao, M.M.; Yang, W.L.; Yang, F.Y.; Zhang, L.; Huang, W.J.; Hou, W.; Fan, C.F.; Jin, R.H.; Feng, Y.M.; Wang, Y.C.; et al. Cathepsin L Plays a Key Role in SARS-CoV-2 Infection in Humans and Humanized Mice and Is a Promising Target for New Drug Development. *Signal Transduct. Target. Ther.* **2021**, *6*, 134. [CrossRef] [PubMed]
224. Hippisley-Cox, J.; Young, D.; Coupland, C.; Channon, K.M.; Tan, P.S.; Harrison, D.A.; Rowan, K.; Aveyard, P.; Pavord, I.D.; Watkinson, P.J. Risk of Severe COVID-19 Disease with ACE Inhibitors and Angiotensin Receptor Blockers: Cohort Study Including 8.3 million People. *Heart* **2020**, *106*, 1503–1511. [CrossRef] [PubMed]
225. Vaduganathan, M.; Vardeny, O.; Michel, T.; McMurray, J.J.V.; Pfeffer, M.A.; Solomon, S.D. Renin–Angiotensin–Aldosterone System Inhibitors in Patients with COVID-19. *N. Engl. J. Med.* **2020**, *382*, 1653–1659. [CrossRef]
226. Bombardini, T.; Picano, E. Angiotensin-Converting Enzyme 2 as the Molecular Bridge Between Epidemiologic and Clinical Features of COVID-19. *Can. J. Cardiol.* **2020**, *36*, 784.e1–784.e2. [CrossRef]
227. Bosch, B.J.; van der Zee, R.; de Haan, C.A.M.; Rottier, P.J.M. The Coronavirus Spike Protein Is a Class I Virus Fusion Protein: Structural and Functional Characterization of the Fusion Core Complex. *J. Virol.* **2003**, *77*, 8801–8811. [CrossRef]
228. Zhang, H.; Penninger, J.M.; Li, Y.; Zhong, N.; Slutsky, A.S. Angiotensin-Converting Enzyme 2 (ACE2) as a SARS-CoV-2 Receptor: Molecular Mechanisms and Potential Therapeutic Target. *Intensive Care Med.* **2020**, *46*, 586–590. [CrossRef]
229. Chatterjee, B.; Thakur, S.S. ACE2 as a Potential Therapeutic Target for Pandemic COVID-19. *RSC Adv.* **2020**, *10*, 39808–39813. [CrossRef]
230. Li, Y.; Zeng, Z.; Li, Y.; Huang, W.; Zhou, M.; Zhang, X.; Jiang, W. Angiotensin-Converting Enzyme Inhibition Attenuates Lipopolysaccharide-Induced Lung Injury by Regulating the Balance between Angiotensin-Converting Enzyme and Angiotensin-Converting Enzyme 2 and Inhibiting Mitogen-Activated Protein Kinase Activation. *Shock* **2015**, *43*, 395–404. [CrossRef]
231. Soler, M.J.; Ye, M.; Wysocki, J.; William, J.; Lloveras, J.; Batlle, D. Localization of ACE2 in the Renal Vasculature: Amplification by Angiotensin II Type 1 Receptor Blockade Using Telmisartan. *Am. J. Physiol. Renal Physiol.* **2009**, *296*, 398–405. [CrossRef]
232. Wang, X.; Ye, Y.; Gong, H.; Wu, J.; Yuan, J.; Wang, S.; Yin, P.; Ding, Z.; Kang, L.; Jiang, Q.; et al. The Effects of Different Angiotensin II Type 1 Receptor Blockers on the Regulation of the ACE-AngII-AT1 and ACE2-Ang(1–7)-Mas Axes in Pressure Overload-Induced Cardiac Remodeling in Male Mice. *J. Mol. Cell Cardiol.* **2016**, *97*, 180–190. [CrossRef] [PubMed]
233. Ahmad, I.; Pawara, R.; Surana, S.; Patel, H. The Repurposed ACE2 Inhibitors: SARS-CoV-2 Entry Blockers of COVID-19. *Top. Curr. Chem.* **2021**, *379*, 40. [CrossRef] [PubMed]
234. Pant, S.; Singh, M.; Ravichandiran, V.; Murty, U.S.N.; Srivastava, H.K. Peptide-like and Small-Molecule Inhibitors against COVID-19. *J. Biomol. Struct. Dyn.* **2021**, *39*, 1–10. [CrossRef] [PubMed]
235. Veeramachaneni, G.K.; Thunuguntla, V.B.S.C.; Bobbillapati, J.; Bondili, J.S. Structural and Simulation Analysis of Hotspot Residues Interactions of SARS-CoV 2 with Human ACE2 Receptor. *J. Biomol. Struct. Dyn.* **2021**, *39*, 4015–4025. [CrossRef]
236. Lan, J.; Ge, J.; Yu, J.; Shan, S.; Zhou, H.; Fan, S.; Zhang, Q.; Shi, X.; Wang, Q.; Zhang, L.; et al. Structure of the SARS-CoV-2 Spike Receptor-Binding Domain Bound to the ACE2 Receptor. *Nature* **2020**, *581*, 215–220. [CrossRef]
237. Wan, Y.; Shang, J.; Graham, R.; Baric, R.S.; Li, F. Receptor Recognition by the Novel Coronavirus from Wuhan: An Analysis Based on Decade-Long Structural Studies of SARS Coronavirus. *J. Virol.* **2020**, *94*, 127–147. [CrossRef]
238. Han, D.P.; Penn-Nicholson, A.; Cho, M.W. Identification of Critical Determinants on ACE2 for SARS-CoV Entry and Development of a Potent Entry Inhibitor. *Virology* **2006**, *350*, 15–25. [CrossRef]
239. Jani, V.; Koulgi, S.; Uppuladinne, V.N.M.; Sonavane, U.; Joshi, R. Computational Drug Repurposing Studies on the ACE2-Spike (RBD) Interface of SARS-CoV-2. *ChemRxiv* **2020**. [CrossRef]
240. O'Keefe, B.R.; Giomarelli, B.; Barnard, D.L.; Shenoy, S.R.; Chan, P.K.S.; McMahon, J.B.; Palmer, K.E.; Barnett, B.W.; Meyerholz, D.K.; Wohlford-Lenane, C.L.; et al. Broad-Spectrum in Vitro Activity and in Vivo Efficacy of the Antiviral Protein Griffithsin against Emerging Viruses of the Family Coronaviridae. *J. Virol.* **2010**, *84*, 2511–2521. [CrossRef]
241. Mille, J.K.; Whittaker, G.R. Host Cell Entry of Middle East Respiratory Syndrome Coronavirus after Two-Step, Furin-Mediated Activation of the Spike Protein. *Proc. Natl. Acad. Sci. USA* **2014**, *111*, 15214–15219. [CrossRef]
242. Ho, T.Y.; Wu, S.L.; Chen, J.C.; Li, C.C.; Hsiang, C.Y. Emodin Blocks the SARS Coronavirus Spike Protein and Angiotensin-Converting Enzyme 2 Interaction. *Antiviral Res.* **2007**, *74*, 92. [CrossRef] [PubMed]
243. Schwarz, S.; Wang, K.; Yu, W.; Sun, B.; Schwarz, W. Emodin Inhibits Current through SARS-Associated Coronavirus 3a Protein. *Antiviral Res.* **2011**, *90*, 64–69. [CrossRef] [PubMed]

244. Hoffmann, M.; Hofmann-Winkler, H.; Smith, J.C.; Krüger, N.; Arora, P.; Sørensen, L.K.; Søgaard, O.S.; Hasselstrøm, J.B.; Winkler, M.; Hempel, T.; et al. Camostat Mesylate Inhibits SARS-CoV-2 Activation by TMPRSS2-Related Proteases and Its Metabolite GBPA Exerts Antiviral Activity. *EBioMedicine* **2021**, *65*, 103255. [CrossRef]
245. Muhseen, Z.T.; Hameed, A.R.; Al-Hasani, H.M.H.; Tahir ul Qamar, M.; Li, G. Promising Terpenes as SARS-CoV-2 Spike Receptor-Binding Domain (RBD) Attachment Inhibitors to the Human ACE2 Receptor: Integrated Computational Approach. *J. Mol. Liq.* **2020**, *320*, 114493. [CrossRef] [PubMed]
246. Hakami, A.R. Targeting the RBD of Omicron Variant (B.1.1.529) with Medicinal Phytocompounds to Abrogate the Binding of Spike Glycoprotein with the HACE2 Using Computational Molecular Search and Simulation Approach. *Biology* **2022**, *11*, 258. [CrossRef]
247. Dyall, J.; Coleman, C.M.; Hart, B.J.; Venkataraman, T.; Holbrook, M.R.; Kindrachuk, J.; Johnson, R.F.; Olinger, G.G.; Jahrling, P.B.; Laidlaw, M.; et al. Repurposing of Clinically Developed Drugs for Treatment of Middle East Respiratory Syndrome Coronavirus Infection. *Antimicrob. Agents Chemother.* **2014**, *58*, 4885–4893. [CrossRef]
248. Zhao, Y.; Ren, J.; Harlos, K.; Jones, D.M.; Zeltina, A.; Bowden, T.A.; Padilla-Parra, S.; Fry, E.E.; Stuart, D.I. Toremifene Interacts with and Destabilizes the Ebola Virus Glycoprotein. *Nature* **2016**, *535*, 169–172. [CrossRef]
249. Marques-Pereira, C.; Pires, M.N.; Gouveia, R.P.; Pereira, N.N.; Caniceiro, A.B.; Rosário-Ferreira, N.; Moreira, I.S. SARS-CoV-2 Membrane Protein: From Genomic Data to Structural New Insights. *Int. J. Mol. Sci.* **2022**, *23*, 2986. [CrossRef]
250. Kumar, P.; Kumar, A.; Garg, N.; Giri, R. An Insight into SARS-CoV-2 Membrane Protein Interaction with Spike, Envelope, and Nucleocapsid Proteins. *J. Biomol. Struct. Dyn.* **2021**, 1–10. [CrossRef]
251. Gupta, M.K.; Vemula, S.; Donde, R.; Gouda, G.; Behera, L.; Vadde, R. In-Silico Approaches to Detect Inhibitors of the Human Severe Acute Respiratory Syndrome Coronavirus Envelope Protein Ion Channel. *J. Biomol. Struct. Dyn.* **2021**, *39*, 2617–2627. [CrossRef]
252. Liu, X.; Verma, A.; Garcia, G.; Ramage, H.; Lucas, A.; Myers, R.L.; Michaelson, J.J.; Coryell, W.; Kumar, A.; Charney, A.W.; et al. Targeting the Coronavirus Nucleocapsid Protein through GSK-3 Inhibition. *Proc. Natl. Acad. Sci. USA* **2021**, *118*, e2113401118. [CrossRef] [PubMed]
253. Roh, C. A Facile Inhibitor Screening of SARS Coronavirus N Protein Using Nanoparticle-Based RNA Oligonucleotide. *Int. J. Nanomed.* **2012**, *7*, 2173. [CrossRef]
254. Kwarteng, A.; Asiedu, E.; Sakyi, S.A.; Asiedu, S.O. Targeting the SARS-CoV2 Nucleocapsid Protein for Potential Therapeutics Using Immuno-Informatics and Structure-Based Drug Discovery Techniques. *Biomed. Pharmacother.* **2020**, *132*, 110914. [CrossRef]
255. Lavigne, M.; Helynck, O.; Rigolet, P.; Boudria-Souilah, R.; Nowakowski, M.; Baron, B.; Brülé, S.; Hoos, S.; Raynal, B.; Guittat, L.; et al. SARS-CoV-2 Nsp3 Unique Domain SUD Interacts with Guanine Quadruplexes and G4-Ligands Inhibit This Interaction. *Nucleic Acids Res* **2021**, *49*, 7695–7712. [CrossRef]
256. Lim, C.T.; Tan, K.W.; Wu, M.; Ulferts, R.; Armstrong, L.A.; Ozono, E.; Drury, L.S.; Milligan, J.C.; Zeisner, T.U.; Zeng, J.; et al. Identifying SARS-CoV-2 Antiviral Compounds by Screening for Small Molecule Inhibitors of Nsp3 Papain-like Protease. *Biochem. J.* **2021**, *478*, 2517. [CrossRef] [PubMed]
257. Roe, M.K.; Junod, N.A.; Young, A.R.; Beachboard, D.C.; Stobart, C.C. Targeting Novel Structural and Functional Features of Coronavirus Protease Nsp5 (3CL pro, M pro) in the Age of COVID-19. *J. Gen. Virol.* **2021**, *102*, 001558. [CrossRef]
258. Malik, Y.A. Properties of Coronavirus and SARS-CoV-2. *Malays. J. Pathol.* **2020**, *42*, 3–11.
259. Lee, J.; Worrall, L.J.; Vuckovic, M.; Rosell, F.I.; Gentile, F.; Ton, A.T.; Caveney, N.A.; Ban, F.; Cherkasov, A.; Paetzel, M.; et al. Crystallographic Structure of Wild-Type SARS-CoV-2 Main Protease Acyl-Enzyme Intermediate with Physiological C-Terminal Autoprocessing Site. *Nat. Commun.* **2020**, *11*, 5877. [CrossRef]
260. Milligan, J.C.; Zeisner, T.U.; Papageorgiou, G.; Joshi, D.; Soudy, C.; Ulferts, R.; Wu, M.; Lim, C.T.; Tan, K.W.; Weissmann, F.; et al. Identifying SARS-CoV-2 Antiviral Compounds by Screening for Small Molecule Inhibitors of Nsp5 Main Protease. *Biochem. J.* **2021**, *478*, 2499–2515. [CrossRef]
261. Dai, W.; Zhang, B.; Jiang, X.M.; Su, H.; Li, J.; Zhao, Y.; Xie, X.; Jin, Z.; Peng, J.; Liu, F.; et al. Structure-Based Design of Antiviral Drug Candidates Targeting the SARS-CoV-2 Main Protease. *Science* **2020**, *368*, 1331–1335. [CrossRef] [PubMed]
262. Naidu, S.A.G.; Tripathi, Y.B.; Shree, P.; Clemens, R.A.; Naidu, A.S. Phytonutrient Inhibitors of SARS-CoV-2/NSP5-Encoded Main Protease (Mpro) Autocleavage Enzyme Critical for COVID-19 Pathogenesis. *J. Diet. Suppl.* **2021**. *Online ahead of print*. [CrossRef] [PubMed]
263. Chu, C.K.; Gadthula, S.; Chen, X.; Choo, H.; Olgen, S.; Barnard, D.L.; Sidwell, R.W. Antiviral Activity of Nucleoside Analogues against SARS-Coronavirus (SARS-CoV). *Antivir. Chem. Chemother.* **2006**, *17*, 285–289. [CrossRef]
264. Gandhi, Y.; Mishra, S.K.; Rawat, H.; Grewal, J.; Kumar, R.; Shakya, S.K.; Jain, V.K.; Babu, G.; Singh, A.; Singh, R.; et al. Phytomedicines Explored under in Vitro and in Silico Studies against Coronavirus: An Opportunity to Develop Traditional Medicines. *S. Afr. J. Bot.* **2022**, in press. [CrossRef] [PubMed]
265. Pompei, R.; Laconi, S.; Ingianni, A. Antiviral Properties of Glycyrrhizic Acid and Its Semisynthetic Derivatives. *Mini Rev. Med. Chem.* **2009**, *9*, 996–1001. [CrossRef] [PubMed]
266. Hoever, G.; Baltina, L.; Michaelis, M.; Kondratenko, R.; Baltina, L.; Tolstikov, G.A.; Doerr, H.W.; Cinatl, J. Antiviral Activity of Glycyrrhizic Acid Derivatives against SARS-Coronavirus. *J. Med. Chem.* **2005**, *48*, 1256–1259. [CrossRef]

267. Taylor, J.K.; Coleman, C.M.; Postel, S.; Sisk, J.M.; Bernbaum, J.G.; Venkataraman, T.; Sundberg, E.J.; Frieman, M.B. Severe Acute Respiratory Syndrome Coronavirus ORF7a Inhibits Bone Marrow Stromal Antigen 2 Virion Tethering through a Novel Mechanism of Glycosylation Interference. *J. Virol.* **2015**, *89*, 11820–11833. [CrossRef]
268. Narayanan, K.; Huang, C.; Lokugamage, K.; Kamitani, W.; Ikegami, T.; Tseng, C.-T.K.; Makino, S. Severe Acute Respiratory Syndrome Coronavirus Nsp1 Suppresses Host Gene Expression, Including That of Type I Interferon, in Infected Cells. *J. Virol.* **2008**, *82*, 4471–4479. [CrossRef]
269. Forni, D.; Cagliani, R.; Mozzi, A.; Pozzoli, U.; Al-Daghri, N.; Clerici, M.; Sironi, M. Extensive Positive Selection Drives the Evolution of Nonstructural Proteins in Lineage C Betacoronaviruses. *J. Virol.* **2016**, *90*, 3627–3639. [CrossRef] [PubMed]

International Journal of
Molecular Sciences

MDPI

Review

Toll-like Receptor Response to Hepatitis C Virus Infection: A Recent Overview

Mohammad Enamul Hoque Kayesh [1,*], Michinori Kohara [2] and Kyoko Tsukiyama-Kohara [3,*]

1 Department of Microbiology and Public Health, Faculty of Animal Science and Veterinary Medicine, Patuakhali Science and Technology University, Barishal 8210, Bangladesh
2 Department of Microbiology and Cell Biology, Tokyo Metropolitan Institute of Medical Science, Tokyo 156-8506, Japan; kohara-mc@igakuken.or.jp
3 Transboundary Animal Diseases Centre, Joint Faculty of Veterinary Medicine, Kagoshima University, Kagoshima 890-0065, Japan
* Correspondence: mehkayesh@pstu.ac.bd (M.E.H.K.); kkohara@vet.kagoshima-u.ac.jp (K.T.-K.); Tel.: +88-025-506-1677 (M.E.H.K.); +81-99-285-3589 (K.T.-K.)

Abstract: Hepatitis C virus (HCV) infection remains a major global health burden, causing chronic hepatitis, cirrhosis, and hepatocellular carcinoma. Toll-like receptors (TLRs) are evolutionarily conserved pattern recognition receptors that detect pathogen-associated molecular patterns and activate downstream signaling to induce proinflammatory cytokine and chemokine production. An increasing number of studies have suggested the importance of TLR responses in the outcome of HCV infection. However, the exact role of innate immune responses, including TLR response, in controlling chronic HCV infection remains to be established. A proper understanding of the TLR response in HCV infection is essential for devising new therapeutic approaches against HCV infection. In this review, we discuss the progress made in our understanding of the host innate immune response to HCV infection, with a particular focus on the TLR response. In addition, we discuss the mechanisms adopted by HCV to avoid immune surveillance mediated by TLRs.

Keywords: hepatitis C virus; infection; innate immunity; Toll-like receptor; cytokines

Citation: Kayesh, M.E.H.; Kohara, M.; Tsukiyama-Kohara, K. Toll-like Receptor Response to Hepatitis C Virus Infection: A Recent Overview. *Int. J. Mol. Sci.* **2022**, *23*, 5475. https://doi.org/10.3390/ijms23105475

Academic Editors: Oliver Planz and Ralf Kircheis

Received: 21 April 2022
Accepted: 13 May 2022
Published: 13 May 2022

1. Introduction

Hepatitis C virus (HCV) infection is a major global health burden [1,2]. HCV infection frequently causes chronic hepatitis, liver cirrhosis, and hepatocellular carcinoma (HCC) [3]. According to the World Health Organization, 58 million people worldwide are chronically infected with HCV, with approximately 1.5 million new infections occurring each year [4]. HCV is an enveloped, positive-sense, single-stranded RNA virus belonging to the genus *Hepacivirus* and the family *Flaviviridae* [5]. HCV has a $\sim$10-kb long genome, encoding a large polyprotein of approximately 3000 amino acids that is processed by host and viral proteases into three structural (core, E1, and E2) and seven non-structural (NS) proteins (p7, NS2, NS3, NS4A, NS4B, NS5A, and NS5B) [6]. HCV has high genetic diversity resulting in seven major genotypes and more than 60 subtypes [7].

The innate immune system, an essential component of host immunity, plays a key role in the initial detection of invading pathogens, including viruses, and subsequently activates adaptive immunity, thereby playing an important role in the early control of viral infection [8–11]. The host innate immune response is activated upon microbial invasion and detection of evolutionarily conserved structures found on pathogens, called pathogen-associated molecular patterns (PAMPs), by germ-line-encoded pattern recognition receptors (PRRs) [12]. PRRs also recognize molecules released by damaged cells, known as damage-associated molecular patterns [13]. Different PRRs, including Toll-like receptors (TLRs), RIG-I-like receptors (RLRs), NOD-like receptors (NLRs), AIM2-like receptors (ALRs), C-type lectin receptors (CLRs), and intracellular DNA sensors such as cGAS, are key innate immune components that recognize viral components such as viral nucleic acids and proteins [14,15]. However, TLRs are the most widely characterized PRRs, constituting key components of innate immunity, and are involved in the early interaction with PAMPs of invading microbes [16].

TLRs are evolutionarily conserved type I transmembrane proteins that contain a conserved structure of an N-terminal ectodomain of leucine-rich repeats, a single transmembrane domain, and a cytosolic Toll/interleukin (IL)-1 receptor (TIR) domain [11,17]. The cytosolic TIR domain is responsible for the activation of downstream signaling, and TLR signaling pathways are regulated by TIR domain-containing cytosolic adaptor proteins such as myeloid differentiation factor 88 (MyD88), MyD88 adaptor-like (MAL or TIRAP), TIR-domain-containing adaptor protein inducing interferon (IFN)-β (TRIF or TICAM1), TRIF-related adaptor molecule (TRAM or TICAM2), and sterile α- and armadillo-motif-containing protein (SARM) [8,18,19]. The adaptor protein MyD88 is used by nearly all TLR signaling pathways, except TLR3 [20]. Members of each TLR family have similar functions across species [21,22]. TLRs are encoded by a large gene family, and different organisms appear to encode a certain number of TLRs. For example, the TLR family comprises 10 members (TLR1–TLR10) in humans and 12 members (TLR1–TLR9 and TLR11–TLR13) in mice [23]. TLRs can be localized either on the cell surface, such as TLRs 1, 2, 4, 5, 6, and 10, or in intracellular compartments (e.g., the endoplasmic reticulum, endosome, lysosome, or endolysosome) such as TLRs 3, 7, 8, and 9 [23–25]. TLR1, TLR2, TLR4, TLR5, and TLR6 play pivotal roles in viral protein recognition [26]. To recognize viral double-stranded RNA, single-stranded RNA, and DNA, the membrane proteins TLR3, TLR7/8, and TLR9 are used, respectively [27–30].

TLR-induced innate immune responses appear to be a prerequisite for the generation of most adaptive immune responses and play a central role in shaping such responses [25,31]. To provide protection against invading microbes, TLR responses ultimately lead to the induction of IFNs, cytokines, and chemokines via several distinct signaling pathways, which is very important in limiting the spread of infection [14,31,32]. TLR agonists appear to be potential immunomodulators for treating infections and play a significant role in modulating immunotherapeutic effects [33–36]. However, the TLR response may not always be beneficial to the host, but a dysregulated response may lead to immune-mediated pathology rather than protection [37–40]. Therefore, a proper understanding of the TLR response in any infection, including HCV infection, is critical. Against this background, we discuss here the current progress made in our understanding of the host TLR response in HCV infection and the mechanisms adopted by HCV to avoid TLR-mediated immune surveillance, which may help in devising new therapeutic or preventive strategies.

2. TLR Response to HCV Infection

A complex interplay between the IFN system and viral countermeasures exists in HCV infections [41]. During viral replication, HCV PAMPs can be recognized as non-self by PRRs, leading to the activation of innate and adaptive immune responses [42]. TLRs are important PRRs that recognize PAMPs present in HCV [43]. TLRs are key triggering molecules for cytokine production, and TLR signaling pathways provide a link between innate and acquired immunity [15,44]. Different TLRs have been found to interact with HCV proteins and nucleic acids. It has been reported that HCV is sensed by TLR3 through the detection of dsRNA intermediates in infected hepatoma cells, which may activate the TLR3-signaling cascade and lead to the production of type I and II IFNs, expression of interferon-stimulated genes, and proinflammatory cytokines limiting HCV replication [45]. Induction of TLR2 and TLR4 expression and modulation of the proinflammatory response by HCV proteins have been reported in Raji cells and peripheral blood mononuclear cells (PBMCs) [46,47]. Several other studies also found increased expression of TLR2 and TLR4 mRNA in chronic hepatitis patients compared to controls [48,49]. However, TLR4 mRNA expression was downregulated in cirrhotic patients when compared to chronic hepatitis patients [48]. Hypo-responsiveness to TLR ligands has been reported in patients with chronic HCV infection [50]. Differential expression of TLRs has been reported in HCV-infected patients [51,52], where increased expression of tumor necrosis factor (TNF)-α, IL-6, and IL-12 p35 in PBMCs was also reported [52]. It has been observed that NS5A can activate the promoter of the TLR4 gene in both hepatocytes and B-cells, and enhanced TLR4 expression induces the induction of IFN-β and IL-6 production in human B-cells [46].

The HCV core and NS3 proteins play significant roles in HCV pathogenesis; it has been shown that they can trigger proinflammatory cytokine production in monocytes, inhibit myeloid dendritic cell accessory cell functions, and provide immunoinhibitory effects via IL-10 induction [53]. Other studies have reported that HCV viral proteins, including HCV core and NS3, could be recognized by TLR2, triggering the activation of inflammatory cells [54,55]. Impaired recognition of HCV core and NS3 proteins caused by the R753Q SNP in TLR2 has been reported [55]. HCV core and NS3 proteins may enhance the activity of IL-1 receptor-associated kinase (IRAK), phosphorylation of p38, extracellular regulated (ERK), and c-jun N-terminal (JNK) kinases and induce the activation of activator protein 1 (AP-1) [54]. In an in vitro study, Chang et al. reported an association between TLR1 and TLR6 in the TLR2-mediated activation of macrophages by HCV core and NS3 proteins [56]. While the involvement of TLR2 in sensing HCV core protein has been observed previously, infectious virions or enveloped HCV-like particles do not activate TLR2 [57]. However, another study showed that HCV core protein can activate TLR2 with decreased IL-6 production by human antigen-presenting cells after subsequent stimulation with TLR2 and TLR4 ligands [58]. HCV core protein may affect pDCs by reducing TLR9-triggered IFN-α as well as TNF-α and IL-10 production [59].

HCV core and NS3 antigens induce TLR1-, TLR2-, and TLR6-mediated inflammatory responses in corneal epithelial cells [60]. Compared to healthy controls, upregulation of TLR2 and TLR4 expression in peripheral monocytes was also observed in patients with chronic hepatitis C, with or without HIV coinfection [61]. An association between TLR4 signaling and the outcome of acute hepatitis C has also been reported [62]. Activation of the TLR3/TRIF signaling pathway by HCV NS5B, a viral RNA-dependent RNA polymerase, has been reported [63]. In addition to TLR3, HCV RNA can also be recognized by RIG-I, triggering the production of multiple cytokines, including type I IFN [42]. In a murine replicon model, the antiviral role of TLR4 activation in suppressing HCV replication was demonstrated [64]. An association between TLR4 gene polymorphisms and chronic HCV infection has also been reported in a Saudi Arabian population [65], suggesting a putative role of TLR4 in HCV infection. However, a larger genome-wide association study is required to validate these associations.

As HCV has an RNA genome, it is likely that TLR7 plays a role in the immune response against HCV infection. Both TLR3 and TLR7 have been suggested to coordinate protective immunity against HCV infection [45,66,67]. A decreased expression of TLR3 and TLR7 mRNA has been reported in chronic HCV patients with a decreased IFN-α expression compared to healthy controls [68–70]. An earlier study reported significantly elevated expression of TLR3 in individuals who spontaneously cleared the virus [71], suggesting a protective role of TLR3 in HCV genotype 3 infection. TLR7 and TLR8 were also found to be elevated in patients with liver cirrhosis [71]. A previous study reported that impaired TLR3- and TLR7/8-mediated cytokine responses may contribute to aggressive HCV recurrence after liver transplantation [72], also indicating the association of these molecules with HCV infection. However, more extensive in vivo studies are required to understand their use in protective immune responses against HCV infections.

It has been reported that TLR7 can sense HCV RNA in exosomes released from infected hepatocytes, inducing type I IFN response [73,74]. An in vitro study demonstrated that TLR7 can induce HCV immunity not only by IFN induction but also via an IFN-independent mechanism [75]. The antiviral roles of TLR7 and TLR8 have also been suggested in HCV infection [76]. In a previous study, it was shown that HCV could be recognized and inhibited by TLR7 and TLR8 via TNF-α production [77]. HCV genomic RNA-induced TLR7- and TLR8-mediated anti-HCV immune responses have been reported in various antigen-presenting cells [78]. Polymorphisms in TLR7 and/or TLR8 genes have been shown to modulate HCV infection outcomes [79–81], suggesting an association between these molecules and HCV infection. The TLR9 rs187084 C allele was reported to be associated with spontaneous virus clearance in women, suggesting the sex-specific effects of TLR9 promoter variants on spontaneous clearance in HCV infection and implying the role of the DNA sensor TLR9 in natural immunity against HCV infection [82]. In a recent study, an association between TLR9 gene polymorphisms and the outcome of the HCV-specific cell-mediated immune response was reported [83], indicating a putative role of TLR9 in HCV infection. A suitable small animal model is still lacking for HCV infection, which is essential for a proper understanding of the TLR response to HCV infection. Tree shrews appear to be a promising animal model for several important viral infections in humans, including hepatitis C virus [84], and show a higher degree of genetic similarity to primates than to rodents [85,86]. In an earlier study, we showed that HCV could trigger innate immune responses in the livers of chronically infected tree shrews, with significant induction of intrahepatic TLR3, TLR7, and TLR8 mRNA [87]. For simplicity, the findings obtained in different studies are shown in Figure 1, without indication of cell type/system, and highlight that various TLRs are implicated in HCV infection, which may influence viral pathogenesis. Therefore, a complete understanding of the TLR response in HCV infection is critical for designing new and successful therapeutic or preventive interventions.

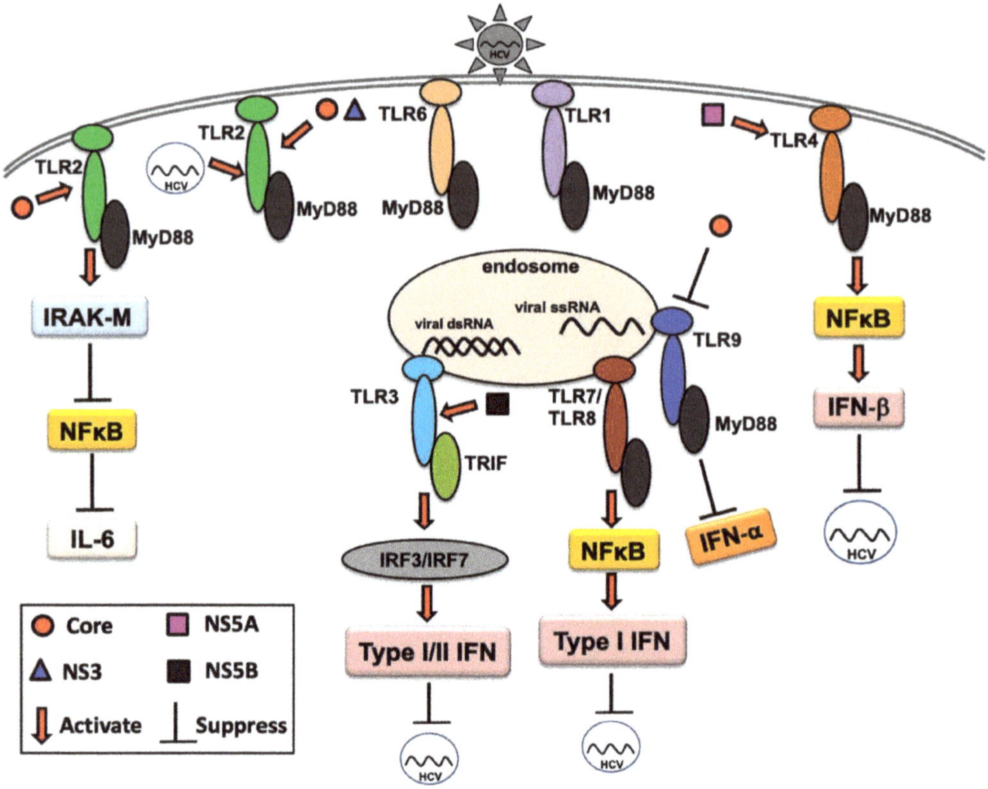

Figure 1. TLR response to HCV infection. Red arrows indicate the induction/activation of components of TLR signaling by HCV or its proteins; black lines indicate the inhibition of the host innate immune response or inhibition of HCV replication, as appropriate.

3. Inhibition of Innate Immune Response by HCV Infection

The host has evolved multifaceted innate immune mechanisms to sense and counteract HCV infection. However, the success of innate IFN response in inhibiting HCV infection remains low. In a large proportion of patients, HCV persistence has been observed for decades despite the expression of hundreds of interferon-stimulated genes, indicating the inability of the IFN system to clear HCV infection [30,88,89]. In addition, HCV has developed multiple strategies for innate immune regulation, including proteolytic cleavage of molecules that play key roles in the induction of the IFN response, changes in IFN-induced effector proteins, interference with the function of CD8+ T cells, and immune escape in T- and B-cell epitopes [90]. HCV encodes several proteins, including core, NS3/4A, NS4B, and NS5A, which play active roles in inhibiting the innate immune response [42,91–93]. HCV frameshift (F) protein, which is expressed by a translational ribosomal frameshift [94], has also been suggested to play a role in the immune evasion mechanism [92].

NS3/4A plays a central role in HCV pathogenesis by cleaving several host proteins [95–97]. HCV NS3/4A has the potential to cleave TRIF and impair the TLR3-dependent signaling pathway [45,93,98]. It has also been reported that NS3/4A protease can disrupt RIG-I signaling by cleavage or delocalization of IFN-β promoter stimulator 1, also known as mitochondrial antiviral signaling protein (MAVS), preventing downstream activation of IRF-3 and IFN-β induction [93,98–102]. Notably, cleavage of MAVS by NS3/4A has been reported in the infected human liver, demonstrating the importance of this cleavage in HCV infection in vivo [101,103]. HCV proteins may interfere with IFN-induced intracellular signaling, which could be an important mechanism for viral persistence and treatment resistance [104]. HCV core may inhibit the IFN-signaling pathway by interfering with the Janus kinase/signal transducer and activator of the transcription pathway [105–107]. Several studies have reported the inhibition or degradation of STAT1 by HCV core protein [104,108–111], which may inhibit the JAK/STAT-signaling pathway of the host response. Reduced levels of STAT2 phosphorylation caused by HCV core proteins have also been reported [112,113].

Several studies have shown the implication of HCV core protein in the activation of the NF-κB pathway for inducing inflammatory response [114–117]; however, an implication of HCV core protein in the suppression of the NF-κB pathway has also been reported [118]. STAT3 has been found to be downregulated in HCV-infected livers and in Huh7 cells [119], which may favor viral replication. Impaired TLR4 signaling in HCV-infected dendritic cells has been previously demonstrated [120]. It has been reported that NS4B can interact with the stimulator of IFN genes (STING), which may cause inhibition of downstream signaling [91,121,122]. HCV NS4A, NS4B, and NS5A may inhibit IFN-β induction, contributing to the persistence of this virus [63]. NS5A can directly bind to MyD88, a major adaptor molecule in TLR signaling, inhibiting the activation of TLR-mediated cytokine production [123]. NS5A can induce IL-8 expression, associated with the interruption of IFN-α [124]. NS5A also blocks the antiviral activity of 2′–5′ oligoadenylate synthetase (2′–5′ OAS) [125]. It has been reported that HCV may utilize NS5A and E2 to inhibit PKR-mediated antiviral defense [126–128]. A previous study also suggested a role for NS5A in the inhibition of the IFN response that is activated by HCV via RIG-I and MDA5 [129]. NS5A may interact with nucleosome assembly protein 1-like 1 (NAP1L1), a nuclear-cytoplasmic chaperone, which may downregulate genes essential for innate immunity, such as RIG-I- and TLR3-mediated responses [130].

Chronic HCV infection results in impaired TLR response in pDCs as well as impaired activation of naive CD4 T cells, with reduced activation marker (HLA-DR) and cytokine (IFN-α) expression upon R-848 stimulation [131]. Samrat et al. reported that HCV core and F protein can induce poor T-cell responses, resulting in low granzyme B expression by CD4+ and CD8+ T cells [92]. It has been reported that HCV p7 protein has an immune evasion function, which may suppress antiviral IFN function by counteracting IFN-inducible protein 6-16 (IFI6-16) [132]. Based on these findings, it is assumed that HCV and its proteins play a crucial role in inhibiting or suppressing the host innate immune response (Figure 2) by different known and unknown mechanisms; therefore, a clear understanding of immune inhibition or evasion is critical for devising new therapeutic and preventive strategies to control HCV infection.

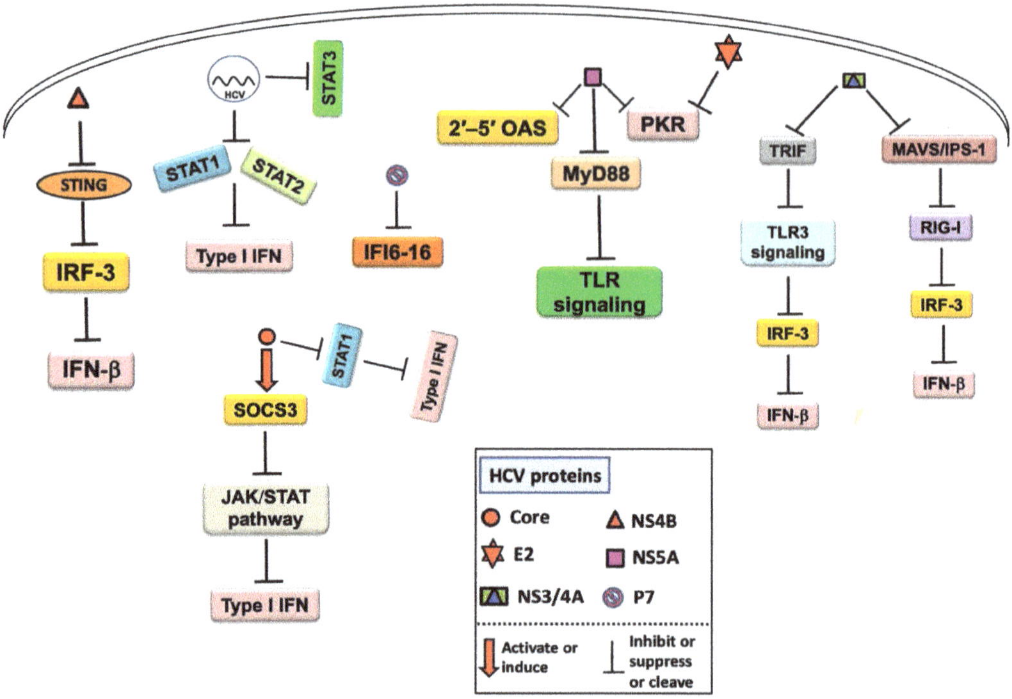

Figure 2. An overview of the mechanism of the host innate immune response inhibition by HCV and its proteins.

4. Potential of TLR Agonists as Immunomodulators

A large number of viruses have been shown to trigger innate immunity via TLRs, which have been found important in the outcome of many viral infections [133–135], suggesting manipulating the TLR response could serve as a therapeutic avenue against viral infections. Although there is a great therapeutic success against HCV infection with newly approved drugs, direct-acting antivirals (DAAs) [136], cirrhosis and HCC have been reported in patients following viral clearance [137]. Moreover, current HCV treatment approaches are not effective in preventing recurrent infections. Notably, high treatment cost has restricted its use in economically weak countries. Therefore, there is an urgent need to develop an alternative therapeutic approach as well as an effective, safe, and affordable vaccine against HCV. There is a growing interest in the use of TLR agonists as vaccine adjuvants [138], which are capable of stimulating innate and adaptive immune responses, thereby improving vaccine efficacy. Recently, TLR agonists have received much attention as immunomodulators with the ability to induce the production of IFN, proinflammatory cytokines, and chemokines and have been found to be promising against many viral infections, including hepatitis B virus and human immunodeficiency virus type 1 (HIV-1) [35,139–141]. Manipulating TLR response has also been found to be promising against SARS-CoV-2 infection. In a ferret model, it has been shown that the injection of TLR2/6 agonist INNA-051 significantly reduced SARS-CoV-2 viral RNA levels in the nose and throat [142]. Additionally, it has been proposed that TLR agonists, including imiquimod, an immune stimulator of TLR7, could serve as an effective therapeutic approach in the early stages of COVID-19 [143].

Chronic HCV infection induces weak cellular immune responses against viral antigens, and viral clearance after acute hepatitis or therapy requires strong and multispecific antiviral CD4+ and CD8+ T-cell responses [144,145]. TLR agonists may play an important role in enhancing immunotherapeutic effects [33–35,138,146]. The use of TLR agonists as immunomodulators to enhance the immune response in chronic HCV infection could be of great interest for the control of HCV infection. Isatoribine, an agonist of TLR7, showed dose-dependent changes in immunologic biomarkers and antiviral effects against HCV infection [147]. The antiviral activity of the synthetic TLR7 agonist was shown to be associated with the stimulation of antiviral genes, such as IRF7, but not with the activation of the IFN-responsive STAT-1 transcription factor [75]. A previous study demonstrated that oral administration of resiquimod, a TLR 7/8 agonist, transiently reduced viral levels but was associated with adverse effects similar to IFN-α [148]. Another study showed that TLR3/4/7/8/9 agonists could induce anti-HCV activity in PBMC supernatants, correlating with IFN-α and the IFN-induced antiviral biomarker 2′,5′-oligoadenylate synthase induction. However, TLR4 and TLR8 agonists induce the proinflammatory cytokines IL-1β and TNF-α at concentrations similar to those inducing antiviral activity, raising concerns regarding adverse side effects [149]. In a randomized clinical trial, oral administration of ANA773, a prodrug of the TLR7 agonist, resulted in an IFN-dependent response leading to a significant decrease in serum HCV RNA levels, with mild to moderate adverse events [150].

It has been reported that TLR7/9 agonists may enhance the inhibition of infectivity and IFN-α production by pDCs, suggesting pDCs could serve as a drug target against HCV infection [151]. It also suggests the possibility of using TLR7/9 agonists in HCV vaccine development. An earlier study showed that although a hepatitis C viral-like particle vaccine adjuvanted with TLR2 agonists, R_4Pam_2Cys and E_8Pam_2Cys, induced higher antibody titers in mice, it did not induce stronger NAb responses compared to vaccines without adjuvants [152], suggesting the usefulness of TLR agonists as vaccine adjuvants for HCV vaccines. Using appropriate adjuvants in HCV vaccine candidates, the induction of a strong T- and B-cell immune response may be enhanced [153]. In an in vitro study with an HCV-infected hepatoma cell line, Huh7.5, Dominguez-Molina et al. showed that TLR agonists can enhance antiviral pDCs function primarily through IFN-α production against HCV infection [151].

5. Conclusions

From the currently available data, it is understood that there is a differential expression of TLRs in HCV infection. Moreover, TLR3 and TLR7 may play a protective role against HCV infection. However, the exact role of the TLR response to HCV infection requires further extensive study, which also requires a suitable small animal model. HCV has developed multiple strategies to inhibit the innate immune response toward establishing a chronic infection, which also requires additional investigation for developing new therapeutic approaches. As there are mixed effects, an extensive study is still required to select the best-suited TLR agonist for use as a vaccine adjuvant for HCV vaccine candidates. Overall, a clear understanding of TLR interactions in HCV infection is critical for providing new therapeutic and preventive approaches to fight the disease, including TLR agonist-adjuvanted HCV vaccines.

Author Contributions: Conceptualization, M.E.H.K., M.K. and K.T.-K.; writing—original draft preparation, M.E.H.K. and K.T.-K.; writing—review and editing, M.E.H.K., M.K. and K.T.-K.; All authors have read and agreed to the published version of the manuscript.

Funding: Grant from the Tokyo Metropolitan Government for drug development targeting liver cirrhosis.

Institutional Review Board Statement: Not applicable.

Informed Consent Statement: Not applicable.

Conflicts of Interest: The authors declare no conflict of interest.

References

1. Mohd Hanafiah, K.; Groeger, J.; Flaxman, A.D.; Wiersma, S.T. Global epidemiology of hepatitis C virus infection: New estimates of age-specific antibody to HCV seroprevalence. *Hepatology* **2013**, *57*, 1333–1342. [CrossRef] [PubMed]
2. Lavanchy, D. The global burden of hepatitis C. *Liver Int.* **2009**, *29* (Suppl. S1), 74–81. [CrossRef] [PubMed]
3. Saito, I.; Miyamura, T.; Ohbayashi, A.; Harada, H.; Katayama, T.; Kikuchi, S.; Watanabe, Y.; Koi, S.; Onji, M.; Ohta, Y.; et al. Hepatitis C virus infection is associated with the development of hepatocellular carcinoma. *Proc. Natl. Acad. Sci. USA* **1990**, *87*, 6547–6549. [CrossRef] [PubMed]
4. World Health Organization (WHO). Hepatitis C. Updated on 27 July 2021. 2021. Available online: https://www.who.int/news-room/fact-sheets/detail/hepatitis-c (accessed on 25 March 2022).
5. Choo, Q.L.; Kuo, G.; Weiner, A.J.; Overby, L.R.; Bradley, D.W.; Houghton, M. Isolation of a cDNA clone derived from a blood-borne non-A, non-B viral hepatitis genome. *Science* **1989**, *244*, 359–362. [CrossRef] [PubMed]
6. Moradpour, D.; Penin, F.; Rice, C.M. Replication of hepatitis C virus. *Nat. Rev. Microbiol.* **2007**, *5*, 453–463. [CrossRef]
7. Smith, D.B.; Bukh, J.; Kuiken, C.; Muerhoff, A.S.; Rice, C.M.; Stapleton, J.T.; Simmonds, P. Expanded classification of hepatitis C virus into 7 genotypes and 67 subtypes: Updated criteria and genotype assignment web resource. *Hepatology* **2014**, *59*, 318–327. [CrossRef]
8. Takeda, K.; Akira, S. Toll-like receptors in innate immunity. *Int. Immunol.* **2005**, *17*, 1–14. [CrossRef]
9. Takeuchi, O.; Akira, S. Innate immunity to virus infection. *Immunol. Rev.* **2009**, *227*, 75–86. [CrossRef]
10. Zuniga, E.I.; Macal, M.; Lewis, G.M.; Harker, J.A. Innate and Adaptive Immune Regulation During Chronic Viral Infections. *Annu. Rev. Virol.* **2015**, *2*, 573–597. [CrossRef]
11. Medzhitov, R.; Janeway, C., Jr. Innate immunity. *N. Engl. J. Med.* **2000**, *343*, 338–344. [CrossRef]
12. Mogensen, T.H. Pathogen recognition and inflammatory signaling in innate immune defenses. *Clin. Microbiol. Rev.* **2009**, *22*, 240–273. [CrossRef] [PubMed]
13. Amarante-Mendes, G.P.; Adjemian, S.; Branco, L.M.; Zanetti, L.C.; Weinlich, R.; Bortoluci, K.R. Pattern Recognition Receptors and the Host Cell Death Molecular Machinery. *Front. Immunol.* **2018**, *9*, 2379. [CrossRef] [PubMed]
14. Kawai, T.; Akira, S. Toll-like receptors and their crosstalk with other innate receptors in infection and immunity. *Immunity* **2011**, *34*, 637–650. [CrossRef] [PubMed]
15. Akira, S.; Uematsu, S.; Takeuchi, O. Pathogen recognition and innate immunity. *Cell* **2006**, *124*, 783–801. [CrossRef] [PubMed]
16. Akira, S.; Takeda, K.; Kaisho, T. Toll-like receptors: Critical proteins linking innate and acquired immunity. *Nat. Immunol.* **2001**, *2*, 675–680. [CrossRef]
17. Fitzgerald, K.A.; Kagan, J.C. Toll-like Receptors and the Control of Immunity. *Cell* **2020**, *180*, 1044–1066. [CrossRef]
18. Narayanan, K.B.; Park, H.H. Toll/interleukin-1 receptor (TIR) domain-mediated cellular signaling pathways. *Apoptosis* **2015**, *20*, 196–209. [CrossRef]
19. O'Neill, L.A.; Bowie, A.G. The family of five: TIR-domain-containing adaptors in Toll-like receptor signalling. *Nat. Rev. Immunol.* **2007**, *7*, 353–364. [CrossRef]
20. Lee, M.S.; Kim, Y.J. Signaling pathways downstream of pattern-recognition receptors and their cross talk. *Annu. Rev. Biochem.* **2007**, *76*, 447–480. [CrossRef]
21. Kaisho, T.; Akira, S. Toll-like receptor function and signaling. *J. Allergy Clin. Immunol.* **2006**, *117*, 979–987. [CrossRef]
22. Akira, S.; Takeda, K. Toll-like receptor signalling. *Nat. Rev. Immunol.* **2004**, *4*, 499–511. [CrossRef] [PubMed]
23. Kawasaki, T.; Kawai, T. Toll-like receptor signaling pathways. *Front. Immunol.* **2014**, *5*, 461. [CrossRef] [PubMed]
24. O'Neill, L.A.; Golenbock, D.; Bowie, A.G. The history of Toll-like receptors-redefining innate immunity. *Nat. Rev. Immunol.* **2013**, *13*, 453–460. [CrossRef] [PubMed]
25. Chaturvedi, A.; Pierce, S.K. How location governs toll-like receptor signaling. *Traffic* **2009**, *10*, 621–628. [CrossRef] [PubMed]
26. Prinz, M.; Heikenwalder, M.; Schwarz, P.; Takeda, K.; Akira, S.; Aguzzi, A. Prion pathogenesis in the absence of Toll-like receptor signalling. *EMBO Rep.* **2003**, *4*, 195–199. [CrossRef] [PubMed]
27. Kawai, T.; Akira, S. Toll-like receptor and RIG-I-like receptor signaling. *Ann. N. Y. Acad. Sci.* **2008**, *1143*, 1–20. [CrossRef] [PubMed]
28. Alexopoulou, L.; Holt, A.C.; Medzhitov, R.; Flavell, R.A. Recognition of double-stranded RNA and activation of NF-kappaB by Toll-like receptor 3. *Nature* **2001**, *413*, 732–738. [CrossRef]
29. Diebold, S.S.; Kaisho, T.; Hemmi, H.; Akira, S.; Reis e Sousa, C. Innate antiviral responses by means of TLR7-mediated recognition of single-stranded RNA. *Science* **2004**, *303*, 1529–1531. [CrossRef]
30. Heim, M.H.; Thimme, R. Innate and adaptive immune responses in HCV infections. *J. Hepatol.* **2014**, *61*, S14–S25. [CrossRef]
31. Takeuchi, O.; Akira, S. Pattern recognition receptors and inflammation. *Cell* **2010**, *140*, 805–820. [CrossRef]
32. Iwasaki, A.; Medzhitov, R. Toll-like receptor control of the adaptive immune responses. *Nat. Immunol.* **2004**, *5*, 987–995. [CrossRef] [PubMed]
33. Mullins, S.R.; Vasilakos, J.P.; Deschler, K.; Grigsby, I.; Gillis, P.; John, J.; Elder, M.J.; Swales, J.; Timosenko, E.; Cooper, Z.; et al. Intratumoral immunotherapy with TLR7/8 agonist MEDI9197 modulates the tumor microenvironment leading to enhanced activity when combined with other immunotherapies. *J. Immunother. Cancer* **2019**, *7*, 244. [CrossRef] [PubMed]
34. Surendran, N.; Simmons, A.; Pichichero, M.E. TLR agonist combinations that stimulate Th type I polarizing responses from human neonates. *Innate Immun.* **2018**, *24*, 240–251. [CrossRef] [PubMed]

35. Kayesh, M.E.H.; Kohara, M.; Tsukiyama-Kohara, K. Toll-Like Receptor Response to Hepatitis B Virus Infection and Potential of TLR Agonists as Immunomodulators for Treating Chronic Hepatitis B: An Overview. *Int. J. Mol. Sci.* **2021**, *22*, 10462. [CrossRef]
36. Owen, A.M.; Fults, J.B.; Patil, N.K.; Hernandez, A.; Bohannon, J.K. TLR Agonists as Mediators of Trained Immunity: Mechanistic Insight and Immunotherapeutic Potential to Combat Infection. *Front. Immunol.* **2020**, *11*, 622614. [CrossRef]
37. Yokota, S.; Okabayashi, T.; Fujii, N. The battle between virus and host: Modulation of Toll-like receptor signaling pathways by virus infection. *Mediat. Inflamm.* **2010**, *2010*, 184328. [CrossRef]
38. Huang, B.; Zhao, J.; Unkeless, J.C.; Feng, Z.H.; Xiong, H. TLR signaling by tumor and immune cells: A double-edged sword. *Oncogene* **2008**, *27*, 218–224. [CrossRef]
39. Ebermeyer, T.; Cognasse, F.; Berthelot, P.; Mismetti, P.; Garraud, O.; Hamzeh-Cognasse, H. Platelet Innate Immune Receptors and TLRs: A Double-Edged Sword. *Int. J. Mol. Sci.* **2021**, *22*, 7894. [CrossRef]
40. Zheng, M.; Karki, R.; Williams, E.P.; Yang, D.; Fitzpatrick, E.; Vogel, P.; Jonsson, C.B.; Kanneganti, T.D. TLR2 senses the SARS-CoV-2 envelope protein to produce inflammatory cytokines. *Nat. Immunol.* **2021**, *22*, 829–838. [CrossRef]
41. Metz, P.; Reuter, A.; Bender, S.; Bartenschlager, R. Interferon-stimulated genes and their role in controlling hepatitis C virus. *J. Hepatol.* **2013**, *59*, 1331–1341. [CrossRef]
42. Horner, S.M.; Gale, M., Jr. Regulation of hepatic innate immunity by hepatitis C virus. *Nat. Med.* **2013**, *19*, 879–888. [CrossRef] [PubMed]
43. Yang, D.R.; Zhu, H.Z. Hepatitis C virus and antiviral innate immunity: Who wins at tug-of-war? *World J. Gastroenterol.* **2015**, *21*, 3786–3800. [CrossRef] [PubMed]
44. Kawai, T.; Akira, S. The role of pattern-recognition receptors in innate immunity: Update on Toll-like receptors. *Nat. Immunol.* **2010**, *11*, 373–384. [CrossRef] [PubMed]
45. Wang, N.; Liang, Y.; Devaraj, S.; Wang, J.; Lemon, S.M.; Li, K. Toll-like receptor 3 mediates establishment of an antiviral state against hepatitis C virus in hepatoma cells. *J. Virol.* **2009**, *83*, 9824–9834. [CrossRef]
46. Machida, K.; Cheng, K.T.; Sung, V.M.; Levine, A.M.; Foung, S.; Lai, M.M. Hepatitis C virus induces toll-like receptor 4 expression, leading to enhanced production of beta interferon and interleukin-6. *J. Virol.* **2006**, *80*, 866–874. [CrossRef]
47. Feldmann, G.; Nischalke, H.D.; Nattermann, J.; Banas, B.; Berg, T.; Teschendorf, C.; Schmiegel, W.; Duhrsen, U.; Halangk, J.; Iwan, A.; et al. Induction of interleukin-6 by hepatitis C virus core protein in hepatitis C-associated mixed cryoglobulinemia and B-cell non-Hodgkin's lymphoma. *Clin. Cancer Res.* **2006**, *12*, 4491–4498. [CrossRef]
48. Shehata, M.A.; Abou El-Enein, A.; El-Sharnouby, G.A. Significance of toll-like receptors 2 and 4 mRNA expression in chronic hepatitis C virus infection. *Egypt J. Immunol.* **2006**, *13*, 141–152.
49. Wang, J.P.; Zhang, Y.; Wei, X.; Li, J.; Nan, X.P.; Yu, H.T.; Li, Y.; Wang, P.Z.; Bai, X.F. Circulating Toll-like receptor (TLR) 2, TLR4, and regulatory T cells in patients with chronic hepatitis C. *APMIS* **2010**, *118*, 261–270. [CrossRef]
50. Chung, H.; Watanabe, T.; Kudo, M.; Chiba, T. Correlation between hyporesponsiveness to Toll-like receptor ligands and liver dysfunction in patients with chronic hepatitis C virus infection. *J. Viral Hepat.* **2011**, *18*, e561–e567. [CrossRef]
51. He, Q.; Graham, C.S.; Durante Mangoni, E.; Koziel, M.J. Differential expression of toll-like receptor mRNA in treatment non-responders and sustained virologic responders at baseline in patients with chronic hepatitis C. *Liver Int.* **2006**, *26*, 1100–1110. [CrossRef]
52. Sato, K.; Ishikawa, T.; Okumura, A.; Yamauchi, T.; Sato, S.; Ayada, M.; Matsumoto, E.; Hotta, N.; Oohashi, T.; Fukuzawa, Y.; et al. Expression of Toll-like receptors in chronic hepatitis C virus infection. *J. Gastroenterol. Hepatol.* **2007**, *22*, 1627–1632. [CrossRef] [PubMed]
53. Dolganiuc, A.; Kodys, K.; Kopasz, A.; Marshall, C.; Do, T.; Romics, L., Jr.; Mandrekar, P.; Zapp, M.; Szabo, G. Hepatitis C virus core and nonstructural protein 3 proteins induce pro- and anti-inflammatory cytokines and inhibit dendritic cell differentiation. *J. Immunol.* **2003**, *170*, 5615–5624. [CrossRef] [PubMed]
54. Dolganiuc, A.; Oak, S.; Kodys, K.; Golenbock, D.T.; Finberg, R.W.; Kurt-Jones, E.; Szabo, G. Hepatitis C core and nonstructural 3 proteins trigger toll-like receptor 2-mediated pathways and inflammatory activation. *Gastroenterology* **2004**, *127*, 1513–1524. [CrossRef]
55. Brown, R.A.; Gralewski, J.H.; Eid, A.J.; Knoll, B.M.; Finberg, R.W.; Razonable, R.R. R753Q single-nucleotide polymorphism impairs toll-like receptor 2 recognition of hepatitis C virus core and nonstructural 3 proteins. *Transplantation* **2010**, *89*, 811–815. [CrossRef] [PubMed]
56. Chang, S.; Dolganiuc, A.; Szabo, G. Toll-like receptors 1 and 6 are involved in TLR2-mediated macrophage activation by hepatitis C virus core and NS3 proteins. *J. Leukoc. Biol.* **2007**, *82*, 479–487. [CrossRef]
57. Hoffmann, M.; Zeisel, M.B.; Jilg, N.; Paranhos-Baccala, G.; Stoll-Keller, F.; Wakita, T.; Hafkemeyer, P.; Blum, H.E.; Barth, H.; Henneke, P.; et al. Toll-like receptor 2 senses hepatitis C virus core protein but not infectious viral particles. *J. Innate Immun.* **2009**, *1*, 446–454. [CrossRef]
58. Chung, H.; Watanabe, T.; Kudo, M.; Chiba, T. Hepatitis C virus core protein induces homotolerance and cross-tolerance to Toll-like receptor ligands by activation of Toll-like receptor 2. *J. Infect. Dis.* **2010**, *202*, 853–861. [CrossRef]
59. Dolganiuc, A.; Chang, S.; Kodys, K.; Mandrekar, P.; Bakis, G.; Cormier, M.; Szabo, G. Hepatitis C virus (HCV) core protein-induced, monocyte-mediated mechanisms of reduced IFN-alpha and plasmacytoid dendritic cell loss in chronic HCV infection. *J. Immunol.* **2006**, *177*, 6758–6768. [CrossRef]

60. Rajalakshmy, A.R.; Malathi, J.; Madhavan, H.N. HCV core and NS3 proteins mediate toll like receptor induced innate immune response in corneal epithelium. *Exp. Eye Res.* **2014**, *128*, 117–128. [CrossRef]
61. Riordan, S.M.; Skinner, N.A.; Kurtovic, J.; Locarnini, S.; McIver, C.J.; Williams, R.; Visvanathan, K. Toll-like receptor expression in chronic hepatitis C: Correlation with pro-inflammatory cytokine levels and liver injury. *Inflamm. Res.* **2006**, *55*, 279–285. [CrossRef]
62. Chen Yi Mei, S.L.; Burchell, J.; Skinner, N.; Millen, R.; Matthews, G.; Hellard, M.; Dore, G.J.; Desmond, P.V.; Sundararajan, V.; Thompson, A.J.; et al. Toll-like Receptor Expression and Signaling in Peripheral Blood Mononuclear Cells Correlate with Clinical Outcomes in Acute Hepatitis C Virus Infection. *J. Infect. Dis.* **2016**, *214*, 739–747. [CrossRef] [PubMed]
63. Moriyama, M.; Kato, N.; Otsuka, M.; Shao, R.X.; Taniguchi, H.; Kawabe, T.; Omata, M. Interferon-beta is activated by hepatitis C virus NS5B and inhibited by NS4A, NS4B, and NS5A. *Hepatol. Int.* **2007**, *1*, 302–310. [CrossRef] [PubMed]
64. Broering, R.; Wu, J.; Meng, Z.; Hilgard, P.; Lu, M.; Trippler, M.; Szczeponek, A.; Gerken, G.; Schlaak, J.F. Toll-like receptor-stimulated non-parenchymal liver cells can regulate hepatitis C virus replication. *J. Hepatol.* **2008**, *48*, 914–922. [CrossRef] [PubMed]
65. Al-Qahtani, A.A.; Al-Anazi, M.R.; Al-Zoghaibi, F.; Abdo, A.A.; Sanai, F.M.; Khan, M.Q.; Albenmousa, A.; Al-Ashgar, H.I.; Al-Ahdal, M.N. The association of toll-like receptor 4 polymorphism with hepatitis C virus infection in Saudi Arabian patients. *Biomed. Res. Int.* **2014**, *2014*, 357062. [CrossRef]
66. Zhang, Y.L.; Guo, Y.J.; Bin, L.; Sun, S.H. Hepatitis C virus single-stranded RNA induces innate immunity via Toll-like receptor 7. *J. Hepatol.* **2009**, *51*, 29–38. [CrossRef]
67. Li, K.; Li, N.L.; Wei, D.; Pfeffer, S.R.; Fan, M.; Pfeffer, L.M. Activation of chemokine and inflammatory cytokine response in hepatitis C virus-infected hepatocytes depends on Toll-like receptor 3 sensing of hepatitis C virus double-stranded RNA intermediates. *Hepatology* **2012**, *55*, 666–675. [CrossRef]
68. Motavaf, M.; Noorbakhsh, F.; Alavian, S.M.; Sharifi, Z. Distinct Toll-like Receptor 3 and 7 Expression in Peripheral Blood Mononuclear Cells from Patients with Chronic Hepatitis C Infection. *Hepat. Mon.* **2014**, *14*, e16421. [CrossRef]
69. Mohammed, K.I.; Adel, L.A.; Ali-Eldin, F.A.; Eladawy, S. Expression of Toll like receptors 3 & 7 in peripheral blood from patients with chronic hepatitis C virus infection and their correlation with interferon-alpha. *Egypt J. Immunol.* **2013**, *20*, 13–22.
70. Suef, R.A.; Mohamed, E.E.M.; Mansour, M.T.M.; Weigand, K.; Farag, M.M.S. Differential expression of viral pathogen-associated molecular pattern receptors mRNA in Egyptian chronic hepatitis C virus patients. *Egypt. J. Med. Hum. Genet.* **2021**, *22*, 13. [CrossRef]
71. Firdaus, R.; Biswas, A.; Saha, K.; Mukherjee, A.; Pal, F.; Chaudhuri, S.; Chandra, A.; Konar, A.; Sadhukhan, P.C. Modulation of TLR 3, 7 and 8 expressions in HCV genotype 3 infected individuals: Potential correlations of pathogenesis and spontaneous clearance. *Biomed. Res. Int.* **2014**, *2014*, 491064. [CrossRef]
72. Howell, J.; Sawhney, R.; Skinner, N.; Gow, P.; Angus, P.; Ratnam, D.; Visvanathan, K. Toll-like receptor 3 and 7/8 function is impaired in hepatitis C rapid fibrosis progression post-liver transplantation. *Am. J. Transplant.* **2013**, *13*, 943–953. [CrossRef] [PubMed]
73. Takahashi, K.; Asabe, S.; Wieland, S.; Garaigorta, U.; Gastaminza, P.; Isogawa, M.; Chisari, F.V. Plasmacytoid dendritic cells sense hepatitis C virus-infected cells, produce interferon, and inhibit infection. *Proc. Natl. Acad. Sci. USA* **2010**, *107*, 7431–7436. [CrossRef] [PubMed]
74. Dreux, M.; Garaigorta, U.; Boyd, B.; Decembre, E.; Chung, J.; Whitten-Bauer, C.; Wieland, S.; Chisari, F.V. Short-range exosomal transfer of viral RNA from infected cells to plasmacytoid dendritic cells triggers innate immunity. *Cell Host Microbe* **2012**, *12*, 558–570. [CrossRef] [PubMed]
75. Lee, J.; Wu, C.C.; Lee, K.J.; Chuang, T.H.; Katakura, K.; Liu, Y.T.; Chan, M.; Tawatao, R.; Chung, M.; Shen, C.; et al. Activation of anti-hepatitis C virus responses via Toll-like receptor 7. *Proc. Natl. Acad. Sci. USA* **2006**, *103*, 1828–1833. [CrossRef]
76. Wang, C.H.; Eng, H.L.; Lin, K.H.; Chang, C.H.; Hsieh, C.A.; Lin, Y.L.; Lin, T.M. TLR7 and TLR8 gene variations and susceptibility to hepatitis C virus infection. *PLoS ONE* **2011**, *6*, e26235. [CrossRef]
77. Lee, J.; Tian, Y.; Chan, S.T.; Kim, J.Y.; Cho, C.; Ou, J.H. TNF-alpha Induced by Hepatitis C Virus via TLR7 and TLR8 in Hepatocytes Supports Interferon Signaling via an Autocrine Mechanism. *PLoS Pathog.* **2015**, *11*, e1004937. [CrossRef]
78. Zhang, Y.; El-Far, M.; Dupuy, F.P.; Abdel-Hakeem, M.S.; He, Z.; Procopio, F.A.; Shi, Y.; Haddad, E.K.; Ancuta, P.; Sekaly, R.P.; et al. HCV RNA Activates APCs via TLR7/TLR8 While Virus Selectively Stimulates Macrophages Without Inducing Antiviral Responses. *Sci. Rep.* **2016**, *6*, 29447. [CrossRef]
79. Fakhir, F.Z.; Lkhider, M.; Badre, W.; Alaoui, R.; Meurs, E.F.; Pineau, P.; Ezzikouri, S.; Benjelloun, S. Genetic variations in toll-like receptors 7 and 8 modulate natural hepatitis C outcomes and liver disease progression. *Liver Int.* **2018**, *38*, 432–442. [CrossRef]
80. Wang, C.H.; Eng, H.L.; Lin, K.H.; Liu, H.C.; Chang, C.H.; Lin, T.M. Functional polymorphisms of TLR8 are associated with hepatitis C virus infection. *Immunology* **2014**, *141*, 540–548. [CrossRef]
81. El-Bendary, M.; Neamatallah, M.; Elalfy, H.; Besheer, T.; Elkholi, A.; El-Diasty, M.; Elsareef, M.; Zahran, M.; El-Aarag, B.; Gomaa, A.; et al. The association of single nucleotide polymorphisms of Toll-like receptor 3, Toll-like receptor 7 and Toll-like receptor 8 genes with the susceptibility to HCV infection. *Br. J. Biomed. Sci.* **2018**, *75*, 175–181. [CrossRef]
82. Fischer, J.; Weber, A.N.R.; Bohm, S.; Dickhofer, S.; El Maadidi, S.; Deichsel, D.; Knop, V.; Klinker, H.; Moller, B.; Rasenack, J.; et al. Sex-specific effects of TLR9 promoter variants on spontaneous clearance of HCV infection. *Gut* **2017**, *66*, 1829–1837. [CrossRef] [PubMed]

83. Abdelwahab, S.F.; Hamdy, S.; Osman, A.M.; Zakaria, Z.A.; Galal, I.; Sobhy, M.; Hashem, M.; Allam, W.R.; Abdel-Samiee, M.; Rewisha, E.; et al. Association of the polymorphism of the Toll-like receptor (TLR)-3 and TLR-9 genes with hepatitis C virus-specific cell-mediated immunity outcomes among Egyptian health-care workers. *Clin. Exp. Immunol.* **2021**, *203*, 3–12. [CrossRef] [PubMed]
84. Kayesh, M.E.H.; Sanada, T.; Kohara, M.; Tsukiyama-Kohara, K. Tree Shrew as an Emerging Small Animal Model for Human Viral Infection: A Recent Overview. *Viruses* **2021**, *13*, 1641. [CrossRef] [PubMed]
85. Fan, Y.; Huang, Z.Y.; Cao, C.C.; Chen, C.S.; Chen, Y.X.; Fan, D.D.; He, J.; Hou, H.L.; Hu, L.; Hu, X.T.; et al. Genome of the Chinese tree shrew. *Nat. Commun.* **2013**, *4*, 1426. [CrossRef]
86. Sanada, T.; Tsukiyama-Kohara, K.; Shin, I.T.; Yamamoto, N.; Kayesh, M.E.H.; Yamane, D.; Takano, J.I.; Shiogama, Y.; Yasutomi, Y.; Ikeo, K.; et al. Construction of complete Tupaia belangeri transcriptome database by whole-genome and comprehensive RNA sequencing. *Sci. Rep.* **2019**, *9*, 12372. [CrossRef]
87. Kayesh, M.E.H.; Ezzikouri, S.; Sanada, T.; Chi, H.; Hayashi, Y.; Rebbani, K.; Kitab, B.; Matsuu, A.; Miyoshi, N.; Hishima, T.; et al. Oxidative Stress and Immune Responses during Hepatitis C Virus Infection in Tupaia belangeri. *Sci. Rep.* **2017**, *7*, 9848. [CrossRef]
88. Sarasin-Filipowicz, M.; Oakeley, E.J.; Duong, F.H.; Christen, V.; Terracciano, L.; Filipowicz, W.; Heim, M.H. Interferon signaling and treatment outcome in chronic hepatitis C. *Proc. Natl. Acad. Sci. USA* **2008**, *105*, 7034–7039. [CrossRef]
89. Asselah, T.; Bieche, I.; Narguet, S.; Sabbagh, A.; Laurendeau, I.; Ripault, M.P.; Boyer, N.; Martinot-Peignoux, M.; Valla, D.; Vidaud, M.; et al. Liver gene expression signature to predict response to pegylated interferon plus ribavirin combination therapy in patients with chronic hepatitis C. *Gut* **2008**, *57*, 516–524. [CrossRef]
90. Thimme, R.; Binder, M.; Bartenschlager, R. Failure of innate and adaptive immune responses in controlling hepatitis C virus infection. *FEMS Microbiol. Rev.* **2012**, *36*, 663–683. [CrossRef]
91. Ding, Q.; Cao, X.; Lu, J.; Huang, B.; Liu, Y.J.; Kato, N.; Shu, H.B.; Zhong, J. Hepatitis C virus NS4B blocks the interaction of STING and TBK1 to evade host innate immunity. *J. Hepatol.* **2013**, *59*, 52–58. [CrossRef]
92. Samrat, S.K.; Vedi, S.; Singh, S.; Li, W.; Kumar, R.; Agrawal, B. Immunization with Recombinant Adenoviral Vectors Expressing HCV Core or F Proteins Leads to T Cells with Reduced Effector Molecules Granzyme B and IFN-gamma: A Potential New Strategy for Immune Evasion in HCV Infection. *Viral Immunol.* **2015**, *28*, 309–324. [CrossRef] [PubMed]
93. Li, K.; Foy, E.; Ferreon, J.C.; Nakamura, M.; Ferreon, A.C.; Ikeda, M.; Ray, S.C.; Gale, M., Jr.; Lemon, S.M. Immune evasion by hepatitis C virus NS3/4A protease-mediated cleavage of the Toll-like receptor 3 adaptor protein TRIF. *Proc. Natl. Acad. Sci. USA* **2005**, *102*, 2992–2997. [CrossRef] [PubMed]
94. Xu, Z.; Choi, J.; Lu, W.; Ou, J.H. Hepatitis C virus f protein is a short-lived protein associated with the endoplasmic reticulum. *J. Virol.* **2003**, *77*, 1578–1583. [CrossRef] [PubMed]
95. Oshiumi, H.; Miyashita, M.; Matsumoto, M.; Seya, T. A distinct role of Riplet-mediated K63-Linked polyubiquitination of the RIG-I repressor domain in human antiviral innate immune responses. *PLoS Pathog.* **2013**, *9*, e1003533. [CrossRef]
96. Morikawa, K.; Lange, C.M.; Gouttenoire, J.; Meylan, E.; Brass, V.; Penin, F.; Moradpour, D. Nonstructural protein 3-4A: The Swiss army knife of hepatitis C virus. *J. Viral Hepat.* **2011**, *18*, 305–315. [CrossRef]
97. Kang, X.; Chen, X.; He, Y.; Guo, D.; Guo, L.; Zhong, J.; Shu, H.B. DDB1 is a cellular substrate of NS3/4A protease and required for hepatitis C virus replication. *Virology* **2013**, *435*, 385–394. [CrossRef]
98. Rehermann, B. Hepatitis C virus versus innate and adaptive immune responses: A tale of coevolution and coexistence. *J. Clin. Investig.* **2009**, *119*, 1745–1754. [CrossRef]
99. Foy, E.; Li, K.; Sumpter, R., Jr.; Loo, Y.M.; Johnson, C.L.; Wang, C.; Fish, P.M.; Yoneyama, M.; Fujita, T.; Lemon, S.M.; et al. Control of antiviral defenses through hepatitis C virus disruption of retinoic acid-inducible gene-I signaling. *Proc. Natl. Acad. Sci. USA* **2005**, *102*, 2986–2991. [CrossRef]
100. Sklan, E.H.; Charuworn, P.; Pang, P.S.; Glenn, J.S. Mechanisms of HCV survival in the host. *Nat. Rev. Gastroenterol. Hepatol.* **2009**, *6*, 217–227. [CrossRef]
101. Loo, Y.M.; Owen, D.M.; Li, K.; Erickson, A.K.; Johnson, C.L.; Fish, P.M.; Carney, D.S.; Wang, T.; Ishida, H.; Yoneyama, M.; et al. Viral and therapeutic control of IFN-beta promoter stimulator 1 during hepatitis C virus infection. *Proc. Natl. Acad. Sci. USA* **2006**, *103*, 6001–6006. [CrossRef]
102. Meylan, E.; Curran, J.; Hofmann, K.; Moradpour, D.; Binder, M.; Bartenschlager, R.; Tschopp, J. Cardif is an adaptor protein in the RIG-I antiviral pathway and is targeted by hepatitis C virus. *Nature* **2005**, *437*, 1167–1172. [CrossRef] [PubMed]
103. Bellecave, P.; Sarasin-Filipowicz, M.; Donze, O.; Kennel, A.; Gouttenoire, J.; Meylan, E.; Terracciano, L.; Tschopp, J.; Sarrazin, C.; Berg, T.; et al. Cleavage of mitochondrial antiviral signaling protein in the liver of patients with chronic hepatitis C correlates with a reduced activation of the endogenous interferon system. *Hepatology* **2010**, *51*, 1127–1136. [CrossRef] [PubMed]
104. Blindenbacher, A.; Duong, F.H.; Hunziker, L.; Stutvoet, S.T.; Wang, X.; Terracciano, L.; Moradpour, D.; Blum, H.E.; Alonzi, T.; Tripodi, M.; et al. Expression of hepatitis c virus proteins inhibits interferon alpha signaling in the liver of transgenic mice. *Gastroenterology* **2003**, *124*, 1465–1475. [CrossRef]
105. Bode, J.G.; Ludwig, S.; Ehrhardt, C.; Albrecht, U.; Erhardt, A.; Schaper, F.; Heinrich, P.C.; Haussinger, D. IFN-alpha antagonistic activity of HCV core protein involves induction of suppressor of cytokine signaling-3. *FASEB J.* **2003**, *17*, 488–490. [CrossRef] [PubMed]

106. Basu, A.; Meyer, K.; Ray, R.B.; Ray, R. Hepatitis C virus core protein modulates the interferon-induced transacting factors of Jak/Stat signaling pathway but does not affect the activation of downstream IRF-1 or 561 gene. *Virology* **2001**, *288*, 379–390. [CrossRef]
107. Stone, A.E.; Mitchell, A.; Brownell, J.; Miklin, D.J.; Golden-Mason, L.; Polyak, S.J.; Gale, M.J., Jr.; Rosen, H.R. Hepatitis C virus core protein inhibits interferon production by a human plasmacytoid dendritic cell line and dysregulates interferon regulatory factor-7 and signal transducer and activator of transcription (STAT) 1 protein expression. *PLoS ONE* **2014**, *9*, e95627. [CrossRef]
108. Lin, W.; Choe, W.H.; Hiasa, Y.; Kamegaya, Y.; Blackard, J.T.; Schmidt, E.V.; Chung, R.T. Hepatitis C virus expression suppresses interferon signaling by degrading STAT1. *Gastroenterology* **2005**, *128*, 1034–1041. [CrossRef]
109. Lin, W.; Kim, S.S.; Yeung, E.; Kamegaya, Y.; Blackard, J.T.; Kim, K.A.; Holtzman, M.J.; Chung, R.T. Hepatitis C virus core protein blocks interferon signaling by interaction with the STAT1 SH2 domain. *J. Virol.* **2006**, *80*, 9226–9235. [CrossRef]
110. Heim, M.H.; Moradpour, D.; Blum, H.E. Expression of hepatitis C virus proteins inhibits signal transduction through the Jak-STAT pathway. *J. Virol.* **1999**, *73*, 8469–8475. [CrossRef]
111. Duong, F.H.; Filipowicz, M.; Tripodi, M.; La Monica, N.; Heim, M.H. Hepatitis C virus inhibits interferon signaling through up-regulation of protein phosphatase 2A. *Gastroenterology* **2004**, *126*, 263–277. [CrossRef]
112. Luquin, E.; Larrea, E.; Civeira, M.P.; Prieto, J.; Aldabe, R. HCV structural proteins interfere with interferon-alpha Jak/STAT signalling pathway. *Antivir. Res.* **2007**, *76*, 194–197. [CrossRef] [PubMed]
113. Hosui, A.; Ohkawa, K.; Ishida, H.; Sato, A.; Nakanishi, F.; Ueda, K.; Takehara, T.; Kasahara, A.; Sasaki, Y.; Hori, M.; et al. Hepatitis C virus core protein differently regulates the JAK-STAT signaling pathway under interleukin-6 and interferon-gamma stimuli. *J. Biol. Chem.* **2003**, *278*, 28562–28571. [CrossRef] [PubMed]
114. Yoshida, H.; Kato, N.; Shiratori, Y.; Otsuka, M.; Maeda, S.; Kato, J.; Omata, M. Hepatitis C virus core protein activates nuclear factor kappa B-dependent signaling through tumor necrosis factor receptor-associated factor. *J. Biol. Chem.* **2001**, *276*, 16399–16405. [CrossRef] [PubMed]
115. You, L.R.; Chen, C.M.; Lee, Y.H. Hepatitis C virus core protein enhances NF-kappaB signal pathway triggering by lymphotoxin-beta receptor ligand and tumor necrosis factor alpha. *J. Virol.* **1999**, *73*, 1672–1681. [CrossRef]
116. Marusawa, H.; Hijikata, M.; Chiba, T.; Shimotohno, K. Hepatitis C virus core protein inhibits Fas-and tumor necrosis factor alpha-mediated apoptosis via NF-kappaB activation. *J. Virol.* **1999**, *73*, 4713–4720. [CrossRef]
117. Kato, N.; Yoshida, H.; Ono-Nita, S.K.; Kato, J.; Goto, T.; Otsuka, M.; Lan, K.; Matsushima, K.; Shiratori, Y.; Omata, M. Activation of intracellular signaling by hepatitis B and C viruses: C-viral core is the most potent signal inducer. *Hepatology* **2000**, *32*, 405–412. [CrossRef]
118. Nguyen, H.; Sankaran, S.; Dandekar, S. Hepatitis C virus core protein induces expression of genes regulating immune evasion and anti-apoptosis in hepatocytes. *Virology* **2006**, *354*, 58–68. [CrossRef]
119. Larrea, E.; Aldabe, R.; Molano, E.; Fernandez-Rodriguez, C.M.; Ametzazurra, A.; Civeira, M.P.; Prieto, J. Altered expression and activation of signal transducers and activators of transcription (STATs) in hepatitis C virus infection: In vivo and in vitro studies. *Gut* **2006**, *55*, 1188–1196. [CrossRef]
120. Agaugue, S.; Perrin-Cocon, L.; Andre, P.; Lotteau, V. Hepatitis C lipo-Viro-particle from chronically infected patients interferes with TLR4 signaling in dendritic cell. *PLoS ONE* **2007**, *2*, e330. [CrossRef]
121. Nitta, S.; Sakamoto, N.; Nakagawa, M.; Kakinuma, S.; Mishima, K.; Kusano-Kitazume, A.; Kiyohashi, K.; Murakawa, M.; Nishimura-Sakurai, Y.; Azuma, S.; et al. Hepatitis C virus NS4B protein targets STING and abrogates RIG-I-mediated type I interferon-dependent innate immunity. *Hepatology* **2013**, *57*, 46–58. [CrossRef]
122. Yi, G.; Wen, Y.; Shu, C.; Han, Q.; Konan, K.V.; Li, P.; Kao, C.C. Hepatitis C Virus NS4B Can Suppress STING Accumulation To Evade Innate Immune Responses. *J. Virol.* **2016**, *90*, 254–265. [CrossRef] [PubMed]
123. Abe, T.; Kaname, Y.; Hamamoto, I.; Tsuda, Y.; Wen, X.; Taguwa, S.; Moriishi, K.; Takeuchi, O.; Kawai, T.; Kanto, T.; et al. Hepatitis C virus nonstructural protein 5A modulates the toll-like receptor-MyD88-dependent signaling pathway in macrophage cell lines. *J. Virol.* **2007**, *81*, 8953–8966. [CrossRef] [PubMed]
124. Polyak, S.J.; Khabar, K.S.; Paschal, D.M.; Ezelle, H.J.; Duverlie, G.; Barber, G.N.; Levy, D.E.; Mukaida, N.; Gretch, D.R. Hepatitis C virus nonstructural 5A protein induces interleukin-8, leading to partial inhibition of the interferon-induced antiviral response. *J. Virol.* **2001**, *75*, 6095–6106. [CrossRef] [PubMed]
125. Horner, S.M.; Gale, M., Jr. Intracellular innate immune cascades and interferon defenses that control hepatitis C virus. *J. Interferon Cytokine Res.* **2009**, *29*, 489–498. [CrossRef]
126. Gale, M.J., Jr.; Korth, M.J.; Tang, N.M.; Tan, S.L.; Hopkins, D.A.; Dever, T.E.; Polyak, S.J.; Gretch, D.R.; Katze, M.G. Evidence that hepatitis C virus resistance to interferon is mediated through repression of the PKR protein kinase by the nonstructural 5A protein. *Virology* **1997**, *230*, 217–227. [CrossRef]
127. Taylor, D.R.; Shi, S.T.; Romano, P.R.; Barber, G.N.; Lai, M.M. Inhibition of the interferon-inducible protein kinase PKR by HCV E2 protein. *Science* **1999**, *285*, 107–110. [CrossRef]
128. Noguchi, T.; Satoh, S.; Noshi, T.; Hatada, E.; Fukuda, R.; Kawai, A.; Ikeda, S.; Hijikata, M.; Shimotohno, K. Effects of mutation in hepatitis C virus nonstructural protein 5A on interferon resistance mediated by inhibition of PKR kinase activity in mammalian cells. *Microbiol. Immunol.* **2001**, *45*, 829–840. [CrossRef]

129. Hiet, M.S.; Bauhofer, O.; Zayas, M.; Roth, H.; Tanaka, Y.; Schirmacher, P.; Willemsen, J.; Grunvogel, O.; Bender, S.; Binder, M.; et al. Control of temporal activation of hepatitis C virus-induced interferon response by domain 2 of nonstructural protein 5A. *J. Hepatol.* **2015**, *63*, 829–837. [CrossRef]
130. Cevik, R.E.; Cesarec, M.; Da Silva Filipe, A.; Licastro, D.; McLauchlan, J.; Marcello, A. Hepatitis C Virus NS5A Targets Nucleosome Assembly Protein NAP1L1 to Control the Innate Cellular Response. *J. Virol.* **2017**, *91*, e00880-17. [CrossRef]
131. Yonkers, N.L.; Rodriguez, B.; Milkovich, K.A.; Asaad, R.; Lederman, M.M.; Heeger, P.S.; Anthony, D.D. TLR ligand-dependent activation of naive CD4 T cells by plasmacytoid dendritic cells is impaired in hepatitis C virus infection. *J. Immunol.* **2007**, *178*, 4436–4444. [CrossRef]
132. Qi, H.; Chu, V.; Wu, N.C.; Chen, Z.; Truong, S.; Brar, G.; Su, S.Y.; Du, Y.; Arumugaswami, V.; Olson, C.A.; et al. Systematic identification of anti-interferon function on hepatitis C virus genome reveals p7 as an immune evasion protein. *Proc. Natl. Acad. Sci. USA* **2017**, *114*, 2018–2023. [CrossRef] [PubMed]
133. Carty, M.; Bowie, A.G. Recent insights into the role of Toll-like receptors in viral infection. *Clin. Exp. Immunol.* **2010**, *161*, 397–406. [CrossRef] [PubMed]
134. Patel, M.C.; Shirey, K.A.; Pletneva, L.M.; Boukhvalova, M.S.; Garzino-Demo, A.; Vogel, S.N.; Blanco, J.C. Novel drugs targeting Toll-like receptors for antiviral therapy. *Future Virol.* **2014**, *9*, 811–829. [CrossRef] [PubMed]
135. Mifsud, E.J.; Tan, A.C.; Jackson, D.C. TLR Agonists as Modulators of the Innate Immune Response and Their Potential as Agents Against Infectious Disease. *Front. Immunol.* **2014**, *5*, 79. [CrossRef]
136. Li, G.; De Clercq, E. Current therapy for chronic hepatitis C: The role of direct-acting antivirals. *Antivir. Res.* **2017**, *142*, 83–122. [CrossRef]
137. Chinchilla-Lopez, P.; Qi, X.; Yoshida, E.M.; Mendez-Sanchez, N. The Direct-Acting Antivirals for Hepatitis C Virus and the Risk for Hepatocellular Carcinoma. *Ann. Hepatol.* **2017**, *16*, 328–330. [CrossRef]
138. Kayesh, M.E.H.; Kohara, M.; Tsukiyama-Kohara, K. An Overview of Recent Insights into the Response of TLR to SARS-CoV-2 Infection and the Potential of TLR Agonists as SARS-CoV-2 Vaccine Adjuvants. *Viruses* **2021**, *13*, 2302. [CrossRef]
139. Lucifora, J.; Bonnin, M.; Aillot, L.; Fusil, F.; Maadadi, S.; Dimier, L.; Michelet, M.; Floriot, O.; Ollivier, A.; Rivoire, M.; et al. Direct antiviral properties of TLR ligands against HBV replication in immune-competent hepatocytes. *Sci. Rep.* **2018**, *8*, 5390. [CrossRef]
140. Niu, C.; Li, L.; Daffis, S.; Lucifora, J.; Bonnin, M.; Maadadi, S.; Salas, E.; Chu, R.; Ramos, H.; Livingston, C.M.; et al. Toll-like receptor 7 agonist GS-9620 induces prolonged inhibition of HBV via a type I interferon-dependent mechanism. *J. Hepatol.* **2018**, *68*, 922–931. [CrossRef]
141. Martinsen, J.T.; Gunst, J.D.; Hojen, J.F.; Tolstrup, M.; Sogaard, O.S. The Use of Toll-Like Receptor Agonists in HIV-1 Cure Strategies. *Front. Immunol.* **2020**, *11*, 1112. [CrossRef]
142. Proud, P.C.; Tsitoura, D.; Watson, R.J.; Chua, B.Y.; Aram, M.J.; Bewley, K.R.; Cavell, B.E.; Cobb, R.; Dowall, S.; Fotheringham, S.A.; et al. Prophylactic intranasal administration of a TLR2/6 agonist reduces upper respiratory tract viral shedding in a SARS-CoV-2 challenge ferret model. *EBioMedicine* **2021**, *63*, 103153. [CrossRef] [PubMed]
143. Angelopoulou, A.; Alexandris, N.; Konstantinou, E.; Mesiakaris, K.; Zanidis, C.; Farsalinos, K.; Poulas, K. Imiquimod—A toll like receptor 7 agonist—Is an ideal option for management of COVID 19. *Environ. Res.* **2020**, *188*, 109858. [CrossRef] [PubMed]
144. Missale, G.; Bertoni, R.; Lamonaca, V.; Valli, A.; Massari, M.; Mori, C.; Rumi, M.G.; Houghton, M.; Fiaccadori, F.; Ferrari, C. Different clinical behaviors of acute hepatitis C virus infection are associated with different vigor of the anti-viral cell-mediated immune response. *J. Clin. Investig.* **1996**, *98*, 706–714. [CrossRef] [PubMed]
145. Thimme, R.; Oldach, D.; Chang, K.M.; Steiger, C.; Ray, S.C.; Chisari, F.V. Determinants of viral clearance and persistence during acute hepatitis C virus infection. *J. Exp. Med.* **2001**, *194*, 1395–1406. [CrossRef]
146. Hennessy, E.J.; Parker, A.E.; O'Neill, L.A. Targeting Toll-like receptors: Emerging therapeutics? *Nat. Rev. Drug Discov.* **2010**, *9*, 293–307. [CrossRef]
147. Horsmans, Y.; Berg, T.; Desager, J.P.; Mueller, T.; Schott, E.; Fletcher, S.P.; Steffy, K.R.; Bauman, L.A.; Kerr, B.M.; Averett, D.R. Isatoribine, an agonist of TLR7, reduces plasma virus concentration in chronic hepatitis C infection. *Hepatology* **2005**, *42*, 724–731. [CrossRef]
148. Pockros, P.J.; Guyader, D.; Patton, H.; Tong, M.J.; Wright, T.; McHutchison, J.G.; Meng, T.C. Oral resiquimod in chronic HCV infection: Safety and efficacy in 2 placebo-controlled, double-blind phase IIa studies. *J. Hepatol.* **2007**, *47*, 174–182. [CrossRef]
149. Thomas, A.; Laxton, C.; Rodman, J.; Myangar, N.; Horscroft, N.; Parkinson, T. Investigating Toll-like receptor agonists for potential to treat hepatitis C virus infection. *Antimicrob. Agents Chemother.* **2007**, *51*, 2969–2978. [CrossRef]
150. Bergmann, J.F.; de Bruijne, J.; Hotho, D.M.; de Knegt, R.J.; Boonstra, A.; Weegink, C.J.; van Vliet, A.A.; van de Wetering, J.; Fletcher, S.P.; Bauman, L.A.; et al. Randomised clinical trial: Anti-viral activity of ANA773, an oral inducer of endogenous interferons acting via TLR7, in chronic HCV. *Aliment. Pharmacol. Ther.* **2011**, *34*, 443–453. [CrossRef]
151. Dominguez-Molina, B.; Machmach, K.; Perales, C.; Tarancon-Diez, L.; Gallego, I.; Sheldon, J.L.; Leal, M.; Domingo, E.; Ruiz-Mateos, E. Toll-Like Receptor 7 (TLR-7) and TLR-9 Agonists Improve Hepatitis C Virus Replication and Infectivity Inhibition by Plasmacytoid Dendritic Cells. *J. Virol.* **2018**, *92*, e01219-18. [CrossRef]
152. Christiansen, D.; Earnest-Silveira, L.; Chua, B.; Boo, I.; Drummer, H.E.; Grubor-Bauk, B.; Gowans, E.J.; Jackson, D.C.; Torresi, J. Antibody Responses to a Quadrivalent Hepatitis C Viral-Like Particle Vaccine Adjuvanted with Toll-Like Receptor 2 Agonists. *Viral Immunol.* **2018**, *31*, 338–343. [CrossRef] [PubMed]
153. Sepulveda-Crespo, D.; Resino, S.; Martinez, I. Innate Immune Response against Hepatitis C Virus: Targets for Vaccine Adjuvants. *Vaccines* **2020**, *8*, 313. [CrossRef] [PubMed]

International Journal of
Molecular Sciences

MDPI

Review

Is There a Role for Immunoregulatory and Antiviral Oligonucleotides Acting in the Extracellular Space? A Review and Hypothesis

Aleksandra Dondalska †, Sandra Axberg Pålsson † and Anna-Lena Spetz *

Department of Molecular Biosciences, The Wenner-Gren Institute, Stockholm University, 10691 Stockholm, Sweden
* Correspondence: anna-lena.spetz@su.se
† These authors contributed equally to this work.

Abstract: Here, we link approved and emerging nucleic acid-based therapies with the expanding universe of small non-coding RNAs (sncRNAs) and the innate immune responses that sense oligonucleotides taken up into endosomes. The Toll-like receptors (TLRs) 3, 7, 8, and 9 are located in endosomes and can detect nucleic acids taken up through endocytic routes. These receptors are key triggers in the defense against viruses and/or bacterial infections, yet they also constitute an Achilles heel towards the discrimination between self- and pathogenic nucleic acids. The compartmentalization of nucleic acids and the activity of nucleases are key components in avoiding autoimmune reactions against nucleic acids, but we still lack knowledge on the plethora of nucleic acids that might be released into the extracellular space upon infections, inflammation, and other stress responses involving increased cell death. We review recent findings that a set of single-stranded oligonucleotides (length of 25–40 nucleotides (nt)) can temporarily block ligands destined for endosomes expressing TLRs in human monocyte-derived dendritic cells. We discuss knowledge gaps and highlight the existence of a pool of RNA with an approximate length of 30–40 nt that may still have unappreciated regulatory functions in physiology and in the defense against viruses as gatekeepers of endosomal uptake through certain routes.

Keywords: oligonucleotide; TLR; sncRNA; endocytosis; broad-spectrum; antiviral agent; nucleolin; virus entry; immunoregulation; RNA therapeutics

Citation: Dondalska, A.; Axberg Pålsson, S.; Spetz, A.-L. Is There a Role for Immunoregulatory and Antiviral Oligonucleotides Acting in the Extracellular Space? A Review and Hypothesis. *Int. J. Mol. Sci.* **2022**, *23*, 14593. https://doi.org/10.3390/ijms232314593

Academic Editors: Oliver Planz and Ralf Kircheis

Received: 30 October 2022
Accepted: 21 November 2022
Published: 23 November 2022

1. Introduction

Recent advances have identified a pool of single-stranded oligonucleotides (ssONs) (either DNA or RNA) with features resulting in the capability to inhibit a broad range of enveloped viruses by binding to, or shielding, viral entry receptors [1–5]. These oligonucleotides do not target a specific sequence (hence, they are not antisense or sequence mimics) and can be administered in vivo without using cell-penetrating peptides or other delivery systems. One such class of ssDNA, oligonucleotides are the nucleic acid polymers (NAPs), which act by facilitating interactions with un-complexed amphipathic alpha helices (Reviewed in [1]). These NAPs are typically 40-mer phosphorothioate oligonucleotides and have been shown to interact with HIV-1 gp41 [6], the surface glycoprotein of lymphocytic choriomeningitis virus [7], prion proteins [8], and hepatitis delta antigen [9] in a sequence-independent manner. However, some NAPs act in other steps of the viral life cycle such as REP 2139, which was reported to inhibit secretion of Hepatitis B surface antigen (HBsAg) [10,11]. The proposed mechanism of action of REP 2139 is that blocked replenishment of HBsAg in the circulation facilitates host-mediated viral clearance [10]. A historical background with more in-depth introduction to the chemistry of nucleic acids therapeutics was recently published [12].

Another set of ssDNA, which are also acting in a sequence-independent manner, was shown to inhibit TLR3 activation [13–15]. It was revealed that this class of ssONs inhibits TLR3 activation by temporarily inhibiting clathrin-mediated endocytosis, thereby preventing the uptake of the TLR3 ligand dsRNA [15]. Notably, the inhibitory concentration $(IC)_{50}$ of ssONs with a capacity to inhibit TLR3 activation, is around 125 nM [15]. In addition, this class of ssONs were shown to inhibit infection of viruses such as respiratory syncytial virus (RSV) [3] and HIV-1 [5] in low nM concentrations. These potent activities of ssONs (in the low nM range) make us wonder whether there could be a role for naturally occurring oligonucleotides in the extracellular space and if so, which are the potential sources of such oligonucleotides? This review will focus on TLR activation requiring endocytic uptake of ligands and mechanisms on how such TLR activation can be regulated. Moreover, we review data showing that certain oligonucleotides possess antiviral activity without requiring delivery systems for intracellular uptake. Lastly, this review will provide information on the development of therapeutic approaches utilizing ssONs that act in the extracellular space.

2. Characteristics of Therapeutic Oligonucleotides

2.1. ASO

Single-stranded antisense oligonucleotides (ASO) are small (approx. 15–30 nt long), synthetic, nucleic acid polymers that often contain various chemical modifications to improve their therapeutic efficacy (such as stability and cellular uptake). ASO sequences are designed to bind to their target RNA by utilizing Watson and Crick base pairing in order to modulate splicing, affect translation initiation, or promote recruitment endonucleases such as RNase H to increase degradation of the disease-causing RNA. The first approved ASO was fomivirsen, which was used to locally treat cytomegalovirus (CMV) infection [12]. There are currently two widely used classes of ASO; the "gapmers" (utilizing RNase H) and the "splice-switching" (steric block) ASO. The current ASO that utilize the enzymatic activity of endogenous RNase H follow a "gapmer" pattern, meaning that a middle region of 6–10 DNA bases is surrounded by RNA bases that promote target binding. The RNase H recognizes the RNA–DNA heteroduplex substrates that are formed when the DNA-based oligonucleotide binds to the cognate mRNA transcript and causes RNA degradation. RNase H can cleave the RNA strand of a DNA–RNA duplex in both the nucleus and the cytoplasm. Hence, the function of gapmers is reliant on intracellular uptake of the ASO. ASOs typically enter cells via endocytosis and after uptake into early endosomes, they have to cross the endosomal lipid barrier to access their target in the cytoplasm and/or nucleus [16]. However, the mechanism for so called "endosomal escape" is poorly understood. Nevertheless, there are currently three different gapmers approved by the FDA and/or EMA mipomersen, inotersen, and volanesorsen, and they are all 20-mers (Table 1).

High affinity, sterically blocking ASO, are designed to bind to their target mRNA thereby masking specific sequences within their target transcript, which causes interference with RNA–RNA or RNA–protein interactions. ASO approaches often modulate alternative splicing, which leads to selective exclusion or retainment of a specific exon(s). Splice correction methods have often been used to restore the translational reading frame in order to salvage the target protein production [17]. So far, five different splice-switching ASO have been approved by the FDA; eteplirsen (30-mer), nusinersen (18-mer), golodirsen (25-mer), viltolarsen (21-mer), casimersen (22-mer) (Table 1).

Table 1. Characteristics of clinically approved therapeutic oligonucleotides.

Trade Name (Name), Company	Class	Chemistry	Indication, (Target), Organ	FDA/EMA Approval Year	Comment
Vitravene (fomivirsen) Ionis Pharma Novartis	ASO	21-mer PS DNA	CMV renitis, (viral *IE2* mRNA), eye	1998	First approved nucleic acid drug [18]. Withdrawn from use due to reduced clinical need. Mechanism unclear.
Macugen, (pegaptanib) NeXstar Pharma Eyetech/Pfizer	Aptamer	28-mer 2′-F/2′-OMe/pegylated RNA [19,20]	Age-related macular degeneration (VEGF-165), eye	2004	Anti-angiogenic, intravitreal injection.
Kynamro (mipomersen) Ionis Pharma, Genzyme Kastle Tx	Gapmer ASO	20-mer PS 2′-MOE [21]	Homozygous familial hypercholesterolaemia, (*APOB* mRNA), liver	2013	RNase H-mediated cleavage of apolipoprotein B mRNA. Subcutaneous (SC) injection.
Defitelio (defibrotide), Jazz Pharma	Mix of DNA isolated from porcine mucosa	Mix of PO- ssDNA and dsDNA	Hepatic veno-occlusive disease [22], (NA), liver	2016	Sequence-independent mechanism of action. Intravenous injection (IV) [23].
Exondys 51 (eteplirsen), Sarepta Tx	ASO	30-mer PMO	Duchenne muscular dystrophy, (*DMD* exon 51), skeletal muscle	2016	Steric block, splice-switching, [24] IV injection.
Spinraza (nusinersen) Ionis Pharma Biogen	ASO	18-mer PS 2′-MOE	Spinal muscular atrophy, (*SMN2* exon 7), CNS	2016	Steric block, splice-switching [25], Intrathecal injection.
Onpattro (patisiran) Alnylam Pharma	siRNA	19+2¤-mer 2′-OMe, ds [26]	Hereditary transthyretin-mediated amyloidosis, (*TTR*), liver	2018	Lipid nanoparticle formulation, IV injection.
Tegsedi (inotersen) Inonis Pharma Akcea Pharma	Gapmer ASO	20-mer PS 2′-MOE [27]	Hereditary transthyretin amyloidosis, (*TTR*), liver	2018	RNaseH mechanism of action, leading to reductions in TTR protein [28], SC injection.
Waylivra (volanesorsen) Ionis Pharma Akcea Pharma	Gapmer ASO	20-mer PS 2′-MOE [29]	Familial chylomicronaemia syndrome, (*APOC3*), liver	2019	Only approved by EMA not FDA. RNaseH mechanism of action, leading to reductions in apoC3 proteins, SC injection.
Givlaari (givosiran) Alnylam Pharma	siRNA	21/23-mer With partial PS, 2′-F, 2′-OMe, ds [30]	Acute hepatic porphyria (*ALAS1*), liver	2019	GalNAc conjugate to target hepatocytes, SC injection.
Vyondys 53 (golodirsen) Sarepta Tx	ASO	25-mer PMO	Duchenne muscular dystrophy, (*DMD* exon 53), skeletal muscle	2019	Splice-switching [31], IV injection.
Viltepso (viltolarsen) NS Pharma	ASO	21-mer PMO	Duchenne muscular dystrophy, (*DMD* exon 53), skeletal muscle	2020	Splice-switching [32], IV injection.
Oxlumo (lumasiran) Alnylam Pharma	siRNA	21/23-mer With partial PS, 2′-F, 2′-OMe, ds [33]	Primary hyperoxaluria type 1 (*HAO1*), liver	2020	GalNAc conjugate to target hepatocytes, SC injection.
Leqvio (inclisiran) Alnylam Pharma Novartis	siRNA	21/23-mer With partial PS, 2′-F, 2′-OMe, ds *	Hypercholesterolemia (*PCSK9*), liver	2021	GalNAc conjugate to target hepatocytes, SC injection.
Amondys 45 (casimersen) Sarepta Tx	ASO	22-mer PMO	Duchenne muscular dystrophy, (*DMD* exon 45), skeletal muscle	2021	Splice-switching, [34] IV injection.
Amvuttra (vutrisiran) Alnylam Pharma	siRNA	21/23-mer With partial PS, 2′-F, 2′-OMe, ds ^	Hereditary transthyretin amyloidosis [35] (*TTR*), liver	2022	GalNAc conjugate to target hepatocytes, SC injection.

Abbreviations: 2′-Ome, 2′-O-methyl; 2′-F, 2′-fluoro; PEG, polyethylene glycol; PO, phosphodiester; PS, phosphorothioate; PMO, phosphorodiamidate morpholino oligomer; 2′-MOE, 2′-O-methoxyethyl; ds, double-stranded; ss, single-stranded; *PCSK9*, proprotein convertase subtilisin kexin type 9; GalNac, N-Acetylgalactosamine; IV, intravenous; SC, subcutaneous. * LEQVIO® (inclisiran) injection, for subcutaneous use (fda.gov). ^ Novel Drug Approvals for 2022 | FDA. ¤ duplex of two 21-mer RNA with 19 complementary bases and terminal 2-nucleotide 3′overhangs.

2.2. siRNA

The antisense (or guide) strand of a double-stranded small interfering RNA (siRNA) also binds its target mRNA by Watson–Crick base pairing and is likewise reliant on uptake into cells. The other strand is designated by the passenger or sense strand. siRNA work by guiding the Argonaute (AGO) 2 protein, as part of the RNA-induced silencing complex (RISC), to the complementary target transcripts. Exact base pairing between the siRNA and the target transcript results in cleavage of the antisense strand, leading to gene silencing [17]. AGO2 protein and the RISC complex have specific structural requirements that the oligonucleotide must possess in order to bind, which limits the extent of chemical modifications that can be introduced and reflects in the stability and cellular uptake [16]. As of October 2022, five siRNAs have received FDA approval; patisiran, givosiran, lumasiran inclisiran, and vutrisiran (Table 1).

2.3. Others

Aptamers are ssONs typically approx. 20–200 nt that are folded into defined secondary structures and act as ligands that bind to target proteins in a similar way as antibodies. They can be generated by using an in vitro evolution methodology called SELEX (systematic evolution of ligands by exponential enrichment) [17]. Currently, the only approved aptamer is pegaptanib, which is an RNA-based aptamer that targets the VEGF-165 vascular endothelial growth factor isoform. Pegaptanib is used for its anti-angiogenic properties for the treatment of neovascular age-related macular degeneration [19,20]. There are currently more than 40 clinical trials listed at https://clinicaltrials.gov (accessed on 20 October 2022) concerning investigations of aptamers.

The approved drug defibrotide is produced from porcine mucosa and is composed of a mixture of ssDNA and dsDNA (hence phosphodiester (PO)-DNA) of different sizes [22]. The mechanism of action is poorly defined as of yet, however, it is not strictly sequence dependent [23]. The use of naturally occurring PO-DNA in a drug like defibrotide, indicates that there is still a considerable amount to be revealed, in terms of how pools of oligonucleotides contribute to the regulation of normal physiological mechanisms.

The approved medicines shown in Table 1 demonstrate the significance of oligonucleotides in current medical practice. The future importance of oligonucleotide-based therapies is further demonstrated by the notion that there are more than 250 clinical trials listed in https://clinicaltrials.gov (accessed on 20 October 2022) utilizing oligonucleotide therapeutics and numerous studies have been performed over the years. The concepts in progress include different types of CpG-oligodeoxynucleotide (ODN), NAPs, micro RNA (miRNA) (using mimics or anti), CRISPR-based technologies, small nucleolar RNA (snoRNA), Ribozymes, long non-coding RNA (lncRNA) (using antagoNATs or small activating RNA), and of course mRNA (vaccines, VEGF cardiac regeneration) approaches that were boosted during COVID-19 with the success of these vaccines [12,16]. Hence, these are exciting times for therapeutic oligonucleotides and there will likely be many more approved within the next years. Yet, there are still many unanswered questions in terms of optimizing delivery systems into cells of different organs and also how to exploit oligonucleotides that act on the cell surface, circumventing the obstacles of intracellular delivery. The field of therapeutic oligonucleotides is naturally directly linked to the identification and functional characterization of new non-coding RNAs and also how the immune system is reacting to oligonucleotides depending on their subcellular and/or extracellular location.

3. The Expanding Identification of Small Non-Coding RNAs

In the last decades, the existence of functional regulatory sncRNA has been revealed in all kingdoms of life, e.g., from bacteria and archaea to various eukaryotes due to transcriptome-wide studies and advances in high-throughput sequencing technologies [36]. Nevertheless, technical challenges still remain to accurately discover and measure the plethora of such sncRNA. There is a substantial amount of information about siRNAs, miRNAs, and PIWI-interacting RNAs (piRNAs) which act by base pairing to their respective RNA and/or DNA targets to exert RNA-silencing effects (such as post-transcriptional mRNA cleavage, decay or translational repression and transcriptional silencing) via AGO or PIWI. PIWI is an abbreviation of P-element Induced WImpy testis in Drosophila and are highly conserved RNA-binding proteins, which are present in both plants and animals. There are also emerging data on other non-canonical sncRNA that are often approx. 13–200 nt and derive from longer RNAs such as transfer RNA (tRNA), ribosomal RNA (rRNA), Y RNA (yRNA), small nuclear RNA (snRNA), snoRNA, vault RNA (vtRNA) and even mRNA as well as lncRNA. Several authors ascribe sncRNAs in the range of approx. 15–50 nt [37–39]. Similar to many non-coding RNAs in history, these emerging sncRNAs were initially considered to be random degradation products without a defined functional role, but increasing evidence shows their functional regulatory role in both health and diseases. There are links to cancer, immunity, viral infection, neurological diseases, stem cells, retrotransposon control, and epigenetic inheritance (recently reviewed in [38]).

The definition of "small" non-coding RNA is relatively subjective in different contexts and is likely to be more streamlined in the future when more insights are gained. Nevertheless, the different sizes in terms of numbers of nt are important to keep track of as certain features are linked to the length, and the current sequencing techniques used to identify sncRNA are often biased based on sizes of captured RNA [38]. For example, many protocols were focusing on sequencing miRNA and siRNA by using a pre-size selection with a cut-off <30 nt RNA by recovering RNA from electrophoresed gels and thereby prevented the discovery of sncRNAs, which were more than 30 nt [38]. The RNA size selection was later extended to approx. 45 nt, which can include PIWI-interacting RNAs (piRNA) and also led to the discovery of tRNA-derived-small RNA (tsRNA) and yRNA-derived-small RNA (ysRNA) found at 30–40 nt (Table 2 nomenclature of fragments in part from [38]). In addition, it was recently realized that many non-canonical sncRNAs carry various RNA modifications, some of which can prevent their detection by traditional RNAseq [38]. Hence, there is a need to develop novel sequencing techniques that can directly sequence RNA and simultaneously identify modifications, which can be as many as 150 types [40].

Another major challenge concerns the subcellular spatial compartmentalization including measurements in the extracellular space of sncRNAs. Great advances have been made to spatially map the transcriptome based on in situ hybridization, either through multiplexed imaging [41], or sequencing [42], and the single-cell resolution is likely to be further improved. However, these methods are optimized for long mRNA, whereas the short length of sncRNAs limits nucleic acid probe design resulting in binding to multiple targets [38]. It is going to be exciting times to follow the development of paradigm-changing tools allowing for the spatial and compartmentalized high-resolution discovery of sncRNAs in various tissues in health and disease. It was reported that the profile of ncRNAs can be changed upon virus infections [43–45]. For instance, RSV infection can induce the upregulation of certain tRNA fragments that enhance the replication of the virus by affecting the antiviral response [43]. It is conceivable that sncRNAs possess key functions both inside and outside of cells if released from dying cells [46], extracellular vesicles (reviewed in [16]), or hypothetically directly exported from cells. Notably, there are highly conserved cytoplasmic non-coding yRNAs, which range in size from 70 to 115 nt, and typically form Ro ribonucleoproteins (RNPs) by binding to Ro60 and La proteins. These RNPs are often targeted by the immune system in autoimmune diseases, pointing towards an Achilles heel to discriminate self-nucleic acids from foreign ones [46,47]. Although Table 2 is very brief and will require updates on definitions of different types of sncRNAs, which is best performed by the sncRNA field of experts, it points out some key differences in terms of length intervals. siRNA, miRNA, piRNA, and also other types of relatively short ssONs are working by base pairing and gene silencing using AGO proteins or PIWI. The somewhat longer yet highly abundant pool of ssONs of approx. 30–40 nt do not have the features to be able to work via AGO or PIWI and may not necessarily act in a sequence-dependent manner.

Notably, fragments derived from lncRNA and mRNA may also fall into the category of 30–40 nt [39].

Table 2. Brief description * of different types of sncRNA.

sncRNA	Length	Comment on Mechanism and/or Function	Ref.
siRNA	20–27 nt	Base pairing and gene silencing with AGO.	[48]
miRNA	21–23 nt	Base pairing and gene silencing with AGO.	[48]
piRNA	21–35 nt	Base pairing and gene silencing with PIWI.	[49]
tsRNA -tRF-1 -tRF-3 -tRF-5 -3′tRNA halves -5′tRNA halves	13–40 nt 13–30 nt 13–30 nt 13–30 nt 30–40 nt 30–40 nt	Gene silencing not always sequence dependent. Tumor suppression, T cell inhibition, affect virus replication and influence stress responses. Transfer RNAs are characterized by their typical cloverleaf structure which can be processed into abundant fragments.	[37]
rsRNA -several names of identified fragments.	Multiple lengths	Shorter 19–24 nt involved in gene silencing with AGO. Longer 74–130 nt function unclear. Often derived from 45S, 5S, and 28S rRNAs in PBMCs with lengths of 15–42.	[37,50]
ysRNA	Multiple lengths often 26–40 nt but also 83–112 nt	Overexpression of a 57 nt increased IL-10 production and administration in vivo in rats conferred cardioprotection. Sequencing in human PBMCs revealed abundant ysRNAs of approx. 26–40 nt. Often derived from YRNA-RNY4 and YRNA-RNY1 in PBMCs	[50,51]
snsRNA	Approx. 16–40 nt	SnRNA is as a family of highly conserved ncRNAs located in the nucleus and associated with Sm ribonucleoproteins and other specific proteins, to form small nuclear ribonucleoproteins. The function of fragments is largely unknown.	[39,52]
snosRNA	Often 20–24 nt but also 17–19 nt and 27–33 nt	Some were shown to be similar to miRNA and can use AGO for gene silencing.	[52,53]
vtsRNA	Full-length vRNA is approx. 100 nt and can be processed into approx. 23 nt	RNA components of Vault ribonucleoprotein particles, which are located in the cytoplasm. Control of apoptosis and autophagy, lysosome biogenesis, and function in cancer cells.	[54]

* The list and definition of different sncRNAs will require substantial revisions by the field of experts in the sncRNA field and this table is a very rough overview to put the universe of small (s)-derived non-coding RNA into perspective. With improved sequencing including multiple lengths, the picture may change. Peripheral blood mononuclear cells: PBMC.

4. Endosomal Immune Receptors Recognizing Nucleic Acids

The innate immune system employs an array of pattern recognition receptors (PRRs) to recognize potentially dangerous microbes and particles. Accordingly, these groups of receptors are specific for highly conserved features of foreign invaders termed pathogen-associated molecular patterns (PAMPs) or host-derived molecules called danger-associated molecular patterns (DAMPs). Infection and tissue damage can induce a rapid inflammatory response due to the activation of the innate immune system via PRRs which recognize different PAMPs and DAMPs. PRRs are found in particular locations in the cell corresponding to their ligand specificity; hence, the receptors that detect extracellular PAMPs are expressed on the surface of the cell, while PRRs found in the cytosol, sense microbial infections and include NOD-like receptors, RIG-like receptors (RLRs, (RIG-1, MDA-5, and DExD/H-box helicases)), and cytosolic DNA sensors [55–57]. Other PRRs, including certain TLRs, are expressed in endosomal compartments and can recognize foreign nucleic acids or microbes that have been endocytosed by the cell [56,57]. Recognition and binding to cognate PAMP or DAMP ligands by a specific PRR lead to the activation of a signaling cascade that culminates in a coordinated intracellular innate immune response designed to control infection or heal a stress response; this includes type I and III interferons and pro-inflammatory cytokines and chemokines, as well as factors that modulate the expression of innate genes that promote an antiviral cellular state. These genes encode factors with direct antiviral action or genes that modulate the metabolic or cell cycle state, which leads to restricted viral production. Secreted interferons and cytokines will augment the innate response locally and recruit immune cells to the site, which altogether results in the expression of a plethora of interferon-stimulated genes (ISGs), facilitating additional antiviral activities [57].

TLRs belong to a conserved family of transmembrane glycoprotein receptors, for which humans possess 10 genes. All TLRs are similarly organized with an extracellular leucine-rich-repeat domain, a transmembrane domain, and a cytosolic Toll-IL1R (TIR) domain that extends into the cytosol and mediates downstream signaling after receptor activation [58]. TLRs are highly expressed in many immune cells and endothelial cells as well as epithelial cells, and keratinocytes. However, each cell type expresses a distinct repertoire of TLRs [59]. As shown in more detail in Figure 1, the activation of TLRs triggers a critical immune response for host defense that is specific for the respective TLR and induces signaling that leads to the generation of different effector molecules including interferons and pro-inflammatory cytokines. However, dysregulation of signaling or ligand recognition by TLRs is associated with the pathogenesis of inflammatory and autoimmune diseases [60]. Hence, there is a distinct requirement to downregulate the innate response to avoid chronic inflammatory responses [56,61]. Yet, the response has a crucial role in not only responding to the pathogens per se, but also taking part in the healing response and clearance of DAMPs.

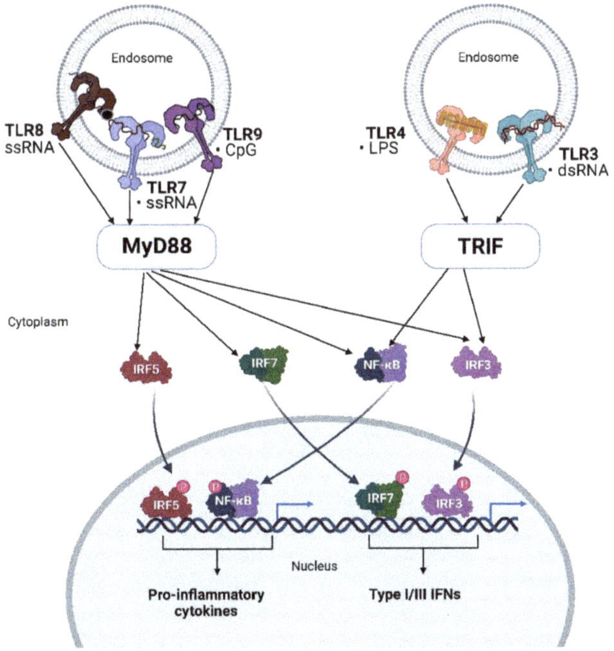

Figure 1. Schematic illustration of TLRs localized to endosomes and the downstream signaling pathways modified from [57]. The expression pattern of endosomal TLRs differs between different cell types and there are also species variations as well as differential up- and downregulation in response to danger signals. TLR4 can localize both to the plasma membrane and the endosomes depending on the cell type in response to ligands such as LPS. TLR3 recognizes dsRNA, TLR7 recognizes ssRNA on one binding site combined with binding to guanosine on the other site, while TLR8 recognizes ssRNA combined with binding to uridine. TLR9 responds to unmethylated CpG motifs that are present at high frequency in bacteria. TLR signaling begins once ligand binding has induced receptor dimerization, which is followed by the engagement of the TIR domain-containing adaptor proteins TRIF (right) or MyD88 (left). The engagement of adaptor proteins will initiate downstream signaling cascades which will lead to phosphorylation and nuclear translocation of transcription factors encoding for proinflammatory cytokines or type I and III interferons. Created with BioRender.com.

Foreign nucleic acids are specifically recognized by RLRs, cytosolic DNA sensors, and a subgroup of TLRs [57]. The subgroup of TLRs that can recognize nucleic acids consists of TLR3, 7, 8, and 9, which all primarily reside in endosomal compartments unlike other nucleic acid sensors, which are found in the cytosol. In general, these TLRs can signal from endosomes by binding to structures that are only accessible once they are taken up and degraded or part of a virus utilizing a certain endocytic pathway to enter the cell. Notably, each endosomal TLR recognizes a specific type of nucleic acid [57].

TLR3 recognizes dsRNA after uptake through clathrin-mediated endocytosis [62]. The sources of dsRNA recognized by TLR3 have been reported to be long dsRNA [63,64] but also ssRNA with stem-loop structures [65]. Although TLR3 seems to have limited sequence specificity, the affinity increases proportionately to dsRNA length [63,64]. The dsRNA has been suggested to bind through electrostatic interactions and hydrogen bonds and to require a minimum length of 40–50 bp for TLR3 activation [63,64,66,67].

TLR7 and TLR8 detect ssRNA but also require binding of RNA degradation products for signaling [68–72]. Hence, TLR7 and TLR8 have two ligand binding sites that have to be occupied for activation and the presence of both products will activate TLR7 or TLR8 signaling. TLR7 has been shown to have a preference for binding to guanosine at the first binding site and recognizes a 3-mer motif embedded within long stretches of polyU-ssRNA at the second site, in which the critical residue is uridine [68,69]. TLR8, on the other hand, has a preference for binding to uridine at the first site and has been suggested to recognize GU-rich ssRNA at the second site [70,72].

TLR9 responds to unmethylated CpG motifs, which consist of a central unmethylated CG dimer flanked by 5′ purines and 3′ pyrimidines, present at high frequency in bacteria [55]. TLR9 is highly expressed in plasmacytoid dendritic cells but can also be detected in B cells and keratinocytes [73,74]. Notably, there are differences between species in TLR9 expression as rodents express TLR9 also in macrophages and myeloid-derived dendritic cells. Therefore, studies in mice may overestimate the activity of TLR9. Four distinct classes of CpG have been identified and used in multiple clinical trials as adjuvants in vaccines or cancer treatment, recently reviewed in [73]. The K-type ODNs (also referred to as B-type) contain 1–5 CpG motifs and have a PS backbone. The D-type ODNs (also referred to as A-type) typically express a single CpG motif flanked by palindromic sequences enabling the formation of a stem-loop structure. The central nucleotides in the D-types are phosphodiester (PO) while the ends are capped with PS (polyG motifs at 5′, 3′or both ends).

The activation of nucleic acid sensing TLRs triggers the induction of numerous molecules involved in the innate immune response. These include pro-inflammatory cytokines such as IL-6 and TNF-α that are highly produced by macrophages, conventional dendritic cells, and B cells, and type I interferons (IFNs), which are secreted in high amounts by plasmacytoid dendritic cells [75]. As shown in more detail in Figure 1, the binding of agonists to all TLRs, except TLR3, results in a signaling cascade depending at least in part on adaptor myeloid differentiation primary response 88 (Myd88) [58,75]. While TLR7 and TLR9 downstream signaling is completely reliant on MyD88, TLR3 solely utilizes the TIR domain-containing adaptor protein-inducing interferon-beta (TRIF). The compartmentalization of nucleic-acid sensing TLRs to endosomal compartments is crucial to their specialization as it allows the innate immune system to detect non-self, internalized nucleic acids, and additionally assures protection from their activation by self-nucleic acids and subsequent induction of autoimmunity [61].

For therapeutic oligonucleotide design, it is preferred to evade immune recognition since unmodified oligonucleotides can activate RIG-I and PKR, which detect dsRNA in the cytoplasm. However, it has been shown that certain chemical modifications such as 2′-OMe-modifications of uridine and guanidine residues in siRNA aid in immune evasion [16]. Nevertheless, there is still a knowledge gap in our understanding of immunogenic properties when designing therapeutic nucleic acids. It can be noted that oligonucleotides with neutral backbones have so far not been implicated in immune activation [16].

TLR3 is expressed in the endosomal compartments of diverse leukocytes such as myeloid dendritic cells, natural killer cells, and macrophages as well as non-immune cells including fibroblasts, endothelial cells, epithelial cells, keratinocytes, and neurons [14,59,74,76–79]. Due to the notion that TLR3 recognizes dsRNA, which can be found in some viral genomes and is produced during viral replication cycles, it constitutes a key component in the recognition and clearance of certain viral infections [80,81]. The activation of TLR3 ensues after suitable ligand binding and leads to the interaction and dimerization of TLR3 extracellular domains [60]. This is followed by signal transduction which leads to the activation of nuclear factor kappa-light-chain-enhancer of activated B cells (NF-κB) and the production of pro-inflammatory cytokines. Moreover, TRIF-mediated signaling can lead to the production of type 1 IFNs through the phosphorylation and activation of interferon regulatory factor 3 (IRF3). [60,82]. The release of pro-inflammatory cytokines and type I IFN by activated or infected cells is crucial in initiating inflammatory and adaptive antiviral responses [60].

TLR3 activation is not limited to agonists of viral origin and can be induced by dsRNA released by damaged tissue or the widely used synthetic dsRNA analog, polyI:C. While TLR3 activation is important for an efficacious immune response towards viral infection, in the course of which dsRNA accumulates throughout viral replication, it has the potential to initiate undesirable effects. In principle, endogenous nucleic acids can trigger TLR3-dependent immune responses, thereby contributing to inflammatory pathologies and autoimmunity [83–85]. Further, excessive TLR3 activation has been shown to increase pathology in some viral infections [86–88] and it has been implicated in undesirable outcomes such as virus-induced asthma [89]. However, damage of the lung does not necessarily have to be due to infection, but can also be enhanced by sterile cell death, e.g., during oxygen treatment of patients with acute respiratory distress syndrome [90], organ transplant complications, cardiovascular disease and diabetes or autoimmune reactions such as arthritis [76,77,91–93]. All nucleic acid sensing endosomal TLRs have been implicated in asthma, wherein TLR3 has been associated with induction, and TLR7, TLR8 and TLR9 are associated in disease exacerbation [94,95]. Human TLR3-mediated immunity, at least in some individuals, is pivotal for protection against HSV-1 infection in the CNS [81] and there is an increasing number of reports showing augmented disease severity in patients with TLR3 deficiency [96–98], otherwise, TLR3 signaling is remarkably redundant.

Altogether, there is extensive support in the literature that over-reactivities involving the recognition of nucleic acids, which are taken up into endosomes and trigger TLRs, have a key role in the pathogenesis of several autoimmune and allergic reactions [14,46,47,80,99].

5. Discovery of SOMIE

We previously showed that a 35 mer CpG ssON (B-type) could inhibit TLR3 signaling in primary human monocyte derived cells (moDC) that express TLR3/4/8, but lack TLR7/9 [14]. This was a serendipity finding discovered when we were combining different TLR agonists and measured moDC activation in order to disclose combinations with additive and/or synergistic effects. As the moDC lacked TLR9 expression, we realized that the inhibitory effect was not dependent on TLR9. Consequently, removal of the CpG motifs and replacement with another nucleotide did not alter the inhibitory effect [14]. We provided evidence that polyI:C–induced DC maturation was inhibited by ssON on all levels investigated: upregulation of maturation markers, production of type I interferons and proinflammatory cytokines, and phosphorylation of vital transcription factors (NFkB, IRF-3). We also showed that polyI:C–induced cytokine production in the airways of cynomolgus macaques was significantly blocked by ssON [14].

As a next step, we further elucidated mechanisms involved in the inhibition of TLR3. It was previously found that ssONs containing TTAGGG motifs to mimic telomeric DNA had a general immunosuppressive effect because of inhibition of STAT signaling [100], while ssONs with polyG motifs inhibited TLR9 activation (and in some cases also TLR7 and TLR8), with the suggested mechanism of competitive antagonism [101–103]. However, the ssONs discovered in [14] did not have such motifs.

We synthesized a panel (varying sequences, modifications and length) of ssON to identify the requirements for the inhibition of dsRNA-mediated activation [15]. We discovered that ssON not only inhibited TLR3 activation, but also inhibited the activation of TLR7 in PBMC making it unlikely that it was a direct TLR3 antagonistic mechanism. Further, we found that ssON modulated TLR4 activation that was dependent on endosomal uptake, while leaving TLR4 signaling from the plasma membrane unaffected in human moDC [15]. Extracellular cargo destined for TLR3/4/7 signaling endosomes is taken up by endocytosis. We therefore investigated whether the inhibition of TLR activation was due to decreased uptake of ligands into endosomes. Indeed, we provided evidence that certain ssON temporarily downregulate clathrin-mediated endocytic activity, thereby revealing a gate keeping mechanism for TLR3/4/7 activation. We termed the ssON-mediated interference of endocytosis SOMIE and showed that both single-stranded RNA and DNA, but not dsDNA, have this capability [15] (Figure 2).

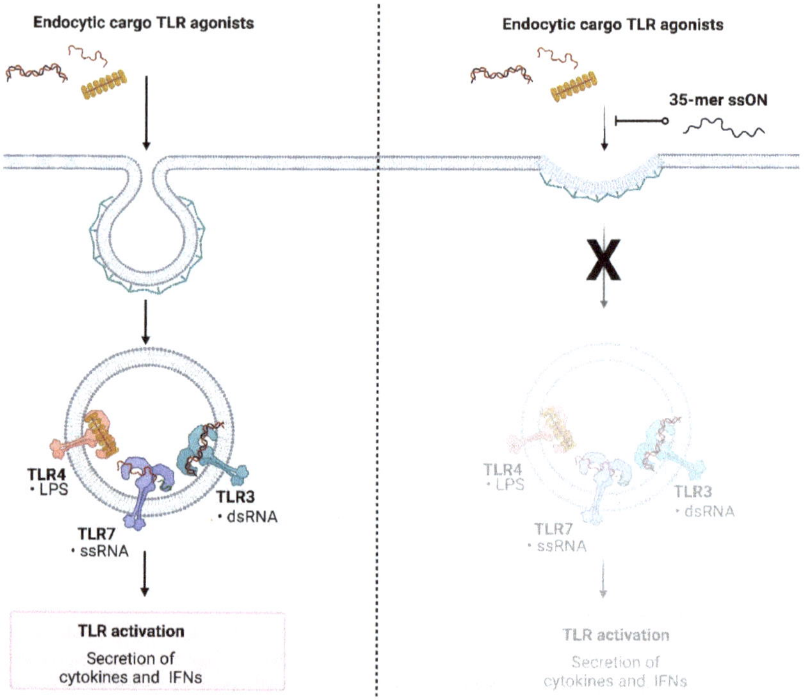

Figure 2. Schematic picture of ssON-mediated interference of endocytosis (SOMIE) modified from [15]. TLR ligands LPS, dsRNA and ssRNA are taken up through clathrin-mediated endocytosis into endosomes expressing TLR4, TLR3, and TLR7 in cells such as human moDC, which will trigger TLR activation and secretion of cytokines and IFNs (**left**). In the presence of 35-mer ssON in the extracellular space, clathrin-mediated endocytosis is temporarily inhibited until ssON is degraded into smaller pieces and consequently endosomal TLR 4/3/7 activation is blocked in moDC (**right**). Created with BioRender.com.

The endocytic inhibition was concentration dependent and not strictly sequence dependent. However, there was a length requirement of at least 25 nt. It is conceivable that not all cell types respond to the SOMIE effect, but we have shown inhibition of TLR3 activation in moDC, keratinocytes, epithelial cells and fibroblasts [14]. Whole cell proteomic and transcriptomic analyses revealed that there were remarkably few cellular changes occurring in moDC after SOMIE [15]. We further provided evidence that ssON modulate

TLR3 activation in vivo in macaques [15]. This opens possible novel therapeutic avenues for autoimmunity involving endocytosis-dependent TLR3/4/7 activation. Our findings that either ssDNA or ssRNA of at least 25 nt have the capacity to temporarily shut down clathrin-mediated endocytosis opens up intriguing questions in host–viral interactions, RNA biology, and autoimmunity. Our data have implications for questions in several fields such as (1) viral endocytosis, (2) cellular uptake of oligonucleotide therapeutics and endogenous sncRNA (13–24 nt), (3) role of stabilizing modifications of endogenous 25–40 nt pool of RNA, (4) immune regulatory function of non-coding RNAs, and (5) development of ssON-based therapeutics acting in the extracellular space.

6. ssONs Acting as Attachment/Entry Inhibitors of Viruses

Clathrin-mediated endocytosis is a key step for cellular entry of many viruses that can lead to the triggering of TLRs located in endosomes [104]. The development of entry inhibitors aligns with a WHO incentive to target virus-associated host factors, which are theoretically less prone to the development of resistance. As these host factors are employed by multiple viruses, there is also the possibility to achieve a broad antiviral coverage. Targeting virus entry is appealing [105] as it blocks the first step in the viral life cycle and shuts off subsequent replication and pathogenic processes. We therefore reasoned that it would be prudent to investigate whether ssONs with the capacity to inhibit endocytosis could function as an antiviral agent. We indeed obtained data showing reduced influenza A (H1N1) infection in vitro in human cells and in a murine in vivo challenge model after ssON treatment [2]. We next investigated the capacity of ssONs to inhibit RSV infection [3,4]. We found that the ability to block RSV infection was dependent on the length of the oligonucleotide, but not strictly sequence dependent as a selection of ssONs between 25–35 nt have the ability to inhibit RSV infection. Furthermore, we discovered that inhibitory ssONs of either DNA or RNA origin effectively inhibited RSV. We found that the effect to inhibit RSV was not dependent on the PS-modification and PO-ssON could also inhibit RSV infection, although not as efficiently as the stabilized version. Notably, we synthesized a selection of six different sncRNAs derived from yRNA, rRNA, or tRNA, which were identified in the bronchoalveolar lavage of healthy individuals. These sequences were randomly selected from a published data set [106], but they were all in the range of 30–40 nt long. We found that these stabilized sncRNAs also possess the capacity to inhibit RSV infection in a similar low nM range as the "parent" 35-mer ssON [4]. These findings further highlighted the importance of oligonucleotide length, but not necessarily an exact sequence or chemical modification, and revealed that this inhibitory effect could be a naturally occurring phenomenon. We also demonstrated that the likely mode-of action governing the inhibitory effect was acting at the viral entry step by preventing the viral binding to nucleolin [3,4].

7. Nucleolin Is a Binding Partner for ssONs

Nucleolin is a nucleic acid-binding protein that exists abundantly in the cell nucleus, but it is also present in the cytoplasm and the cell membrane [107]. Nucleolin is a multifunctional protein and plays key functions in processes such as chromatin remodeling, transcription of rRNA, rRNA maturation, ribosome assembly, and ribosome biogenesis [107]. Nucleolin can bind to either DNA or RNA and was also reported to encompass DNA and RNA helicase activity [107]. It has been indicated as a shuttling protein present in endosomes and to be responsible for transporting proteins between the nucleus, cytoplasm, and the cell surface [108–110]. Notably, a recent study showed that nucleolin located on the surface of murine DC bound directly to A-type CpG ODN, B-type CpG ODN, and polyI:C and promoted their internalization. In human DCs, nucleolin also contributed to the binding and internalization of both type A and B CpG ODNs [111]. Intriguingly, nucleolin has been reported to be involved in viral attachment and/or entry of not only RSV [112,113], but also other viruses such as HIV-1, HSV-2, influenza, parainfluenza type 3, enterovirus 71, Crimean–Congo hemorrhagic fever virus, adeno-associated virus type 2, coxsackie B virus, Seneca Valley virus [107,114–116].

Nucleolin is currently the only candidate that has fulfilled all requirements to be defined as a functional receptor for RSV. It was demonstrated that pre-treating cells with a nucleolin antibody reduced the infection and pre-treating the virus with soluble nucleolin before infection similarly reduced infection. Furthermore, silencing nucleolin expression by using siRNA, significantly reduced the RSV infection. Additional support for nucleolin as a functional RSV receptor was obtained after the induction of nucleolin expression in normally non-permissive cells, which enabled RSV infection [112,113]. Further studies have shown that by utilizing the DNA aptamer AS1411, which binds nucleolin located on the cell surface, RSV infection can be inhibited in epithelial cell lines and in vivo in mice and cotton rats [117]. The RSV F protein has been shown to be the interacting partner of nucleolin [112] and a recent study showed that the F protein binds specifically to the RBD1,2 (RNA binding domain 1,2) binding site of nucleolin [113].

Several studies indicate that nucleolin is part of a multicomplex consisting of several proteins, which participate in the binding and entry of RSV [113,118]. Furthermore, studies have suggested that RSV facilitates its own uptake by triggering nucleolin to translocate to the surface of cells upon viral binding [119]. A recent report showed that the interaction of the RSV F protein with IGF1R triggered the activation of protein kinase C zeta, which resulted in the recruitment of nucleolin to the cell surface, whereby nucleolin aided in viral entry [120] (Figure 3, left). There is still a lack in our knowledge on which viral-cell binding partners are involved in triggering translocation of nucleolin to the cell surface upon contact with other viruses [116]. It also remains to be further established as to what extent nucleolin is required for the viral entry of other viruses. It is conceivable that there might be differential susceptibility to entry inhibitors targeting nucleolin depending on the virus. Our finding that ssONs with the length of 30–40 nt have the capacity to interfere with nucleolin is intriguing and opens up for additional testing of other viral families reliant on nucleolin during the attachment and/or entry step (Figure 3, right). Notably, we were able to reduce RSV infection in vivo in a murine model [3]. We also studied immune responses occurring locally in the lungs in RSV-challenged mice with or without ssON treatment and found that the ssON treated mice displayed a more profound upregulation of ISGs using the Nanostring technology.

We used qPCR to validate that ssON treatment indeed upregulated expression of ISGs such as *Stat 1*, *Stat 2*, *Ccl2*, and *Cxcl10* [3]. Hence, it is conceivable that blocking one viral entry pathway may force viruses to enter via another uptake route, which hypothetically can trigger cytoplasmatic nucleic acid sensors, which are effective inducers of ISGs. Future studies are required to investigate the cellular and molecular responses occurring locally in the lungs upon ssON treatment, especially as it possesses both antiviral and immunomodulatory properties. This is especially pertinent in an in vivo context as several cell types are involved, both resident and recently infiltrating cells, in a finely tuned and orchestrated immune response.

The finding that we may naturally have sncRNAs in, for example, bronchoalveolar lavage with antiviral potential [4] hints towards a naturally occurring mechanism, although it is conceivable that the PS-modifications may strengthen and prolong the effects.

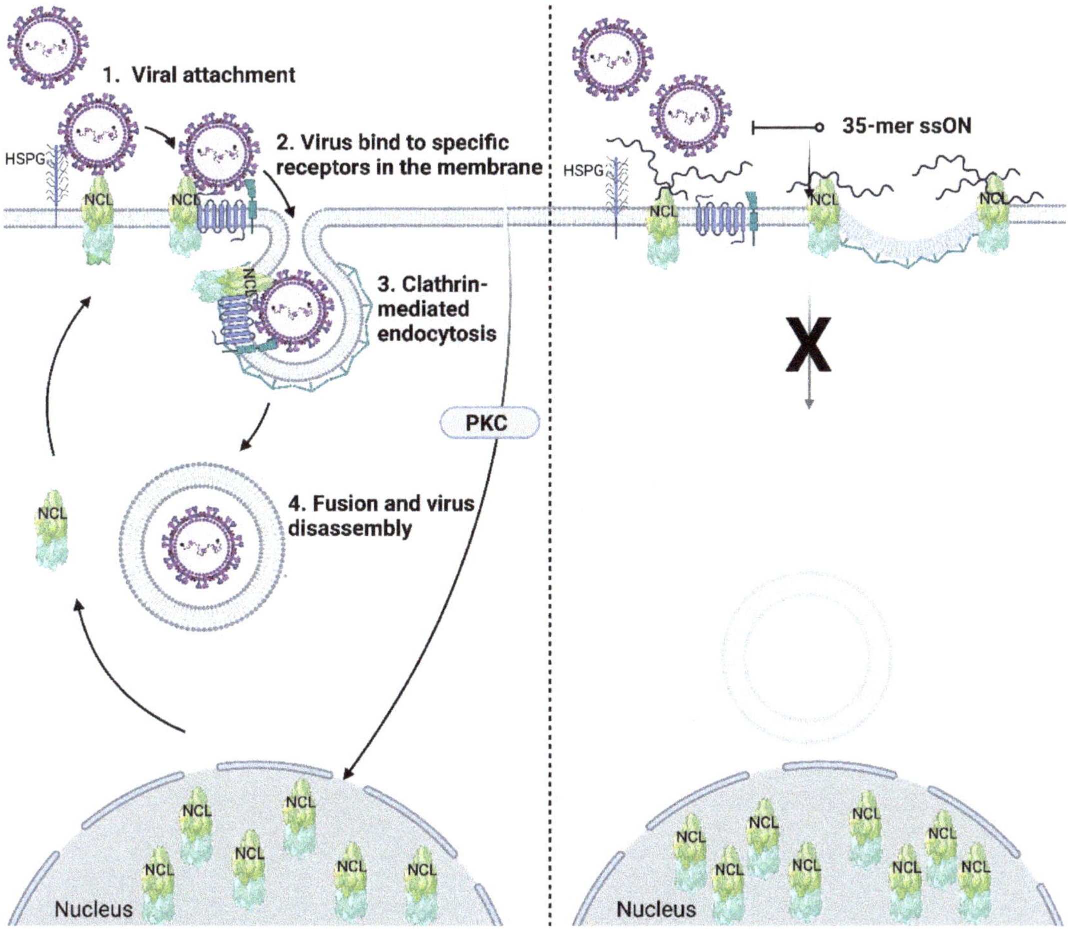

Figure 3. Schematic illustration of nucleolin (NCL) as an attachment receptor for virus and participation in clathrin-mediated endocytosis (**left**), which can be inhibited by 35-mer ssON (**right**). Initially, certain viruses attach to the cellular membrane using heparan sulfate proteoglycan (HSPG) and/or NCL. Next, viruses bind to specific receptors involved in viral entry, and may also trigger intracellular signaling. Hence, viral glycoproteins, such as the RSV F-protein, binds to cellular receptors (IGF1R for RSV) which initiates signaling (PKCζ for IGF1R) [120]. The signaling leads to translocation of NCL from the nucleus to the cell surface. Cell surface NCL also interacts with the C-terminal residues of clathrin light chain A, which is a key component in clathrin-dependent endocytosis. Upon specific receptor binding, the viral entry process is initiated and often occurs through clathrin-mediated endocytosis. Subsequently, viral fusion and virus disassembly may occur. Nucleolin regulates viral attachment and binds nucleic acids as well as aptamer AS1411 through binding of the C-terminal glycine/arginine-rich domain (yellow-green). However, viral proteins (such as RSV F-protein) may also interact with NCL RBD for internalization [113]. The 35-mer ssON but not 15-mer ssON can inhibit viral attachment by shielding NCL [3,4]. We hypothesize that the 35-mer ssONs confer steric hindrance for the endocytosis to occur, while the shorter ssONs are unable to block the molecular movements and are instead readily taken up via clathrin-mediated endocytosis. Created with BioRender.com.

8. Therapeutic Approaches of ssONs Acting in the Extracellular Space

TLR3 is a key receptor for the recognition of dsRNA and the initiation of immune responses against viral infections. However, hyperactive responses can have adverse effects, such as virus-induced asthma but also in other diseases. An over-reactive TLR3 signaling driven by viral, or endogenous dsRNA from dying cells, has been implicated as a common driver in several diseases including viral infections [86–88], acute respiratory distress syndrome [90], asthma, organ transplant complications, cardiovascular disease, and diabetes or autoimmune reactions such as arthritis [76,77,89,91–93]. Experimental evidence supporting TLR3 involvement was often demonstrated by a diminished response in TLR3 knock-out mice and/or exacerbation of the inflammatory condition by polyI:C. As TLR3 is primarily located in the endosomes and is activated (in an acidic environment) upon the binding of dsRNA taken up from dying cells or virus infected cells, it is not an easy target for small molecule inhibitors or antibody-based therapies. New strategies for drug development may therefore be required to prevent TLR3-mediated pathology.

Our results demonstrate that TLR3-triggered immune activation can be modulated by the 35-mer ssON and provide evidence of dampening proinflammatory cytokine release in the airways of cynomolgus macaques and locally in the skin. Injection in the skin led to a reduction in IL-6 (pro-inflammatory) production and induction of IL-10 ("anti-inflammatory") [15]. In addition, the 35-mer ssON ameliorates certain itch in vivo in mice and reduces mast cell degranulation [121]. These findings may open novel perspectives for clinical strategies to prevent or treat inflammatory conditions exacerbated by TLR3 signaling.

We used well-known PS-modifications of the ssONs in the antiviral and immunoregulatory experiments, which enhance the stability of nucleic acid drugs by nucleus-mediated degradation (recent review on different modifications [122]). Additionally, the PS backbone confers protein-binding properties enhancing binding to plasma proteins and cell-surface proteins involved in cellular attachment and/or facilitating uptake into cells [123]. Numerous clinical trials have been conducted with PS-stabilized oligonucleotides (PS-ONs) over the past two decades from which conserved class behaviors have been established. After administration by intravenous infusion or subcutaneous injection, PS-ONs are rapidly cleared from the blood (half-life <1 h), concomitant with accumulation in peripheral organs, mostly liver and kidney. They are relatively resistant (compared with PO-ONs) to nuclease-mediated degradation, but degrade slowly over time with the primary route of elimination via the kidney [124]. Thus, further drug development of PS-stabilized ssON as a broad-spectrum antiviral agent active against RSV [3,4], influenza A [2], and HIV-1 [5], and hypothetically blocking nucleolin-dependent entry driven by various glycoproteins of other pathogenic viruses for which there is still a significant medical need, is highly warranted. It is conceivable that the antiviral mechanism acting in vivo is more complex than inhibition of viral binding to cellular surfaces as differential immune responses may also be evoked that may take part in the defense [3]. Hence, careful pharmacokinetic and formulation studies are required to design future animal studies in combination with mechanistic studies to reveal how antiviral 35-mer ssON´s act in vivo.

9. Conclusions and Future Prospects

Based on recent advances and progress in the field, there is little doubt that additional nucleic acid-based therapeutics will be approved within the coming years. Currently, the majority of oligonucleotide-based therapeutics are centered on base pairing to target a specific sequence to achieve, for example, gene silencing (Table 1). Nevertheless, the expanding field of non-coding RNA biology also teaches us about the plethora of different RNA fragments of which some may exert functions that are non-sequence specific. It is conceivable that they may have a secondary structure required for RNA–protein binding and modifications to increase stability and/or increase affinity. There is a need to develop techniques that can include sequencing of RNA modifications, and preferentially high-resolution spatial techniques, which can reveal the subcellular and/or extracellular location of the non-coding fragments.

There is currently a lack in our understanding about the functions of the pool of non-coding RNA with the approximate length of 30–40 nt. Recent years have explored the functions of shorter fragments such as miRNA, siRNA, and piRNA but there is a still much to reveal concerning the abundant pool of 30–40 nt that can be found to be derived from different RNA such as tRNA, Y RNA, rRNA, mRNA, and lncRNA. One hypothetical view of the 30–40 nt pool is that they confer a "buffering" system to regulate the uptake of endocytic cargo into cells lining (keratinocytes, epithelial, and endocytic cells) and patrolling (classical DC) the body for the detection of pathogenic intruders.

We have provided evidence that some ssONs have the capacity to temporarily inhibit certain endocytic pathways [15] and have provided evidence of their immunoregulatory functions [121] in experimental conditions based on endocytic uptake and triggering of endosomal TLR3 [14,15]. In addition, this pool of 30–40 nt RNA may also have a role in preventing entry of certain viruses [2–4]. There is still a gap in our understanding to what extent they exert broad-spectrum antiviral capacities and if additional immunological properties are associated with their antiviral effects. More knowledge on how the immune system is dealing with extracellular oligonucleotides of different lengths will aid in the design of ASO therapeutics and may also open up the field for a new type of immunoregulatory and/or antiviral therapies.

Author Contributions: Conceptualization, A.-L.S.; writing—original draft preparation, A.-L.S.; writing—review and editing, A.D. and S.A.P. All authors have read and agreed to the published version of the manuscript.

Funding: This project has received funding through the Swedish Research Council (2020-01730) and the European Union's Horizon 2020 research and innovation program under grant agreement No 101003555 (Fight-nCoV) to A.-L.S.

Institutional Review Board Statement: Not applicable.

Informed Consent Statement: Not applicable.

Data Availability Statement: Not applicable.

Conflicts of Interest: A.-L.S. and A.D. declare ownership in TIRmed Pharma having IPR related to ssON. S.A.P. declares to have no competing interest.

References

1. Vaillant, A. Oligonucleotide-Based Therapies for Chronic HBV Infection: A Primer on Biochemistry, Mechanisms and Antiviral Effects. *Viruses* **2022**, *14*, 2052. [CrossRef]
2. Poux, C.; Dondalska, A.; Bergenstrahle, J.; Palsson, S.; Contreras, V.; Arasa, C.; Jarver, P.; Albert, J.; Busse, D.C.; LeGrand, R.; et al. A Single-Stranded Oligonucleotide Inhibits Toll-Like Receptor 3 Activation and Reduces Influenza A (H1N1) Infection. *Front. Immunol.* **2019**, *10*, 2161. [CrossRef] [PubMed]
3. Palsson, S.A.; Dondalska, A.; Bergenstrahle, J.; Rolfes, C.; Bjork, A.; Sedano, L.; Power, U.F.; Rameix-Welti, M.A.; Lundeberg, J.; Wahren-Herlenius, M.; et al. Single-Stranded Oligonucleotide-Mediated Inhibition of Respiratory Syncytial Virus Infection. *Front. Immunol.* **2020**, *11*, 580547. [CrossRef] [PubMed]
4. Palsson, S.A.; Sekar, V.; Kutter, C.; Friedlander, M.R.; Spetz, A.L. Inhibition of Respiratory Syncytial Virus Infection by Small Non-Coding RNA Fragments. *Int. J. Mol. Sci.* **2022**, *23*, 5990. [CrossRef] [PubMed]
5. Cena-Diez, R.; Singh, K.; Spetz, A.L.; Sonnerborg, A. Novel Naturally Occurring Dipeptides and Single-Stranded Oligonucleotide Act as Entry Inhibitors and Exhibit a Strong Synergistic Anti-HIV-1 Profile. *Infect. Dis. Ther.* **2022**, *11*, 1103–1116. [CrossRef] [PubMed]
6. Vaillant, A.; Juteau, J.M.; Lu, H.; Liu, S.; Lackman-Smith, C.; Ptak, R.; Jiang, S. Phosphorothioate oligonucleotides inhibit human immunodeficiency virus type 1 fusion by blocking gp41 core formation. *Antimicrob. Agents Chemother.* **2006**, *50*, 1393–1401. [CrossRef] [PubMed]
7. Lee, A.M.; Rojek, J.M.; Gundersen, A.; Stroher, U.; Juteau, J.M.; Vaillant, A.; Kunz, S. Inhibition of cellular entry of lymphocytic choriomeningitis virus by amphipathic DNA polymers. *Virology* **2008**, *372*, 107–117. [CrossRef] [PubMed]

8. Kocisko, D.A.; Vaillant, A.; Lee, K.S.; Arnold, K.M.; Bertholet, N.; Race, R.E.; Olsen, E.A.; Juteau, J.M.; Caughey, B. Potent antiscrapie activities of degenerate phosphorothioate oligonucleotides. *Antimicrob. Agents Chemother.* **2006**, *50*, 1034–1044. [CrossRef] [PubMed]
9. Beilstein, F.; Blanchet, M.; Vaillant, A.; Sureau, C. Nucleic Acid Polymers Are Active against Hepatitis Delta Virus Infection In Vitro. *J. Virol.* **2018**, *92*, e01416–e01417. [CrossRef] [PubMed]
10. Vaillant, A. REP 2139: Antiviral Mechanisms and Applications in Achieving Functional Control of HBV and HDV Infection. *ACS Infect. Dis.* **2019**, *5*, 675–687. [CrossRef]
11. Boulon, R.; Blanchet, M.; Lemasson, M.; Vaillant, A.; Labonte, P. Characterization of the antiviral effects of REP 2139 on the HBV lifecycle in vitro. *Antiviral. Res.* **2020**, *183*, 104853. [CrossRef] [PubMed]
12. Gait, M.J.; Agrawal, S. Introduction and History of the Chemistry of Nucleic Acids Therapeutics. *Methods Mol. Biol.* **2022**, *2434*, 3–31. [CrossRef] [PubMed]
13. Ranjith-Kumar, C.T.; Duffy, K.E.; Jordan, J.L.; Eaton-Bassiri, A.; Vaughan, R.; Hoose, S.A.; Lamb, R.J.; Sarisky, R.T.; Kao, C.C. Single-stranded oligonucleotides can inhibit cytokine production induced by human toll-like receptor 3. *Mol. Cell. Biol.* **2008**, *28*, 4507–4519. [CrossRef] [PubMed]
14. Skold, A.E.; Hasan, M.; Vargas, L.; Saidi, H.; Bosquet, N.; Le Grand, R.; Smith, C.I.; Spetz, A.L. Single-stranded DNA oligonucleotides inhibit TLR3-mediated responses in human monocyte-derived dendritic cells and in vivo in cynomolgus macaques. *Blood* **2012**, *120*, 768–777. [CrossRef] [PubMed]
15. Jarver, P.; Dondalska, A.; Poux, C.; Sandberg, A.; Bergenstrahle, J.; Skold, A.E.; Dereuddre-Bosquet, N.; Martinon, F.; Palsson, S.; Zaghloul, E.; et al. Single-Stranded Nucleic Acids Regulate TLR3/4/7 Activation through Interference with Clathrin-Mediated Endocytosis. *Sci. Rep.* **2018**, *8*, 15841. [CrossRef] [PubMed]
16. Bost, J.P.; Barriga, H.; Holme, M.N.; Gallud, A.; Maugeri, M.; Gupta, D.; Lehto, T.; Valadi, H.; Esbjorner, E.K.; Stevens, M.M.; et al. Delivery of Oligonucleotide Therapeutics: Chemical Modifications, Lipid Nanoparticles, and Extracellular Vesicles. *ACS Nano* **2021**, *15*, 13993–14021. [CrossRef]
17. Roberts, T.C.; Langer, R.; Wood, M.J.A. Advances in oligonucleotide drug delivery. *Nat. Rev. Drug Discov.* **2020**, *19*, 673–694. [CrossRef]
18. Roehr, B. Fomivirsen approved for CMV retinitis. *J. Int. Assoc. Physicians AIDS Care* **1998**, *4*, 14–16.
19. Ruckman, J.; Green, L.S.; Beeson, J.; Waugh, S.; Gillette, W.L.; Henninger, D.D.; Claesson-Welsh, L.; Janjic, N. 2′-Fluoropyrimidine RNA-based aptamers to the 165-amino acid form of vascular endothelial growth factor (VEGF165). Inhibition of receptor binding and VEGF-induced vascular permeability through interactions requiring the exon 7-encoded domain. *J. Biol. Chem.* **1998**, *273*, 20556–20567. [CrossRef] [PubMed]
20. Gragoudas, E.S.; Adamis, A.P.; Cunningham, E.T., Jr.; Feinsod, M.; Guyer, D.R.; Group, V.I.S.i.O.N.C.T. Pegaptanib for neovascular age-related macular degeneration. *N. Engl. J. Med.* **2004**, *351*, 2805–2816. [CrossRef]
21. Crooke, R.M.; Graham, M.J.; Lemonidis, K.M.; Whipple, C.P.; Koo, S.; Perera, R.J. An apolipoprotein B antisense oligonucleotide lowers LDL cholesterol in hyperlipidemic mice without causing hepatic steatosis. *J. Lipid. Res.* **2005**, *46*, 872–884. [CrossRef] [PubMed]
22. Richardson, P.; Aggarwal, S.; Topaloglu, O.; Villa, K.F.; Corbacioglu, S. Systematic review of defibrotide studies in the treatment of veno-occlusive disease/sinusoidal obstruction syndrome (VOD/SOS). *Bone Marrow Transpl.* **2019**, *54*, 1951–1962. [CrossRef]
23. Richardson, P.G.; Riches, M.L.; Kernan, N.A.; Brochstein, J.A.; Mineishi, S.; Termuhlen, A.M.; Arai, S.; Grupp, S.A.; Guinan, E.C.; Martin, P.L.; et al. Phase 3 trial of defibrotide for the treatment of severe veno-occlusive disease and multi-organ failure. *Blood* **2016**, *127*, 1656–1665. [CrossRef] [PubMed]
24. Lim, K.R.; Maruyama, R.; Yokota, T. Eteplirsen in the treatment of Duchenne muscular dystrophy. *Drug Des. Dev. Ther.* **2017**, *11*, 533–545. [CrossRef] [PubMed]
25. Hua, Y.; Sahashi, K.; Hung, G.; Rigo, F.; Passini, M.A.; Bennett, C.F.; Krainer, A.R. Antisense correction of SMN2 splicing in the CNS rescues necrosis in a type III SMA mouse model. *Genes Dev.* **2010**, *24*, 1634–1644. [CrossRef] [PubMed]
26. Adams, D.; Gonzalez-Duarte, A.; O'Riordan, W.D.; Yang, C.C.; Ueda, M.; Kristen, A.V.; Tournev, I.; Schmidt, H.H.; Coelho, T.; Berk, J.L.; et al. Patisiran, an RNAi Therapeutic, for Hereditary Transthyretin Amyloidosis. *N. Engl. J. Med.* **2018**, *379*, 11–21. [CrossRef] [PubMed]
27. Ackermann, E.J.; Guo, S.; Benson, M.D.; Booten, S.; Freier, S.; Hughes, S.G.; Kim, T.W.; Jesse Kwoh, T.; Matson, J.; Norris, D.; et al. Suppressing transthyretin production in mice, monkeys and humans using 2nd-Generation antisense oligonucleotides. *Amyloid* **2016**, *23*, 148–157. [CrossRef]
28. Benson, M.D.; Waddington-Cruz, M.; Berk, J.L.; Polydefkis, M.; Dyck, P.J.; Wang, A.K.; Plante-Bordeneuve, V.; Barroso, F.A.; Merlini, G.; Obici, L.; et al. Inotersen Treatment for Patients with Hereditary Transthyretin Amyloidosis. *N. Engl. J. Med.* **2018**, *379*, 22–31. [CrossRef]
29. Witztum, J.L.; Gaudet, D.; Freedman, S.D.; Alexander, V.J.; Digenio, A.; Williams, K.R.; Yang, Q.; Hughes, S.G.; Geary, R.S.; Arca, M.; et al. Volanesorsen and Triglyceride Levels in Familial Chylomicronemia Syndrome. *N. Engl. J. Med.* **2019**, *381*, 531–542. [CrossRef]
30. Balwani, M.; Sardh, E.; Ventura, P.; Peiro, P.A.; Rees, D.C.; Stolzel, U.; Bissell, D.M.; Bonkovsky, H.L.; Windyga, J.; Anderson, K.E.; et al. Phase 3 Trial of RNAi Therapeutic Givosiran for Acute Intermittent Porphyria. *N. Engl. J. Med.* **2020**, *382*, 2289–2301. [CrossRef]

31. Servais, L.; Mercuri, E.; Straub, V.; Guglieri, M.; Seferian, A.M.; Scoto, M.; Leone, D.; Koenig, E.; Khan, N.; Dugar, A.; et al. Long-Term Safety and Efficacy Data of Golodirsen in Ambulatory Patients with Duchenne Muscular Dystrophy Amenable to Exon 53 Skipping: A First-in-human, Multicenter, Two-Part, Open-Label, Phase 1/2 Trial. *Nucleic Acid Ther.* **2022**, *32*, 29–39. [CrossRef] [PubMed]
32. Komaki, H.; Nagata, T.; Saito, T.; Masuda, S.; Takeshita, E.; Sasaki, M.; Tachimori, H.; Nakamura, H.; Aoki, Y.; Takeda, S. Systemic administration of the antisense oligonucleotide NS-065/NCNP-01 for skipping of exon 53 in patients with Duchenne muscular dystrophy. *Sci. Transl. Med.* **2018**, *10*, eaan0713. [CrossRef] [PubMed]
33. Al Musaimi, O.; Al Shaer, D.; Albericio, F.; de la Torre, B.G. 2020 FDA TIDES (Peptides and Oligonucleotides) Harvest. *Pharmaceuticals* **2021**, *14*, 145. [CrossRef] [PubMed]
34. Wagner, K.R.; Kuntz, N.L.; Koenig, E.; East, L.; Upadhyay, S.; Han, B.; Shieh, P.B. Safety, tolerability, and pharmacokinetics of casimersen in patients with Duchenne muscular dystrophy amenable to exon 45 skipping: A randomized, double-blind, placebo-controlled, dose-titration trial. *Muscle Nerve* **2021**, *64*, 285–292. [CrossRef] [PubMed]
35. Adams, D.; Tournev, I.L.; Taylor, M.S.; Coelho, T.; Plante-Bordeneuve, V.; Berk, J.L.; Gonzalez-Duarte, A.; Gillmore, J.D.; Low, S.C.; Sekijima, Y.; et al. Efficacy and safety of vutrisiran for patients with hereditary transthyretin-mediated amyloidosis with polyneuropathy: A randomized clinical trial. *Amyloid* **2022**, 1–9. [CrossRef]
36. Cech, T.R.; Steitz, J.A. The noncoding RNA revolution-trashing old rules to forge new ones. *Cell* **2014**, *157*, 77–94. [CrossRef] [PubMed]
37. Rosace, D.; Lopez, J.; Blanco, S. Emerging roles of novel small non-coding regulatory RNAs in immunity and cancer. *RNA Biol.* **2020**, *17*, 1196–1213. [CrossRef]
38. Shi, J.; Zhou, T.; Chen, Q. Exploring the expanding universe of small RNAs. *Nat. Cell Biol.* **2022**, *24*, 415–423. [CrossRef]
39. Wang, H.; Huang, R.; Li, L.; Zhu, J.; Li, Z.; Peng, C.; Zhuang, X.; Lin, H.; Shi, S.; Huang, P. CPA-seq reveals small ncRNAs with methylated nucleosides and diverse termini. *Cell Discov.* **2021**, *7*, 25. [CrossRef]
40. Alfonzo, J.D.; Brown, J.A.; Byers, P.H.; Cheung, V.G.; Maraia, R.J.; Ross, R.L. A call for direct sequencing of full-length RNAs to identify all modifications. *Nat. Genet.* **2021**, *53*, 1113–1116. [CrossRef]
41. Zhuang, X. Spatially resolved single-cell genomics and transcriptomics by imaging. *Nat. Methods* **2021**, *18*, 18–22. [CrossRef] [PubMed]
42. Larsson, L.; Frisen, J.; Lundeberg, J. Spatially resolved transcriptomics adds a new dimension to genomics. *Nat. Methods* **2021**, *18*, 15–18. [CrossRef] [PubMed]
43. Deng, J.; Ptashkin, R.N.; Chen, Y.; Cheng, Z.; Liu, G.; Phan, T.; Deng, X.; Zhou, J.; Lee, I.; Lee, Y.S.; et al. Respiratory Syncytial Virus Utilizes a tRNA Fragment to Suppress Antiviral Responses Through a Novel Targeting Mechanism. *Mol. Ther.* **2015**, *23*, 1622–1629. [CrossRef] [PubMed]
44. Hancock, M.H.; Skalsky, R.L. Roles of Non-coding RNAs During Herpesvirus Infection. *Curr. Top. Microbiol. Immunol.* **2018**, *419*, 243–280. [CrossRef] [PubMed]
45. Henzinger, H.; Barth, D.A.; Klec, C.; Pichler, M. Non-Coding RNAs and SARS-Related Coronaviruses. *Viruses* **2020**, *12*, 1374. [CrossRef] [PubMed]
46. Pelka, K.; Shibata, T.; Miyake, K.; Latz, E. Nucleic acid-sensing TLRs and autoimmunity: Novel insights from structural and cell biology. *Immunol. Rev.* **2016**, *269*, 60–75. [CrossRef] [PubMed]
47. Marshak-Rothstein, A. Toll-like receptors in systemic autoimmune disease. *Nat. Rev. Immunol.* **2006**, *6*, 823–835. [CrossRef]
48. Carthew, R.W.; Sontheimer, E.J. Origins and Mechanisms of miRNAs and siRNAs. *Cell* **2009**, *136*, 642–655. [CrossRef]
49. Ozata, D.M.; Gainetdinov, I.; Zoch, A.; O'Carroll, D.; Zamore, P.D. PIWI-interacting RNAs: Small RNAs with big functions. *Nat. Rev. Genet.* **2019**, *20*, 89–108. [CrossRef]
50. Gu, W.; Shi, J.; Liu, H.; Zhang, X.; Zhou, J.J.; Li, M.; Zhou, D.; Li, R.; Lv, J.; Wen, G.; et al. Peripheral blood non-canonical small non-coding RNAs as novel biomarkers in lung cancer. *Mol. Cancer* **2020**, *19*, 159. [CrossRef]
51. Cambier, L.; de Couto, G.; Ibrahim, A.; Echavez, A.K.; Valle, J.; Liu, W.; Kreke, M.; Smith, R.R.; Marban, L.; Marban, E. Y RNA fragment in extracellular vesicles confers cardioprotection via modulation of IL-10 expression and secretion. *EMBO Mol. Med.* **2017**, *9*, 337–352. [CrossRef] [PubMed]
52. Chen, C.J.; Heard, E. Small RNAs derived from structural non-coding RNAs. *Methods* **2013**, *63*, 76–84. [CrossRef] [PubMed]
53. Taft, R.J.; Glazov, E.A.; Lassmann, T.; Hayashizaki, Y.; Carninci, P.; Mattick, J.S. Small RNAs derived from snoRNAs. *RNA* **2009**, *15*, 1233–1240. [CrossRef]
54. Gallo, S.; Kong, E.; Ferro, I.; Polacek, N. Small but Powerful: The Human Vault RNAs as Multifaceted Modulators of Pro-Survival Characteristics and Tumorigenesis. *Cancers* **2022**, *14*, 2787. [CrossRef] [PubMed]
55. Paludan, S.R.; Bowie, A.G. Immune sensing of DNA. *Immunity* **2013**, *38*, 870–880. [CrossRef] [PubMed]
56. Roers, A.; Hiller, B.; Hornung, V. Recognition of Endogenous Nucleic Acids by the Innate Immune System. *Immunity* **2016**, *44*, 739–754. [CrossRef]
57. Chow, K.T.; Gale, M., Jr.; Loo, Y.M. RIG-I and Other RNA Sensors in Antiviral Immunity. *Annu. Rev. Immunol* **2018**, *36*, 667–694. [CrossRef]
58. Akira, S.; Takeda, K. Toll-like receptor signalling. *Nat. Rev. Immunol.* **2004**, *4*, 499–511. [CrossRef]
59. Takeda, K.; Kaisho, T.; Akira, S. Toll-like receptors. *Annu. Rev. Immunol.* **2003**, *21*, 335–376. [CrossRef]

60. Kawai, T.; Akira, S. The role of pattern-recognition receptors in innate immunity: Update on Toll-like receptors. *Nat. Immunol.* **2010**, *11*, 373–384. [CrossRef]
61. Lind, N.A.; Rael, V.E.; Pestal, K.; Liu, B.; Barton, G.M. Regulation of the nucleic acid-sensing Toll-like receptors. *Nat. Rev. Immunol.* **2022**, *22*, 224–235. [CrossRef] [PubMed]
62. Watanabe, A.; Funami, K.; Seya, T.; Matsumoto, M. The clathrin-mediated endocytic pathway participates in dsRNA-induced IFN-beta production. *J. Immunol.* **2008**, *181*, 5522–5529. [CrossRef]
63. Leonard, J.N.; Ghirlando, R.; Askins, J.; Bell, J.K.; Margulies, D.H.; Davies, D.R.; Segal, D.M. The TLR3 signaling complex forms by cooperative receptor dimerization. *Proc. Natl. Acad. Sci. USA* **2008**, *105*, 258–263. [CrossRef] [PubMed]
64. Fukuda, K.; Watanabe, T.; Tokisue, T.; Tsujita, T.; Nishikawa, S.; Hasegawa, T.; Seya, T.; Matsumoto, M. Modulation of double-stranded RNA recognition by the N-terminal histidine-rich region of the human toll-like receptor 3. *J. Biol. Chem.* **2008**, *283*, 22787–22794. [CrossRef] [PubMed]
65. Tatematsu, M.; Nishikawa, F.; Seya, T.; Matsumoto, M. Toll-like receptor 3 recognizes incomplete stem structures in single-stranded viral RNA. *Nat. Commun.* **2013**, *4*, 1833. [CrossRef]
66. Bell, J.K.; Botos, I.; Hall, P.R.; Askins, J.; Shiloach, J.; Segal, D.M.; Davies, D.R. The molecular structure of the Toll-like receptor 3 ligand-binding domain. *Proc. Natl. Acad. Sci. USA* **2005**, *102*, 10976–10980. [CrossRef]
67. Liu, L.; Botos, I.; Wang, Y.; Leonard, J.N.; Shiloach, J.; Segal, D.M.; Davies, D.R. Structural basis of toll-like receptor 3 signaling with double-stranded RNA. *Science* **2008**, *320*, 379–381. [CrossRef]
68. Lund, J.M.; Alexopoulou, L.; Sato, A.; Karow, M.; Adams, N.C.; Gale, N.W.; Iwasaki, A.; Flavell, R.A. Recognition of single-stranded RNA viruses by Toll-like receptor 7. *Proc. Natl. Acad. Sci. USA* **2004**, *101*, 5598–5603. [CrossRef]
69. Diebold, S.S.; Kaisho, T.; Hemmi, H.; Akira, S.; Reis e Sousa, C. Innate antiviral responses by means of TLR7-mediated recognition of single-stranded RNA. *Science* **2004**, *303*, 1529–1531. [CrossRef]
70. Heil, F.; Hemmi, H.; Hochrein, H.; Ampenberger, F.; Kirschning, C.; Akira, S.; Lipford, G.; Wagner, H.; Bauer, S. Species-specific recognition of single-stranded RNA via toll-like receptor 7 and 8. *Science* **2004**, *303*, 1526–1529. [CrossRef]
71. Lehmann, S.M.; Kruger, C.; Park, B.; Derkow, K.; Rosenberger, K.; Baumgart, J.; Trimbuch, T.; Eom, G.; Hinz, M.; Kaul, D.; et al. An unconventional role for miRNA: Let-7 activates Toll-like receptor 7 and causes neurodegeneration. *Nat. Neurosci.* **2012**, *15*, 827–835. [CrossRef] [PubMed]
72. Tanji, H.; Ohto, U.; Shibata, T.; Taoka, M.; Yamauchi, Y.; Isobe, T.; Miyake, K.; Shimizu, T. Toll-like receptor 8 senses degradation products of single-stranded RNA. *Nat. Struct. Mol. Biol.* **2015**, *22*, 109–115. [CrossRef] [PubMed]
73. Kayraklioglu, N.; Horuluoglu, B.; Klinman, D.M. CpG Oligonucleotides as Vaccine Adjuvants. *Methods Mol. Biol.* **2021**, *2197*, 51–85. [CrossRef] [PubMed]
74. Lebre, M.C.; van der Aar, A.M.; van Baarsen, L.; van Capel, T.M.; Schuitemaker, J.H.; Kapsenberg, M.L.; de Jong, E.C. Human keratinocytes express functional Toll-like receptor 3, 4, 5, and 9. *J. Investig. Dermatol.* **2007**, *127*, 331–341. [CrossRef]
75. Blasius, A.L.; Beutler, B. Intracellular toll-like receptors. *Immunity* **2010**, *32*, 305–315. [CrossRef]
76. Zhao, J.; Huang, X.; McLeod, P.; Jiang, J.; Liu, W.; Haig, A.; Jevnikar, A.M.; Jiang, Z.; Zhang, Z.X. Toll-like receptor 3 is an endogenous sensor of cell death and a potential target for induction of long-term cardiac transplant survival. *Am. J. Transplant.* **2021**, *21*, 3268–3279. [CrossRef]
77. Zhuang, C.; Chen, R.; Zheng, Z.; Lu, J.; Hong, C. Toll-Like Receptor 3 in Cardiovascular Diseases. *Heart Lung Circ.* **2022**, *31*, e93–e109. [CrossRef]
78. Tatematsu, M.; Seya, T.; Matsumoto, M. Beyond dsRNA: Toll-like receptor 3 signalling in RNA-induced immune responses. *Biochem. J.* **2014**, *458*, 195–201. [CrossRef]
79. Gao, D.; Ciancanelli, M.J.; Zhang, P.; Harschnitz, O.; Bondet, V.; Hasek, M.; Chen, J.; Mu, X.; Itan, Y.; Cobat, A.; et al. TLR3 controls constitutive IFN-beta antiviral immunity in human fibroblasts and cortical neurons. *J. Clin. Investig.* **2021**, *131*, e134529. [CrossRef]
80. Wang, Y.; Swiecki, M.; McCartney, S.A.; Colonna, M. dsRNA sensors and plasmacytoid dendritic cells in host defense and autoimmunity. *Immunol. Rev.* **2011**, *243*, 74–90. [CrossRef]
81. Zhang, S.Y.; Herman, M.; Ciancanelli, M.J.; Perez de Diego, R.; Sancho-Shimizu, V.; Abel, L.; Casanova, J.L. TLR3 immunity to infection in mice and humans. *Curr. Opin. Immunol.* **2013**, *25*, 19–33. [CrossRef] [PubMed]
82. Hacker, H.; Redecke, V.; Blagoev, B.; Kratchmarova, I.; Hsu, L.C.; Wang, G.G.; Kamps, M.P.; Raz, E.; Wagner, H.; Hacker, G.; et al. Specificity in Toll-like receptor signalling through distinct effector functions of TRAF3 and TRAF6. *Nature* **2006**, *439*, 204–207. [CrossRef] [PubMed]
83. Kariko, K.; Ni, H.; Capodici, J.; Lamphier, M.; Weissman, D. mRNA is an endogenous ligand for Toll-like receptor 3. *J. Biol. Chem.* **2004**, *279*, 12542–12550. [CrossRef] [PubMed]
84. Cavassani, K.A.; Ishii, M.; Wen, H.; Schaller, M.A.; Lincoln, P.M.; Lukacs, N.W.; Hogaboam, C.M.; Kunkel, S.L. TLR3 is an endogenous sensor of tissue necrosis during acute inflammatory events. *J. Exp. Med.* **2008**, *205*, 2609–2621. [CrossRef] [PubMed]
85. Amarante, M.K.; Watanabe, M.A. Toll-like receptor 3: Involvement with exogenous and endogenous RNA. *Int. Rev. Immunol.* **2010**, *29*, 557–573. [CrossRef] [PubMed]
86. Iwakiri, D.; Zhou, L.; Samanta, M.; Matsumoto, M.; Ebihara, T.; Seya, T.; Imai, S.; Fujieda, M.; Kawa, K.; Takada, K. Epstein-Barr virus (EBV)-encoded small RNA is released from EBV-infected cells and activates signaling from Toll-like receptor 3. *J. Exp. Med.* **2009**, *206*, 2091–2099. [CrossRef]

87. Kindberg, E.; Vene, S.; Mickiene, A.; Lundkvist, A.; Lindquist, L.; Svensson, L. A functional Toll-like receptor 3 gene (TLR3) may be a risk factor for tick-borne encephalitis virus (TBEV) infection. *J. Infect. Dis.* **2011**, *203*, 523–528. [CrossRef]
88. Le Goffic, R.; Balloy, V.; Lagranderie, M.; Alexopoulou, L.; Escriou, N.; Flavell, R.; Chignard, M.; Si-Tahar, M. Detrimental contribution of the Toll-like receptor (TLR)3 to influenza A virus-induced acute pneumonia. *PLoS Pathog.* **2006**, *2*, e53. [CrossRef]
89. Wang, Q.; Miller, D.J.; Bowman, E.R.; Nagarkar, D.R.; Schneider, D.; Zhao, Y.; Linn, M.J.; Goldsmith, A.M.; Bentley, J.K.; Sajjan, U.S.; et al. MDA5 and TLR3 initiate pro-inflammatory signaling pathways leading to rhinovirus-induced airways inflammation and hyperresponsiveness. *PLoS Pathog.* **2011**, *7*, e1002070. [CrossRef]
90. Murray, L.A.; Knight, D.A.; McAlonan, L.; Argentieri, R.; Joshi, A.; Shaheen, F.; Cunningham, M.; Alexopolou, L.; Flavell, R.A.; Sarisky, R.T.; et al. Deleterious role of TLR3 during hyperoxia-induced acute lung injury. *Am. J. Respir. Crit. Care Med.* **2008**, *178*, 1227–1237. [CrossRef]
91. Citores, M.J.; Banos, I.; Noblejas, A.; Rosado, S.; Castejon, R.; Cuervas-Mons, V. Toll-like receptor 3 L412F polymorphism may protect against acute graft rejection in adult patients undergoing liver transplantation for hepatitis C-related cirrhosis. *Transplant. Proc.* **2011**, *43*, 2224–2226. [CrossRef] [PubMed]
92. Strodthoff, D.; Ma, Z.; Wirstrom, T.; Strawbridge, R.J.; Ketelhuth, D.F.; Engel, D.; Clarke, R.; Falkmer, S.; Hamsten, A.; Hansson, G.K.; et al. Toll-Like Receptor 3 Influences Glucose Homeostasis and beta-Cell Insulin Secretion. *Diabetes* **2015**, *64*, 3425–3438. [CrossRef] [PubMed]
93. Brentano, F.; Schorr, O.; Gay, R.E.; Gay, S.; Kyburz, D. RNA released from necrotic synovial fluid cells activates rheumatoid arthritis synovial fibroblasts via Toll-like receptor 3. *Arthritis Rheum.* **2005**, *52*, 2656–2665. [CrossRef] [PubMed]
94. Torres, D.; Dieudonne, A.; Ryffel, B.; Vilain, E.; Si-Tahar, M.; Pichavant, M.; Lassalle, P.; Trottein, F.; Gosset, P. Double-stranded RNA exacerbates pulmonary allergic reaction through TLR3: Implication of airway epithelium and dendritic cells. *J. Immunol.* **2010**, *185*, 451–459. [CrossRef]
95. Papaioannou, A.I.; Spathis, A.; Kostikas, K.; Karakitsos, P.; Papiris, S.; Rossios, C. The role of endosomal toll-like receptors in asthma. *Eur. J. Pharmacol.* **2017**, *808*, 14–20. [CrossRef]
96. Kuo, C.Y.; Ku, C.L.; Lim, H.K.; Hsia, S.H.; Lin, J.J.; Lo, C.C.; Ding, J.Y.; Kuo, R.L.; Casanova, J.L.; Zhang, S.Y.; et al. Life-Threatening Enterovirus 71 Encephalitis in Unrelated Children with Autosomal Dominant TLR3 Deficiency. *J. Clin. Immunol.* **2022**, *42*, 606–617. [CrossRef]
97. Partanen, T.; Chen, J.; Lehtonen, J.; Kuismin, O.; Rusanen, H.; Vapalahti, O.; Vaheri, A.; Anttila, V.J.; Bode, M.; Hautala, N.; et al. Heterozygous TLR3 Mutation in Patients with Hantavirus Encephalitis. *J. Clin. Immunol.* **2020**, *40*, 1156–1162. [CrossRef]
98. Lim, H.K.; Huang, S.X.L.; Chen, J.; Kerner, G.; Gilliaux, O.; Bastard, P.; Dobbs, K.; Hernandez, N.; Goudin, N.; Hasek, M.L.; et al. Severe influenza pneumonitis in children with inherited TLR3 deficiency. *J. Exp. Med.* **2019**, *216*, 2038–2056. [CrossRef]
99. Krieg, A.M.; Vollmer, J. Toll-like receptors 7, 8, and 9: Linking innate immunity to autoimmunity. *Immunol. Rev.* **2007**, *220*, 251–269. [CrossRef]
100. Klinman, D.M.; Tross, D.; Klaschik, S.; Shirota, H.; Sato, T. Therapeutic applications and mechanisms underlying the activity of immunosuppressive oligonucleotides. *Ann. N. Y. Acad. Sci.* **2009**, *1175*, 80–88. [CrossRef]
101. Latz, E.; Verma, A.; Visintin, A.; Gong, M.; Sirois, C.M.; Klein, D.C.; Monks, B.G.; McKnight, C.J.; Lamphier, M.S.; Duprex, W.P.; et al. Ligand-induced conformational changes allosterically activate Toll-like receptor 9. *Nat. Immunol.* **2007**, *8*, 772–779. [CrossRef]
102. Wagner, H. The sweetness of the DNA backbone drives Toll-like receptor 9. *Curr. Opin. Immunol.* **2008**, *20*, 396–400. [CrossRef]
103. Kuznik, A.; Panter, G.; Jerala, R. Recognition of nucleic acids by Toll-like receptors and development of immunomodulatory drugs. *Curr. Med. Chem.* **2010**, *17*, 1899–1914. [CrossRef]
104. Cossart, P.; Helenius, A. Endocytosis of viruses and bacteria. *Cold Spring Harb. Perspect. Biol.* **2014**, *6*, a016972. [CrossRef] [PubMed]
105. Vanderlinden, E.; Naesens, L. Emerging antiviral strategies to interfere with influenza virus entry. *Med. Res. Rev.* **2014**, *34*, 301–339. [CrossRef] [PubMed]
106. Francisco-Garcia, A.S.; Garrido-Martin, E.M.; Rupani, H.; Lau, L.C.K.; Martinez-Nunez, R.T.; Howarth, P.H.; Sanchez-Elsner, T. Small RNA Species and microRNA Profiles are Altered in Severe Asthma Nanovesicles from Broncho Alveolar Lavage and Associate with Impaired Lung Function and Inflammation. *Noncoding RNA* **2019**, *5*, 51. [CrossRef] [PubMed]
107. Abdelmohsen, K.; Gorospe, M. RNA-binding protein nucleolin in disease. *RNA Biol.* **2012**, *9*, 799–808. [CrossRef]
108. Hovanessian, A.G.; Soundaramourty, C.; El Khoury, D.; Nondier, I.; Svab, J.; Krust, B. Surface expressed nucleolin is constantly induced in tumor cells to mediate calcium-dependent ligand internalization. *PLoS ONE* **2010**, *5*, e15787. [CrossRef] [PubMed]
109. Fu, X.; Liang, C.; Li, F.; Wang, L.; Wu, X.; Lu, A.; Xiao, G.; Zhang, G. The Rules and Functions of Nucleocytoplasmic Shuttling Proteins. *Int. J. Mol. Sci.* **2018**, *19*, 1445. [CrossRef] [PubMed]
110. Song, N.; Ding, Y.; Zhuo, W.; He, T.; Fu, Z.; Chen, Y.; Song, X.; Fu, Y.; Luo, Y. The nuclear translocation of endostatin is mediated by its receptor nucleolin in endothelial cells. *Angiogenesis* **2012**, *15*, 697–711. [CrossRef] [PubMed]
111. Kitagawa, S.; Matsuda, T.; Washizaki, A.; Murakami, H.; Yamamoto, T.; Yoshioka, Y. Elucidation of the role of nucleolin as a cell surface receptor for nucleic acid-based adjuvants. *NPJ Vaccines* **2022**, *7*, 115. [CrossRef]
112. Tayyari, F.; Marchant, D.; Moraes, T.J.; Duan, W.; Mastrangelo, P.; Hegele, R.G. Identification of nucleolin as a cellular receptor for human respiratory syncytial virus. *Nat. Med.* **2011**, *17*, 1132–1135. [CrossRef]

113. Mastrangelo, P.; Chin, A.A.; Tan, S.; Jeon, A.H.; Ackerley, C.A.; Siu, K.K.; Lee, J.E.; Hegele, R.G. Identification of RSV Fusion Protein Interaction Domains on the Virus Receptor, Nucleolin. *Viruses* **2021**, *13*, 261. [CrossRef]
114. Battles, M.B.; McLellan, J.S. Respiratory syncytial virus entry and how to block it. *Nat. Rev. Microbiol.* **2019**, *17*, 233–245. [CrossRef]
115. Song, J.; Quan, R.; Wang, D.; Liu, J. Seneca Valley Virus 3C(pro) Mediates Cleavage and Redistribution of Nucleolin to Facilitate Viral Replication. *Microbiol. Spectr.* **2022**, *10*, e0030422. [CrossRef]
116. Tonello, F.; Massimino, M.L.; Peggion, C. Nucleolin: A cell portal for viruses, bacteria, and toxins. *Cell Mol. Life Sci.* **2022**, *79*, 271. [CrossRef]
117. Mastrangelo, P.; Norris, M.J.; Duan, W.; Barrett, E.G.; Moraes, T.J.; Hegele, R.G. Targeting Host Cell Surface Nucleolin for RSV Therapy: Challenges and Opportunities. *Vaccines* **2017**, *5*, 27. [CrossRef]
118. Mastrangelo, P.; Hegele, R.G. The RSV fusion receptor: Not what everyone expected it to be. *Microbes Infect.* **2012**, *14*, 1205–1210. [CrossRef]
119. Anderson, C.S.; Chirkova, T.; Slaunwhite, C.G.; Qiu, X.; Walsh, E.E.; Anderson, L.J.; Mariani, T.J. CX3CR1 Engagement by Respiratory Syncytial Virus Leads to Induction of Nucleolin and Dysregulation of Cilia-related Genes. *J. Virol.* **2021**, *95*, e00095-21. [CrossRef]
120. Griffiths, C.D.; Bilawchuk, L.M.; McDonough, J.E.; Jamieson, K.C.; Elawar, F.; Cen, Y.; Duan, W.; Lin, C.; Song, H.; Casanova, J.L.; et al. IGF1R is an entry receptor for respiratory syncytial virus. *Nature* **2020**, *583*, 615–619. [CrossRef]
121. Dondalska, A.; Ronnberg, E.; Ma, H.; Palsson, S.A.; Magnusdottir, E.; Gao, T.; Adam, L.; Lerner, E.A.; Nilsson, G.; Lagerstrom, M.; et al. Amelioration of Compound 48/80-Mediated Itch and LL-37-Induced Inflammation by a Single-Stranded Oligonucleotide. *Front. Immunol.* **2020**, *11*, 559589. [CrossRef] [PubMed]
122. Gokirmak, T.; Nikan, M.; Wiechmann, S.; Prakash, T.P.; Tanowitz, M.; Seth, P.P. Overcoming the challenges of tissue delivery for oligonucleotide therapeutics. *Trends Pharmacol. Sci.* **2021**, *42*, 588–604. [CrossRef] [PubMed]
123. Crooke, S.T.; Vickers, T.A.; Liang, X.H. Phosphorothioate modified oligonucleotide-protein interactions. *Nucleic Acids Res.* **2020**, *48*, 5235–5253. [CrossRef] [PubMed]
124. Iannitti, T.; Morales-Medina, J.C.; Palmieri, B. Phosphorothioate oligonucleotides: Effectiveness and toxicity. *Curr. Drug Targets* **2014**, *15*, 663–673. [CrossRef]

International Journal of
Molecular Sciences

MDPI

Article

Investigation of TLR2 and TLR4 Polymorphisms and Sepsis Susceptibility: Computational and Experimental Approaches

Mohammed Y. Behairy [1], Ali A. Abdelrahman [2], Eman A. Toraih [3,4], Emad El-Deen A. Ibrahim [5], Marwa M. Azab [2,*], Anwar A. Sayed [6,7,*] and Hany R. Hashem [8]

1 Department of Microbiology and Immunology, Faculty of Pharmacy, University of Sadat City, Sadat City 32958, Egypt
2 Department of Microbiology and Immunology, Faculty of Pharmacy, Suez Canal University, Ismailia 41522, Egypt
3 Department of Surgery, School of Medicine, Tulane University, New Orleans, LA 70112, USA
4 Genetics Unit, Department of Histology & Cell Biology, Faculty of Medicine, Suez Canal University, Ismailia 41522, Egypt
5 Department of Anesthesia, Intensive Care and Pain Management, Faculty of Medicine, Suez Canal University, Ismailia 41522, Egypt
6 Department of Medical Microbiology and Immunology, Taibah University, Madinah 42353, Saudi Arabia
7 Department of Surgery and Cancer, Imperial College London, London SW7 2BX, UK
8 Department of Microbiology and Immunology, Faculty of Pharmacy, Fayoum University, Fayoum 63514, Egypt
* Correspondence: marwaazab2515@yahoo.com (M.M.A.); dsayed@taibahu.edu.sa (A.A.S.); Tel.: +20-10-2429-9630 (M.M.A.); +966-14-861-8888 (A.A.S.)

Citation: Behairy, M.Y.; Abdelrahman, A.A.; Toraih, E.A.; Ibrahim, E.E.-D.A.; Azab, M.M.; Sayed, A.A.; Hashem, H.R. Investigation of TLR2 and TLR4 Polymorphisms and Sepsis Susceptibility: Computational and Experimental Approaches. *Int. J. Mol. Sci.* **2022**, *23*, 10982. https://doi.org/10.3390/ijms231810982

Academic Editors: Oliver Planz and Ralf Kircheis

Received: 23 August 2022
Accepted: 13 September 2022
Published: 19 September 2022

Abstract: Toll-like receptors (TLR) play an eminent role in the regulation of immune responses to invading pathogens during sepsis. TLR genetic variants might influence individual susceptibility to developing sepsis. The current study aimed to investigate the association of genetic polymorphisms of the *TLR2* and *TLR4* with the risk of developing sepsis with both a pilot study and in silico tools. Different in silico tools were used to predict the impact of our SNPs on protein structure, stability, and function. Furthermore, in our prospective study, all patients matching the inclusion criteria in the intensive care units (ICU) were included and followed up, and DNA samples were genotyped using real-time polymerase chain reaction (RT-PCR) technology. There was a significant association between *TLR2* Arg753Gln polymorphisms and sepsis under the over-dominant model ($p = 0.043$). In contrast, we did not find a significant difference with the *TLR4* Asp299Gly polymorphism with sepsis. However, there was a significant association between *TLR4* Asp299Gly polymorphisms and *Acinetobacter baumannii* infection which is quite a virulent organism in ICU ($p = 0.001$) and post-surgical cohorts ($p = 0.033$). Our results conclude that the *TLR2* genotype may be a risk factor for sepsis in adult patients.

Keywords: TLR; polymorphism; infection; sepsis; septic shock

1. Introduction

Infection is one of the prominent causes of human morbidity and mortality, especially in patients requiring critical care [1,2]. Moreover, in intensive care units (ICUs), a serious complication of infection is sepsis and its maximal manifestation, septic shock [3]. Sepsis is an infection-induced life-threatening organ dysfunction with mortality rates reaching 20–70% [4,5].

Infectious diseases have been found to be a major selective pressure [6]. Despite the ambiguity of the precise etiology of sepsis, numerous studies have shown that gene polymorphisms have an important role in affecting individual susceptibility to sepsis [7]. Some polymorphisms of the innate immune system are supposed to mediate a predisposition to infectious complications including the outcome of patients with sepsis [8]. The innate immune system is of crucial importance for both the direct defense against micro-organisms and the activation of the adaptive immune system [9]. The innate immunity system is the main mediator of inflammation, and it recruits specific pattern recognition receptors (PRRs) capable of recognizing micro-organisms through identifying conserved pathogen-associated molecular patterns.

Toll-like receptors (TLR) are the most studied subtypes of pattern recognition receptors with their critical importance in the immune system [10,11]. Among the members of the TLR family, TLR2 and TLR4 are considered the most important PRRs that cover a wide range of antigenic determinants [12]. TLR4 has a distinctive ability to recognize a very wide range of microorganisms including Gram-negative bacteria through Lipopolysaccharide (LPS), in addition to many viruses and Fungi. Meanwhile, TLR2 is regarded as a key molecule in regulating our immune system with a crucial role in the recognition of Peptidoglycans of Gram-positive bacteria, in addition to different ligands of yeast, fungi, viruses, and parasites [12,13].

One of the most studied innate immunity polymorphisms is the *TLR4* Asp299Gly (rs4986790) polymorphism, which interferes with TLR4 signal transduction; thus, it is supposed to affect host susceptibility to infections and microbial invasions [14]. Moreover, structural analysis of *TLR4* Asp299Gly has revealed evidence of a resulted impairment in TLR4 binding to its ligands [15]. Meanwhile, one of the most important polymorphisms of *TLR2* is Arg753Gln; the presence of this SNP was found to impair the signaling pathway of this key receptor [16], thus suggesting increased susceptibility to infections and sepsis.

Consequently, many studies have been conducted all over the world to reveal the prevalence of these SNPs and their impact on infection and sepsis susceptibility, but a varied pattern of prevalence was found for both *TLR4* and *TLR2* SNPs among different populations [13,17,18]. In addition, conflicting results were found regarding their impact on infection and sepsis susceptibility in different populations [19–21]. Therefore, a need was felt for further investigation on these issues. The usage of computational approaches in studying SNPs' impact has gained momentum and importance in recent years [22–25]. Integrating the in silico approach with the experimental one provides great accuracy and depth to the analysis.

In this study, we aimed to investigate the possible role of *TLR2* and *TLR4* polymorphisms in affecting sepsis susceptibility and survival in critically ill patients in the Egyptian population using both in silico analysis and experimental methods.

2. Results

The study involved both a pilot study and in silico analysis. A scheme illustrating the layout of the study plan is shown in Figure 1.

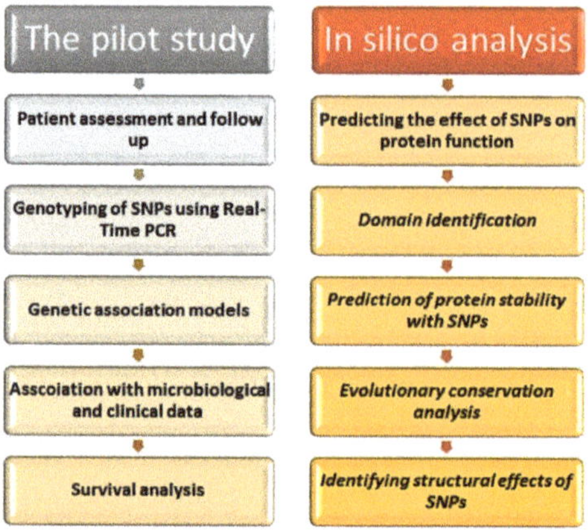

Figure 1. Scheme illustrating the outline of the study plan.

2.1. *In Silico Analysis*

2.1.1. General Information: TLR2

TLR2 gene (ENSG00000137462) is a protein-coding gene located on 4q31.3. It is composed of five exons with a length of 26,564 nucleotides. It is located on Chromosome 4: 153684080-153710643 according to the Genome Reference Consortium Human Build 38 patch release 13 (GRCh38.p13) with NCBI Reference Sequence (NC_000004.12) (https://www.ncbi.nlm.nih.gov/gene/7097 (accessed on 29 August 2021)). There are eight transcripts for this gene (ensemble.org). This gene encodes Toll-like receptor 2 protein, a member of the Toll-like receptor (TLR) family. Figure 2A shows the subcellular localization of TLR2. The predicted network of protein–protein interactions of the TLR2 protein is shown in Figure 2B and its Gene Coexpression matrix (Figure 2C) shows coexpression with CD14, CLEC7A, and LY96 with scores of 0.611, 0.281, and 0.130, respectively (https://string-db.org (accessed on 29 August 2021)). Rs5743708 is an SNP located at Chromosome 4, position: 153705165 (forward strand) with two alleles (G and A). G is the ancestral allele and the minor allele frequency for A equals 0.01. This is a missense variant that causes the replacement of amino acid Arginine with amino acid Glutamine at position 753.

2.1.2. General Information: TLR4

TLR4 gene (ENSG00000136869) is a protein-coding gene located on 9q33.1. It is composed of four exons with a length of 20,333 nucleotides. It is located on Chromosome 9: 117704403-117724735 according to the Genome Reference Consortium Human Build 38 patch release 13 (GRCh38.p13) with NCBI Reference Sequence (NC_000009.12) (https://www.ncbi.nlm.nih.gov/gene/7099 (accessed on 29 August 2021)). There are four transcripts for this gene (ensemble.org). This gene encodes the Toll-like receptor 4 protein, a member of the Toll-like receptor (TLR) family as well. Figure 2D shows the subcellular localization of TLR4. The predicted network of protein–protein interactions of TLR4 protein is shown in Figure 2E and its Gene Coexpression matrix (Figure 2F) shows coexpression with LY86, CD14, and LY96 with scores of 0.301, 0.264, and 0.176, respectively (https://string-db.org (accessed on 29 August 2021)). Rs4986790 is an SNP located at Chromosome 9, position: 117713024 (forward strand) with three alleles (A, G, and T). A is the ancestral allele and the minor allele frequency for G equals 0.06. This is a missense mutation that causes the replacement of amino acid Aspartic acid with amino acid Glycine at position 299.

2.1.3. Predicting the Effect of SNPs on Protein Function

Five bioinformatics tools were used to predict the impact of rs4986790 and rs5743708 on the TLR4 and TLR2 proteins, respectively, to increase the accuracy of the results. For TLR4, all used bioinformatics tools predicted this variation to be neutral or benign as shown in Table 1. While for TLR2, all tools predicted this SNP to be damaging except SNPs and GO which predicted it to be neutral (Table 1). Figure 3 shows the structural and functional effects of SNPs.

Table 1. Predicting the effect of SNPs on protein function using bioinformatics tools.

SNP	Amino Acid Change	SIFT	Polyphen2	PANTHER	PROVEAN	SNPs and GO
rs4986790	D299G	Tolerated	Benign	probably benign	Neutral	Neutral
rs5743708	R753Q	Deleterious	Probably Damaging	probably damaging	Deleterious	Neutral

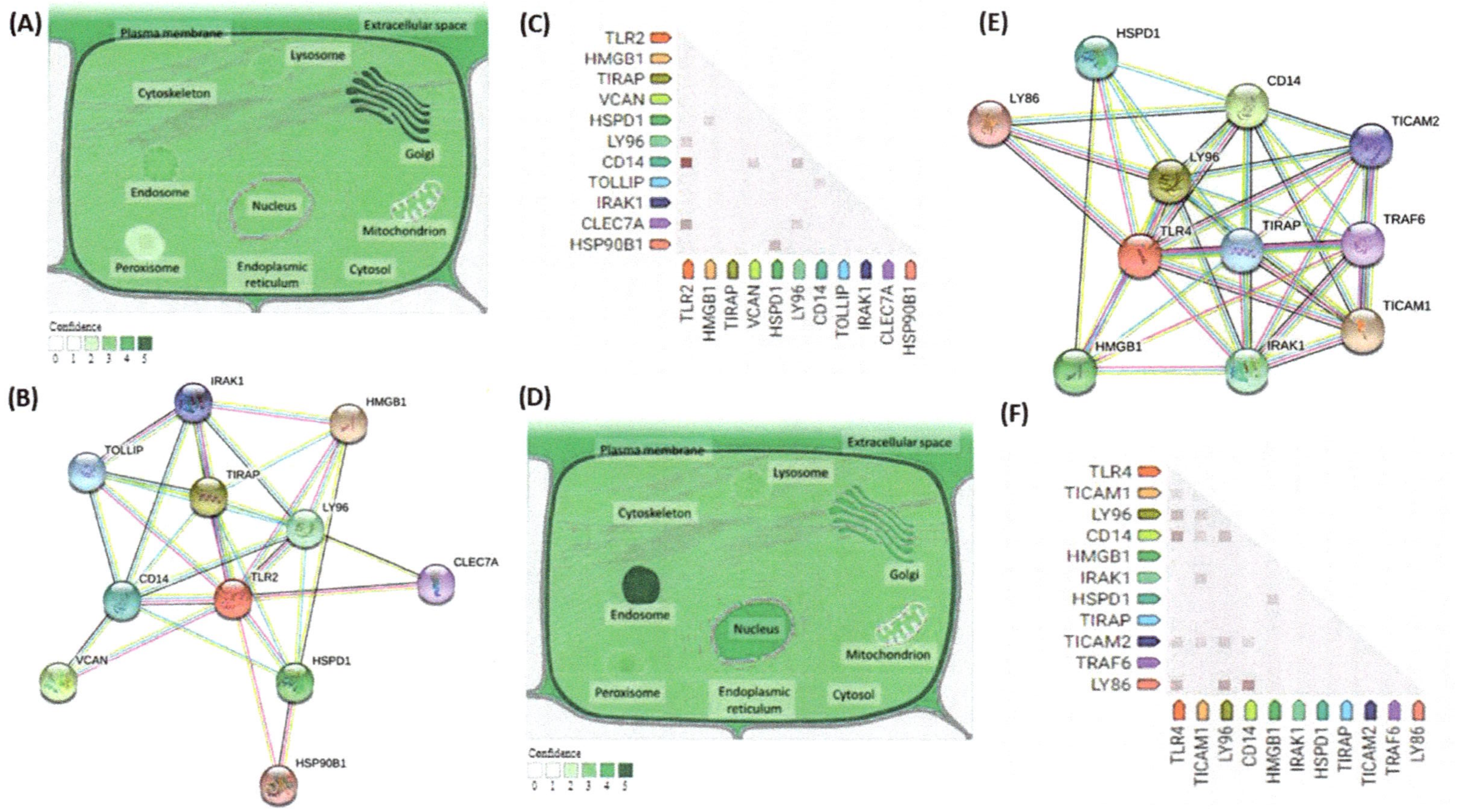

Figure 2. Functional Analysis of TLR2 and TLR4 proteins. (**A**) Subcellular localization of the TLR2 protein. The gradient of green color indicates the degree of confidence (genecards.org) with (compartments.jensenlab.org) as the source of the image. (**B**) Predicted network of protein–protein interactions of the TLR2 protein.

Proteins are represented by nodes, while predicted associations are represented by edges that could be drawn with 7 colored lines that indicate different types of evidence. Redline fusion evidence, Light blue line—database evidence. Green line—neighborhood evidence. Blue line—co-occurrence evidence. Purple line—experimental evidence. Black line—coexpression evidence. Yellow line—text mining evidence. HMGB1: High mobility group protein B1, TIRAP: Toll/interleukin-1 receptor domain-containing adapter protein, VCAN: Versican core protein, HSPD1: 60 kDa heat shock protein, LY96: Lymphocyte antigen 96, CD14: Monocyte differentiation antigen CD14, TOLLIP: Toll-interacting protein, IRAK1: Interleukin-1 receptor-associated kinase 1, CLEC7A: C-type lectin domain family 7 member A, HSP90B1: Endoplasmin. STRING analysis (version 11.5). (**C**) *TLR2* Gene Coexpression matrix. Predict association between protein functions, Color intensity shows the confidence level in the association between protein functions. TLR2 shows coexpression with CD14, CLEC7A, and LY96 with scores of 0.611, 0.281, and 0.130, respectively (https://string-db.org (accessed on 29 August 2021)). (**D**) Subcellular localization of TLR4 protein (genecards.org) with (compartments.jensenlab.org) as the source of the image. (compartments.jensenlab.org). (**E**) 1B predicted a network of protein–protein interactions of the TLR4 protein. TICAM1: TIR domain-containing adapter molecule 1, TICAM2: TIR domain-containing adapter molecule 2, TRAF6: TNF receptor-associated factor 6, LY86: Lymphocyte antigen 86. STRING analysis (version 11.5). (**F**) *TLR4* Gene Coexpression matrix. TLR4 shows coexpression with LY86, CD14, and LY96 with scores of 0.301, 0.264, and 0.176, respectively (https://string-db.org (accessed on 29 August 2021)).

2.1.4. Identifying SNP Location on Protein Domains

Using InterPro revealed that rs5743708 was found to be located on the Toll/Interleukin-1 Receptor Homology (TIR) Domain (InterPro entry: IPR000157) in the TLR2 protein which is an essential domain for protein function, while rs4986790 location was found to belong to a superfamily called Leucine-rich repeat domain superfamily (InterPro entry: IPR032675).

2.1.5. Prediction of Protein Stability with SNPs

I-Mutant 2.0 web server analyzed the effects of rs5743708 and rs4986790 SNPs on the stability of TLR2 and TLR4 proteins, respectively, by calculating free energy change values (DDG) and the Reliability Index value (RI). For TLR4, rs4986790 was found to decrease stability with RI = 3 and DDG = 0.38 Kcal/mol. While for TLR2, rs5743708 was found to decrease stability with RI = 8 and DDG = −0.71 Kcal/mol.

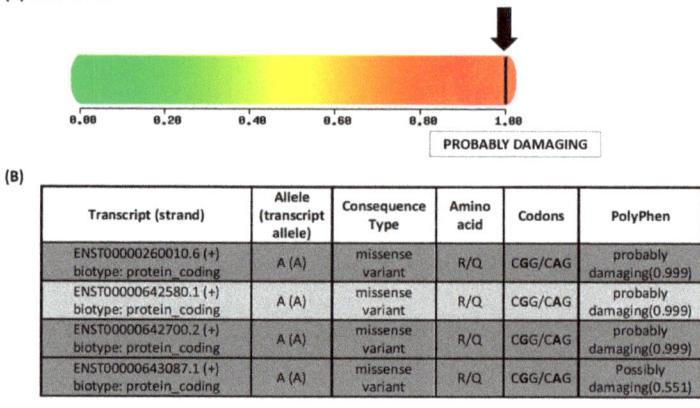

(A) rs5743708

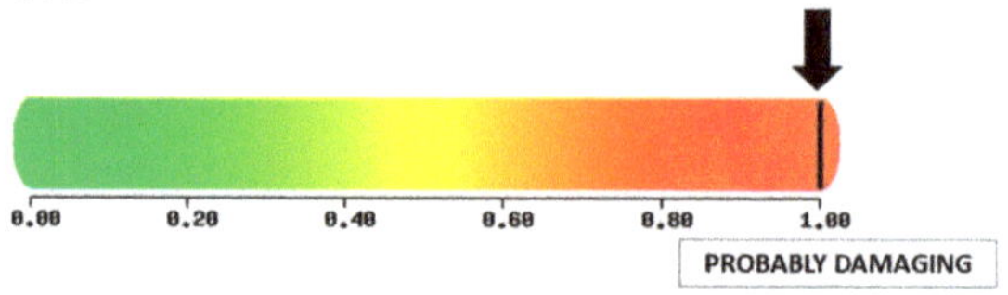

(B)

Transcript (strand)	Allele (transcript allele)	Consequence Type	Amino acid	Codons	PolyPhen
ENST00000260010.6 (+) biotype: protein_coding	A (A)	missense variant	R/Q	CGG/CAG	probably damaging(0.999)
ENST00000642580.1 (+) biotype: protein_coding	A (A)	missense variant	R/Q	CGG/CAG	probably damaging(0.999)
ENST00000642700.2 (+) biotype: protein_coding	A (A)	missense variant	R/Q	CGG/CAG	probably damaging(0.999)
ENST00000643087.1 (+) biotype: protein_coding	A (A)	missense variant	R/Q	CGG/CAG	Possibly damaging(0.551)

(C) rs4986790

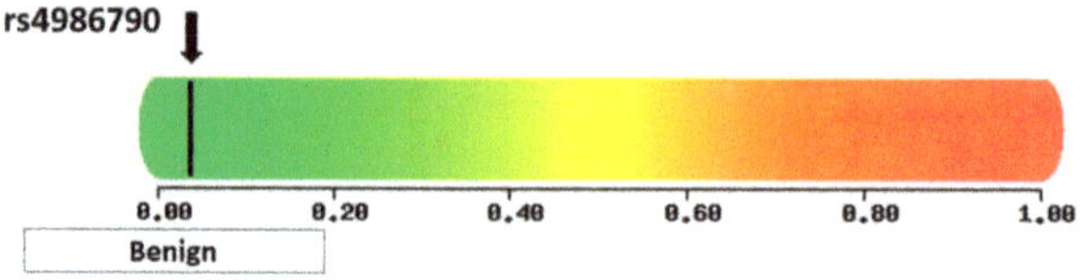

(D)

Transcript (strand)	Allele (transcript allele)	Consequence Type	Amino acid	Codons	PolyPhen
ENST00000355622.8 (+) biotype: protein_coding	T (T)	missense variant	D/V	GAT/GTT	Benign (0.054)
ENST00000394487.5 (+) biotype: protein_coding	T (T)	missense variant	D/V	GAT/GTT	Benign (0.054)
ENST00000472304.2 (+) biotype: protein_coding	T (T)	3 prime UTR variant	-	-	-
ENST00000355622.8 (+) biotype: protein_coding	G (G)	missense variant	D/G	GAT/GGT	Benign (0.033)
ENST00000394487.5 (+) biotype: protein_coding	G (G)	missense variant	D/G	GAT/GGT	Benign (0.033)
ENST00000472304.2 (+) biotype: protein_coding	G (G)	3 prime UTR variant	-	-	-

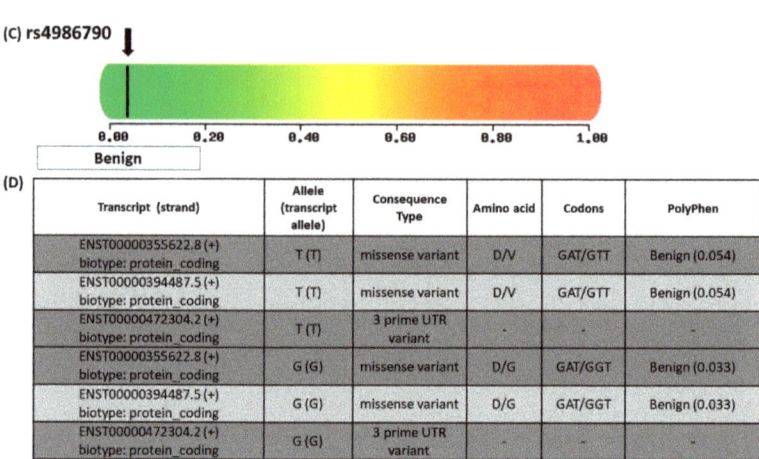

Figure 3. Functional and structural consequences of SNPs (**A**). Predicting the impact of rs5743708 on TLR2 function—the score ranged from benign (0) to damaging (1) (**B**). Table showing the transcripts of rs5743708, allele (transcript allele), consequence type, amino acid fate, codons, and PolyPhen score. R: Arginine, Q: Glutamine (ensemble.org). (**C**) Predicting the impact of rs4986790 on TLR4 function the score ranges from benign (0) to damaging (1) (**D**). Table showing the transcripts of rs4986790, allele (transcript allele), consequence type, amino acid fate, codons, and PolyPhen score. D: Aspartate, V: valine, G: Glycine (ensemble.org).

2.1.6. Conservation Analysis

TLR2 and TLR4 proteins were analyzed by the ConSurf server to perform an evolutionary conservation analysis of their amino acid positions (Figures 4 and 5), respectively. In TLR2, position 753 (R753) was found to be an exposed and functional residue with high conservation. While in TLR4, position 299 (D299) was found to be an exposed and variable residue.

Figure 4. Evolutionary conservation analysis of TLR2 by Consurf.

Figure 5. Evolutionary conservation analysis of TLR4 by Consurf.

2.1.7. Identifying the Structural Effects of SNPs

Using Project HOPE to analyze rs4986790 in TLR4, the new Glycine residue was found to differ in size and charge from the wild residue (Aspartic Acid) which could lead to a loss of interactions (Figure 6A). There was a difference in hydrophobicity too, which could cause loss of hydrogen bonds with possible disturbance of correct folding. Moreover, this replacement leads to an inability to form a Cysteine Bridge with its importance to protein stability, thus affecting the 3D structure of the protein and protein stability. In addition, Glycine flexibility affects the needed stability at that position. Meanwhile, analyzing rs5743708 in TLR2 revealed differences in size and charge between wild and mutant amino acids which could cause a loss of interactions (Figure 6B). Moreover, the different properties could lead to disturbance and elimination of (TIR) Domain function with its importance for protein function.

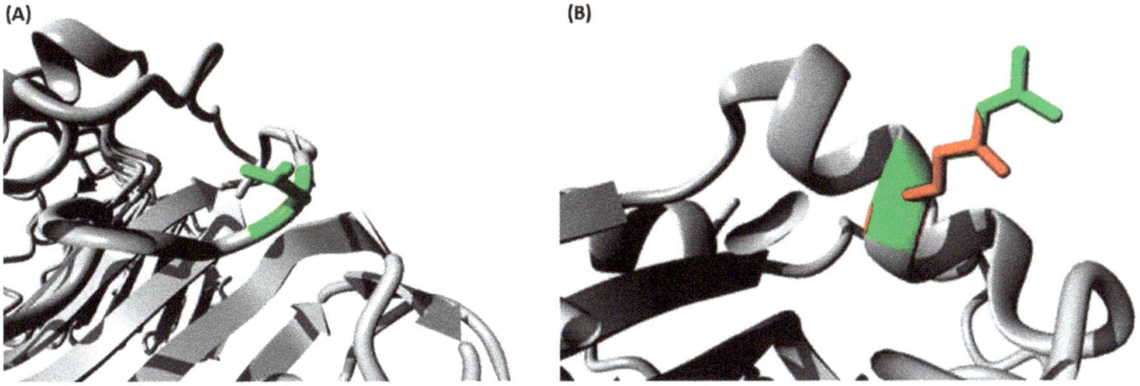

Figure 6. HOPE illustration of mutation structural impacts (**A**). Project HOPE illustration of the structural replacement of Aspartic acid with Glycine at position 299 in the TLR4 protein (colored with grey). The side chain of the wild type is colored in green while Glycine only has a hydrogen atom in its side chain (**B**). Project HOPE illustration of the structural replacement of Arginine with Glutamine at position 753 in the TLR2 protein (colored with grey). The side chain of the wild type is colored green while the side chain of the mutant type is colored red.

2.2. Demographic and Microbiological Data

A total of seventy-five Egyptian unrelated patients were included in the study. All participants had developed an infection. The patients were followed up to assess sepsis and septic shock, and the demographic features and the clinical characteristics of ICU-admitted patients according to developing sepsis are presented in Table 2. The two groups had significant differences in age factor, APACHE score at admission, and some categories of admissions. Causative organisms are listed (Table 2). There was no statistically significant difference between any of the causative organisms and developing sepsis.

Table 2. Demographic and clinical features of ICU admitted patients with and without sepsis.

Variables		All	No Sepsis	Sepsis	*p*-Value	OR (95% CI)
		Demographic Characteristics				
Number		75 (100%)	48 (64.0%)	27 (36.0%)		
Age, years	Mean ± SD	60.0 ± 17.6	55.5 ± 18.9	68.0 ± 11.5	0.003	
	≤40 years	14 (18.7%)	14 (29.2%)	0 (0.0%)	0.002	Reference
	≤60 years	20 (26.7%)	14 (29.2%)	6 (22.2%)		13.0 (0.67–252.6)
	>60 years	41 (54.7%)	20 (41.6%)	21 (77.8%)		30.41 (1.7–543.6)

Table 2. *Cont.*

Variables		**All**	**No Sepsis**	**Sepsis**	***p*-Value**	**OR (95% CI)**
Sex	Male	47 (63.0%)	31 (64.6%)	16 (59.3%)	0.64	Reference
	Female	28 (37.0%)	17 (35.4%)	11 (40.7%)		1.25 (0.48–3.30)
Vital signs	HR	100.6 ± 21.3	101.2 ± 20.8	99.5 ± 22.4	0.75	
	MAP	81.7 ± 25	85.4 ± 27.4	75.2 ± 18.7	0.09	
Concomitant diseases						
Diabetes	Positive	29 (38.7)	17 (35.4%)	12 (44.4%)	0.44	1.46 (0.56–3.82)
Hypertension	Positive	42 (56.0%)	26 (54.2%)	16 (59.3%)	0.67	1.23 (0.47–3.20)
Vascular disease	Positive	27 (36.0%)	15 (31.3%)	12 (44.4%)	0.25	1.76 (0.66–4.66)
Chronic lung disease	Positive	6 (8.0%)	5 (10.4%)	1 (3.7%)	0.41	0.33 (0.04–2.99)
Chronic liver disease	Positive	7 (9.3%)	2 (4.2%)	5 (18.5%)	0.09	5.23 (0.94–29.10)
Chronic renal disease	Positive	17 (22.7%)	10 (20.8%)	7 (25.9%)	0.61	1.33 (0.44–4.02)
ICU assessment						
APACHE score	Mean ± SD	17.4 ± 8.3	15.8 ± 6.0	20.3 ± 10.8	0.024	
Glasgow scale	Mean ± SD	9.8 ± 4.4	9.4 ± 4.1	10.5 ± 4.8	0.29	
Length of stay, days	Mean ± SD	19.7 ± 15.9	17.6 ± 11.8	23.4 ± 21.1	0.13	
Consequence	Discharge	27 (36.0%)	19 (39.58%)	8 (29.63%)	0.69	Reference
	Transferred	5 (6.7%)	3 (6.25%)	2 (7.40%)		1.58 (0.22–11.3)
	Death	43 (57.3%)	26 (54.17%)	17 (62.96%)		1.55 (0.56–4.34)
OS, days	Mean ± SD	19.6 ± 17.2	17.5 ± 13.4	22.8 ± 22	0.33	
Admission category						
Renal	Positive	2 (2.7%)	1 (2.1%)	1 (3.7%)	0.67	1.81 (0.11–30.1)
Cardiovascular	Positive	3 (4%)	2 (4.2%)	1 (3.7%)	0.92	0.88 (0.08–10.2)
Infection	Positive	21 (28%)	7 (14.6%)	14 (51.8%)	0.001	7.23 (2.42–21.6)
Neurology	Positive	20 (26.7%)	17 (35.4%)	3 (11.1%)	0.022	0.27 (0.07–1.03)
Post-surgical	Positive	11 (14.7%)	6 (12.5%)	5 (18.5%)	0.47	1.59 (0.44–5.80)
Respiratory	Positive	10 (13.3%)	10 (20.8%)	0 (0.0%)	0.011	0.07 (0.00–1.19)
Trauma	Positive	3 (4%)	3 (6.3%)	0 (0.0%)	0.54	0.24 (0.01–4.75)
Other causes	Positive	5 (6.7%)	2 (4.2%)	3 (11.1%)	0.24	2.88 (0.45–18.4)
Variables		All	No sepsis	Sepsis	*p*-value	OR (95% CI)
Causative organism in culture						
Enterobacter spp.	Positive	6 (6.3%)	3 (6.3%)	3 (11.1%)	0.45	1.88 (0.35–10.01)
Acinetobacter baumannii	Positive	11 (11.5%)	7 (14.6%)	4 (14.8%)	0.97	1.02 (0.27–3.85)
Candida albicans	Positive	3 (3.1%)	1 (2.1%)	2 (7.4%)	0.25	3.76 (0.32–43.53)
Escherichia coli	Positive	15 (15.8%)	11(22.9%)	4 (14.8%)	0.40	0.59 (0.17–2.06)
Gram negative bacilli	Positive	3 (3.1%)	2 (4.2%)	1 (3.7%)	0.92	0.90 (0.08–10.45)
Klebsiella pneumoniae	Positive	20 (21.1%)	13 (27.1%)	7 (25.9%)	0.91	0.94 (0.32–2.75)
Pseudomonas aeruginosa	Positive	12 (12.6%)	6 (12.5%)	6 (22.2%)	0.27	2.00 (0.57–6.96)
Staph spp.	Positive	17 (17.9%)	12 (25.0%)	5(18.5%)	0.52	0.68 (0.21–2.20)
Streptococcus spp.	Positive	4 (4.2%)	3 (6.25%)	1 (3.7%)	0.63	0.58 (0.06–5.84)
Aeromonas hydephila	Positive	1 (1.1%)	1 (2.1%)	0 (0%)	1.00	0.58 (0.02–14.6)
Proteus spp.	Positive	1 (1.1%)	1 (2.1%)	0 (0%)	1.00	0.58 (0.02–14.6)
Citrobacter spp.	Positive	1 (1.1%)	1 (2.1%)	0 (0%)	1.00	0.58 (0.02–14.6)
Serratia spp.	Positive	1 (1.1%)	1(2.1%)	0 (0%)	1.00	0.58 (0.02–14.6)

Data are shown as a number (percentage) or number ± standard deviation. HR: heart rate in beats per minute; MAP: mean arterial pressure in mmHg, OS: Overall survival. Chi-square (χ^2) or Fisher's exact tests were used for qualitative variables and student's *t*-test was used for quantitative attributes. OR (95% CI), odds ratio, and confidence interval. Statistical analysis at *p*-value < 0.05.

2.3. Allele Frequencies of TLR2 and TLR4 Genes in the Study Population

Genotype and allele frequencies for TLR2 and TLR4 were detailed in Table 3. For TLR4, the frequency of wild-type genotype AA was 91%, while the heterozygous genotype AG was 8%, and the mutant genotype GG was 1%. The genotype frequencies followed the genotype frequencies expected by Hardy–Weinberg equilibrium ($p > 0.05$). For TLR2 the frequency of wild-type genotype GG was 92%, while the heterozygous genotype GA was 5%, and the mutant genotype AA was 3%. The genotype frequencies did not follow the genotype frequencies expected by Hardy–Weinberg equilibrium ($p < 0.05$).

Table 3. Genotype and allele frequencies of *TLR2* and *TLR4* genes in the study population according to developing or not developing sepsis.

Variables	*TLR2* (rs5743708)				*TLR4* (rs4986790)			
	All	Non-Septic	Septic	*p*-Value	All	Non-Septic	Septic	*p*-Value
				Genotype frequencies				
A/A	2 (3)	1 (2)	1 (4)	0.22	68 (91)	45 (94)	23 (85)	0.20
G/A	4 (5)	1 (2)	3 (11)		6 (8)	2 (4)	4 (15)	
G/G	69 (92)	46 (96)	23 (85)		1 (1)	1 (2)	0 (0)	
				Allele frequencies				
A	8 (5)	3 (3)	5 (9)	0.10	142 (95)	92 (96)	50 (93)	0.39
G	142 (95)	93 (97)	49 (91)		8 (5)	4 (4)	4 (7)	
P_{HWE}	0.009	0.032	0.180		0.180	0.063	1.00	

Data are shown as a number (percentage). Fisher's Exact tests were performed. Statistical analysis at p value < 0.05.

Genotype association models for the risk of sepsis were analyzed and a significant association was found between TLR2 Arg753Gln SNP and sepsis under the over dominant model ($p = 0.043$), but in the TLR4 polymorphism this difference did not reach statistical significance (Table 4).

Table 4. Genotype association models for sepsis risk assessment.

Model	Genotype	Non-Septic	Septic	Adjusted OR (95% CI) [a]	*p*-Value
TLR2					
Codominant [b]	G/G	46 (95.8%)	23 (85.2%)	Reference	
	A/G	1 (2.1%)	3 (11.1%)	11.42 (0.84–155.32)	0.12
	A/A	1 (2.1%)	1 (3.7%)	1.65 (0.09–29.49)	
Dominant	G/G	46 (95.8%)	23 (85.2%)	Reference	0.07
	A/G-A/A	2 (4.2%)	4 (14.8%)	5.34 (0.77–36.96)	
Recessive	G/G-A/G	47 (97.9%)	26 (96.3%)	Reference	0.79
	A/A	1 (2.1%)	1 (3.7%)	1.48 (0.09–25.46)	
Over-dominant	G/G-A/A	47 (97.9%)	24 (88.9%)	Reference	0.043
	A/G	1 (2.1%)	3 (11.1%)	11.27 (0.83–152.94)	
Log-additive	—	—	—	2.42 (0.61–9.56)	0.18
TLR4					

Table 4. *Cont.*

Model	Genotype	Non-Septic	Septic	Adjusted OR (95% CI) [a]	p-Value
Codominant [b]	A/A	45 (93.8%)	23 (85.2%)	Reference	0.11
	A/G	2 (4.2%)	4 (14.8%)	7.23 (0.77–67.86)	
	G/G	1 (2.1%)	0 (0%)	0.00 (0.00-NA)	
Dominant	A/A	45 (93.8%)	23 (85.2%)	Reference	0.16
	A/G-G/G	3 (6.2%)	4 (14.8%)	3.68 (0.57–23.57)	
Recessive	A/A-A/G	47 (97.9%)	27 (100%)	Reference	0.36
	G/G	1 (2.1%)	0 (0%)	0.00 (0.00-NA)	
Over-dominant	A/A-G/G	46 (95.8%)	23 (85.2%)	Reference	0.06
	A/G	2 (4.2%)	4 (14.8%)	7.49 (0.79–71.02)	
Log-additive	—	—	—	1.90 (0.45–8.04)	0.37

Values are shown as numbers (%). Chi-square (χ^2) or Fisher's exact tests were used. OR (95% CI), odds ratio, and confidence interval. [a] adjusted for confounding factors (age and sex). [b] represented both heterozygote and homozygote comparison models.

2.4. *TLR2* and *TLR4* Polymorphisms in Relation to Clinical and Laboratory Data

The association of single nucleotide polymorphisms (SNPs) with clinical and laboratory characteristics data is studied in Table 5. There was a statistically significant association between the *TLR4* polymorphism (rs4986790) and infection with *Acinetobacter baumannii* (p = 0.001) and infection with undetermined Gram (−) bacilli. Moreover, a statistically significant association was found between the *TLR4* polymorphism (rs4986790) and post-surgical patients' admission category referred to ICU (p = 0.033). In addition, the *TLR4* polymorphism (rs4986790) had a significant association with the selection of Azithromycin as an empirical antibiotic (p = 0.003), and Imipenem antibiotic (p = 0.024), while the *TLR2* polymorphism (rs5743708) had an association with the selection of Teicoplanin (p < 0.001) and with Ampicillin + Sulbactam (p = 0.022). The selected empirical antibiotic depended on patient status and the severity of infection.

Table 5. Analysis for the association of variants with clinical and laboratory characteristics.

Variables		*TLR2* (rs5743708)	*TLR4* (rs4986790)
		p-Value	p-Value
Demographic	Age, years	0.84	0.99
	Sex	0.27	0.71
Vital signs	HR, beats/min	0.27	0.47
	MAP, mm Hg	0.70	0.84
	SBP, mm Hg	0.70	0.86
	DBP, mm Hg	0.80	0.92
Concomitant diseases	Diabetes	0.80	0.08
	Hypertension	0.95	0.06
	Vascular dis	0.14	0.43
	Chronic lung disease	0.75	0.71
	Chronic liver disease	0.49	0.67
	Chronic renal disease	0.31	0.80

Table 5. *Cont.*

Variables		*TLR2* (rs5743708)	*TLR4* (rs4986790)
		p-Value	*p*-Value
ICU assessment	APACHE score	0.75	0.70
	Glasgow scale	0.89	0.24
	Length of stay	0.84	0.36
	Sepsis	0.22	0.20
	Septic shock	0.74	0.44
	Death	0.75	0.46
	Overall survival	0.52	0.06
Admission category (cause of admission)	Renal	0.91	0.90
	Cardiovascular	0.87	0.25
	Infection	0.07	0.38
	Neurology	0.30	0.77
	Post-surgical	0.70	0.033
	Respiratory	0.22	0.55
	Trauma	0.87	0.85
	Other causes	0.79	0.75
Biochemical data	WBC, $\times 10^3$ cells/µL	0.56	0.16
	HB, g%	0.08	0.31
	Creatinine, mg/dL	0.98	0.24
Causative organism	*Enterobacter* spp.	0.07	0.05
	Acinetobacter spp.	0.70	0.001
	Candida spp.	0.08	0.85
	E. coli	0.44	0.38
	Gram (−) bacilli	0.87	<0.001
	Klebsiella spp.	0.30	0.69
	Pseudomonas spp.	0.73	0.90
	Staph spp.	0.63	0.80
	Streptococcus spp.	0.83	0.80
	Aeromonas spp.	0.95	0.94
	Proteus spp.	0.95	0.94
	Citrobacter spp.	0.95	0.94
	Serratia spp.	0.95	0.94
Variables		*TLR2* (rs5743708)	*TLR4* (rs4986790)
		p-Value	*p*-Value
Type of culture	Blood	0.68	0.76
	Sputum	0.77	0.29
	Urine	0.47	0.41
	Pus	0.35	0.73
	CSF	0.95	0.94

Table 5. *Cont.*

Variables		*TLR2* (rs5743708)	*TLR4* (rs4986790)
		p-Value	*p*-Value
No of infections		0.48	0.94
Empirical antibiotic	No of antibiotics	0.05	0.77
	Cefoperazone	0.91	0.90
	Ceftazidime	0.57	0.90
	Levofloxacin	0.47	0.14
	Cefepime	0.16	0.23
	Ampicillin + sulbactam	0.022	0.70
	Imipenem	0.22	0.024
	Meropenem	0.95	0.94
	Ertapenem	0.75	0.69
	Azithromycin	0.95	0.003
	Rifampicin	0.95	0.94
	Teicoplanin	<0.001	0.94
	Cefotaxime	0.57	0.90
	Piperacillin	0.95	0.94

Chi-square (χ^2) or Fisher's exact tests were used for qualitative variables and student's *t*-test was used for quantitative attributes. Statistical analysis at p value < 0.05.

2.5. Multivariate Analysis in Relation to Developing Sepsis

A multivariate analysis was performed to determine which variable was independently associated with the risk of sepsis (Table 6). Only age was found to be independently associated with the risk of sepsis with a *p*-value of 0.009.

Table 6. Multivariate analysis for the risk of sepsis in ICU-admitted patients.

Risk Factors	OR	95% CI (Lower)	95% CI (Upper)	*p*-Value
Age	0.940	0.897	0.984	0.009
Sex (female)	0.473	0.116	1.925	0.30
HR, beats/min	1.003	0.974	1.033	0.83
MAP, mm Hg	0.898	0.654	1.233	0.51
SBP, mm Hg	1.022	0.915	1.142	0.70
DBP, mm Hg	1.123	0.901	1.400	0.30
WBC, $\times 10^3$ cells/μL	0.926	0.854	1.003	0.06
HB, g%	0.875	0.666	1.150	0.34
Creatinine, mg/dL	0.938	0.724	1.217	0.63
APACHE score	0.942	0.815	1.088	0.42
Glasgow scale	0.962	0.800	1.156	0.68
Length of stay	0.962	0.917	1.008	0.11
TLR2 (A/G)	0.082	0.001	6.474	0.26
TLR2 (G/G)	1.939	0.084	44.584	0.68
TLR4 (A/G)	0.090	0.005	1.785	0.11
TLR4 (G/G)	NA	NA	NA	1.00

OR: odds ratio; CI: confidence interval. Binary logistic regression analysis was performed.

2.6. Survival Analysis

Survival analysis was performed with the usage of Log-rank, Breslow, and Tarone–Ware tests which showed significance only with the length of stay (0.001, 0.001, and 0.001), respectively, and with the post-surgical category of admission with a log-rank test (0.03), as shown in Table 7.

Table 7. Survival analysis in ICU-admitted patients.

Variables		Overall Comparisons		
		Log Rank	**Breslow**	**Tarone–Ware**
Demographic data	Age	0.44	0.36	0.35
	Sex	0.23	0.50	0.36
Vital signs	HR	0.61	0.99	0.84
	MAP	0.86	0.69	0.75
	SBP	0.45	0.46	0.44
	DBP	0.63	0.64	0.64
Concomitant disease	Diabetes	0.87	0.62	0.85
	Hypertension	0.12	0.26	0.17
	Vascular disease	0.39	0.28	0.28
	Chronic liver disease	0.58	0.50	0.50
	Chronic renal disease	0.42	0.55	0.48
ICU assessment	APACHE score	0.81	0.84	0.76
	Glasgow scale	0.51	0.54	0.57
	Length of stay	<0.001	<0.001	<0.001
	Sepsis	0.91	0.82	0.78
	Septic shock	0.94	0.69	0.74
	No empirical drug	0.06	0.09	0.06
Admission category	Renal	0.53	0.54	0.54
	Cardiovascular	0.30	0.56	0.44
	Infection	0.79	0.86	0.82
	Neurology	0.33	0.29	0.31
	Post-surgical	0.030	0.09	0.06
	Respiratory	0.61	0.74	0.62
	Trauma	0.32	0.39	0.37
	Other causes	0.26	0.59	0.40
Lab data	WBC, $\times 10^3$ cells/μL	0.66	0.93	0.84
	HB, g%	0.51	0.69	0.61
	Creatinine, mg/dL	0.71	0.42	0.55
	No of infection	0.48	0.47	0.54
Molecular analysis	TLR2	0.38	0.25	0.27
	TLR4	0.63	0.39	0.43
	Combined	0.12	0.35	0.22

Survival time is shown as mean and standard error, HR: Hazard ratio, CI; confidence interval. Log-rank, Breslow, and Tarone–Ware tests were used to find Kaplan–Meier estimates for survival. Quantitative variables were categorized by their medians.

In addition, Cox regression analysis was applied to the data to determine if any of these variables were independently associated with the duration of survival (Table 8). Hazard risk for TLR2 was 1.89 and hazard risk for TLR4 was 2.25 but these results did not reach significance, so the effect of TLR gene status during time remained constant.

Table 8. Multivariate analysis for the risk of mortality in ICU-admitted patients.

Variables		HR	95% CI	*p*-Value
Demographic data	Age	1.90	(0.47–7.57)	0.36
	Sex	0.42	(0.09–1.93)	0.27
ICU assessment	APACHE score	1.41	(0.31–6.24)	0.65
	Glasgow scale	1.86	(0.38–8.89)	0.44
	Septic shock	0.55	(0.10–2.90)	0.48
	No empirical drug	0.32	(0.04–2.06)	0.62
	No of infection	1.76	(0.19–16.23)	0.23
Molecular analysis	TLR2	1.89	(0.08–43.58)	0.69
	TLR4	2.25	(0.48–10.43)	0.30

HR: hazard risk, CI: confidence interval. Cox Proportional Hazard Regression analysis was performed.

3. Discussion

The remarkable importance of TLR2 and TLR4 in our immune system and in modulating our response to infection suggested potential roles of their important variants, Arg753Gln and Asp299Gly, in increasing susceptibility to infection and sepsis as well.

Different bioinformatics approaches were utilized in our analysis. Investigating the impacts of our variants depended on five various tools with various approaches to achieve a high robustness and effectiveness. While rs4986790 was predicted to possess a benign impact on TLR4 by all tools, rs5743708 was predicted by all tools except SNPs and GO to possess a damaging impact on TLR2. Moreover, the SNPs' positions on the domains of their proteins were determined by InterPro, revealing the presence of rs5743708 on the important TIR domain. The TIR domain has a crucial role in the activation of TLR pathways [26]. Therefore, it is anticipated that this mutation could affect its protein function. In addition, since protein function and structure are critically dependent on its stability [27], the impacts of rs5743708 and rs4986790 on their proteins' stability were investigated revealing how proteins' stability was reduced by these SNPs. Furthermore, concerning the relationship between high scores of conservation and functionally significant residues [28], the conservation analysis was intended to anticipate those SNPs which could affect the significant functions. Rs5743708 of *TLR2* was found to be a functional residue with high conservation. On the contrary, rs4986790 of *TLR4* was found to be a variable residue. In addition, both rs5743708 and rs4986790 were anticipated to induce structural impacts on TLR2 and TLR4, respectively using the HOPE bioinformatics server.

In our prospective study, the genotype frequencies for *TLR4* were in accordance with Hardy–Weinberg equilibrium. On the contrary, the genotype frequencies for *TLR2* were not in accordance with the Hardy–Weinberg equilibrium, and this aberrant result was also found by Saleh et al. in the Egyptian population, in his study about Toll-like receptor-2 polymorphisms and the susceptibility to pulmonary and peritoneal tuberculosis [29] which may require further investigation. The different prevalence of these SNPs between different populations have been steadily observed by different researchers [13,17,18] with obvious differences between Asian, African, and European ethnicities for both SNPs. These different distribution patterns between populations were suspected to be responsible for different susceptibility patterns to infectious diseases and other serious diseases such as coronary artery disease and type 2 Diabetes as well [18,30].

In our study, there was a significant association between TLR2 Arg753Gln polymorphism and sepsis under the over-dominant model ($p = 0.043$), while the TLR4 polymorphism did not show such significance. Some other investigators reached the same results in some populations despite the observed conflict between studies. A meta-analysis study conducted by Gao and colleagues found an association in this study between Arg753Gln SNP and the risk of sepsis among critically ill adult patients in Europe. Meanwhile, this study also shed light on the issue of the conflicting results regarding the TLR2 polymorphism and developing sepsis [21]. The TLR4 polymorphism studies also showed conflicting results; a study conducted in France by Lorenz et al. found that the Asp299Gly and Thr399Ile polymorphisms of TLR4 might potentially be linked to Gram-negative septic shock [19]. On the contrary, some studies showed an absence of association between TLR4 SNP and sepsis; a study conducted by Kumpf et al. found no association between Asp299Gly and Thr399Ile polymorphisms of TLR4 and the incidence of sepsis syndrome or the type of organisms causing surgical infection in German adults [20]. In addition, another study by Shan Xo et al. in Wenzhou found that the Asp299Gly and Thr399Ile polymorphisms may not correlate with susceptibility to sepsis in Chinese Han children [31]. These conflicting results can be seen frequently among different ethnic groups in these types of genetic association studies investigating diseases that depend on several genetic factors [32], as sepsis is believed to be initiated and augmented by multiple genes and there is no full control over sepsis by a single gene [33,34]. Consequently, the various frequency of different SNPs in different ethnic groups, and the difference in the penetration and the effect of SNPs because of other factors such as gender or age variations in different studies could explain these conflicting results among different populations.

Developing infection with *Acinetobacter baumannii* was found to have a statistically significant association with the TLR4 polymorphism ($p = 0.001$). This finding is in agreement with a recent study conducted by Chatzi et al. who found that the Asp299Gly and Thr399Ile polymorphisms of TLR4 could play an essential role in developing multidrug resistance to *Acinetobacter baumannii* in CNS infections [35]. In addition, other researchers have confirmed the role of TLR4 in *Acinetobacter baumannii* infection in vitro and in vivo and found that the production of IL-8 by epithelial A549 cells in the human lung as a response to *Acinetobacter baumannii* required both TLR2 and TLR4 [36]. However, other studies showed that the recognition of *Acinetobacter baumannii* depends on TLR4 rather than TLR2, as TLR4 is the dominant receptor in this type of recognition. Knapp et al. found that TLR4-deficient mice, not TLR2-deficient mice (with intranasal inoculation of *Acinetobacter baumannii* Lipopolysaccharides) showed the impaired production of TNFa in bronchi alveolar lavage fluid and the impaired recruitment of polymorph nuclear cells, compared with Wild Type mice [37]. Moreover, Kim et al. found that the production of *Acinetobacter baumannii*-induced cytokines was impaired with TLR4-deficient bone marrow-derived macrophages or dendritic cells, while it was not the case with TLR2-deficient macrophages [38]. Besides, Erridge et al. found that the activation of human monocytes (resulting from phenol water re-extracted Lipopolysaccharides from *Acinetobacter baumannii*) was the responsibility of the TLR4 signaling pathway [39]. This association between the TLR4 polymorphism and this virulent bacterium could allow proper management and prevention measures where high rates of *Acinetobacter baumannii* infection are found. This could be an important step towards the individualization of host susceptibilities towards virulent microorganisms in intensive care units.

Our study also found a significant association between the *TLR4* polymorphism (rs4986790) and the post-surgical category among patients referred to ICU. This role of the TLR4 polymorphism in post-surgery was investigated by a clinical study conducted by Koch and colleagues who found that the presence of a *TLR4* polymorphism influenced the immune–endocrine stress response which resulted from the systemic inflammation caused by major surgery. They found decreased serum concentrations of ACTH, IL-8, IL-10, and GM-CSF postoperatively in those surgical patients who carried that polymorphism [40]. This might explain this significant association found in our study.

The multivariate analysis was also performed to analyze the effects of our variables on the development of sepsis syndrome, but it was only the age factor that was found to have an independent association with the risk of sepsis in our study group. The age factor is a well-identified risk factor for developing this syndrome [41].

Survival analysis found that the length of stay and the surgical category of admission had a significant association with time of survival in intensive care units. Our results are in agreement with several studies that found an association between prolonged ICU stay and higher hospital mortality as well. Those patients, with an ICU length of stay of 14 days or longer, were found to have a mortality rate of more than 50% [42,43].

Overall, our study is characterized by the usage of both experimental and in silico methods. Our investigation showed promising results regarding the analysis of the role of TLRs variants in infection and sepsis. However, our study had its limitations as in most genetic polymorphism studies, as the number of patients carrying the variant alleles was relatively small due to the small number of these polymorphisms in the general population. As a result, there is a need for multi-center studies conducted on a larger scale to validate these findings.

4. Materials and Methods

4.1. Ethics Statement

The study protocol was approved by Scientific Research Ethics Commission at Suez Canal University (reference No. 201709MH1). All subjects or their next of kin gave informed consent before inclusion in the study.

4.2. In Silico Analysis

4.2.1. General Information

National Center for Biotechnology Information (NCBI) and Ensembl databases were used to retrieve general information about *TLR2* and *TLR4* genes. Subcellular localization was retrieved from compartments.jensenlab.org mainly and genecards.org. Gene coexpression and predicted protein–protein interactions were obtained from the String Biological database. General information about rs5743708 and rs4986790 were brought from the dbSNP and Ensembl databases. (https://web.expasy.org (accessed on 29 August 2021)) was used for retrieving data about the variants' effect on sequences of our proteins with these data gained from UniProtKB/Swiss-Prot databases.

4.2.2. Predicting the Effect of SNPs on Protein Function

Five bioinformatics tools were used to predict the effect of SNPs on protein function to increase the strength and accuracy of results; 1-SIFT (Sorting Intolerant from Tolerant) (https://sift.bii.a-star.edu.sg/ (accessed on 30 August 2021)). SIFT depends on sequence homology in addition to the physical properties of amino acids to predict the effect of missense mutations on protein function [44]. 2-PolyPhen-2 (Polymorphism Phenotyping v2) (http://genetics.bwh.harvard.edu/pph2 (accessed on 30 August 2021)). PolyPhen-2 uses comparative and physical approaches to predict the effect of amino acid substitution [45]. 3-PANTHER (Protein Analysis Trough Evolutionary Relationship) (http://www.pantherdb.org/tools/csnpScoreForm.jsp (accessed on 30 August 2021)). This method depends on calculating the evolutionary preservation of an amino acid to predict the likelihood that a nonsynonymous SNP could cause a functional impact on the protein [46]. 4- PROVEAN (Protein Variation Effect Analyzer) (http://provean.jcvi.org/seq_submit.php (accessed on 30 August 2021)). PROVEAN uses blast hits to calculate the delta alignment score and computes the PROVEAN score finally with a cutoff at −2.5 [47]. 5-SNPs and GO (https://snps.biofold.org/snps-and-go/snps-and-go.html (accessed on 30 August 2021)). SNPs and GO depend on protein functional annotation to predict the impact of variations [48].

4.2.3. The Identification of SNP Location on Protein Domains

The locations of SNPs on conserved domains on TLR2 and TLR4 proteins were identified using the InterPro bioinformatics tool (https://www.ebi.ac.uk/interpro/ (accessed on 30 August 2021)), a bioinformatics tool that could perform functional analysis of protein and identify domains and functional sites [49].

4.2.4. The Prediction of Protein Stability with SNPs

We used I-Mutant 2.0 (https://folding.biofold.org/i-mutant/i-mutant2.0.html (accessed on 30 August 2021)) to predict the stability of the TLR2 and TLR4 proteins with rs5743708 and rs4986790 SNPs, respectively [50]. I-Mutant 2.0 is considered a support vector machine that was tested depending on the ProTherm database which contained the largest experimental data about stability changes with protein mutations [51].

4.2.5. The Identification of Evolutionarily Conserved Positions in a Protein Sequence

This identification was performed using the ConSurf server (https://consurf.tau.ac.il (accessed on 30 August 2021)) which depends on phylogenetic relations between homologous sequences to identify the evolutionary conservation of amino acids in protein sequences [28,52].

4.2.6. The Identification of Structural Effects of SNPs

Structural effects of rs5743708 and rs4986790 SNPs on TLR2 and TLR4, respectively, were analyzed using HOPE (https://www3.cmbi.umcn.nl/hope/ (accessed on 30 August 2021)) which is a mutant analysis server that could analyze the effects of SNPs on protein structure [53].

4.3. The Study Design

This was a prospective observational study that was conducted in intensive care units in Suez Canal University Hospitals, Ismailia, Egypt, for seven months. All ICU Patients who contracted infections with a positive culture or a chest X-ray were included in the study group. All included patients were Egyptian adults of both sexes. Exclusion criteria were patients younger than 18 years old, pregnancy, immune suppression, and patients with radiation therapy or chemotherapy.

Once admitted, general examination and clinical status were assessed for patients; both Acute Physiology and Chronic Health Evaluation (APACHE II) scores and sequential organ failure assessment (SOFA) scores were measured. In addition to vital signs check (blood pressure, heart rate, respiratory rate, central venous pressure, and temperature) and laboratory analyses such as complete blood count, blood sugar, CRP, blood urea nitrogen, serum calcium, potassium, sodium, aspartate aminotransferase, alanine aminotransferase, and arterial blood gas analysis were carried out.

The patients were further followed up to assess infection, sepsis, and septic shock. Routine cultures of sputum, blood, urine, and pus were collected to determine the presence of infection and identify the causing organism. Assessment of sepsis and septic shock was performed by daily evaluation for sepsis or septic shock. Sepsis and septic shock were defined and diagnosed according to "The Third International Consensus Definitions for Sepsis and Septic Shock (Sepsis-3)" [4].

4.4. Samples Collection

Two milliliters of venous blood sample were collected into EDTA tubes from all admitted patients in the study group under complete aseptic conditions and stored at −80 °C until processed for DNA extraction.

4.5. Genotyping

Genomic DNA was extracted from venous blood with a QIAamp DNA Blood Mini kit (Cat. No. 51104; QIAGEN, Hilden, Germany) according to the manufacturer's protocol. The measurement of both the concentration and purity of the extracted DNA was performed by NanoDrop ND-1000 (NanoDrop Tech., Inc., Wilmington, DE, USA).

Genotyping for the *TLR4* gene polymorphism (Asp299Gly; rs4986790) and *TLR2* gene polymorphism (Arg753Gln; rs5743708) was performed using real-time polymerase chain reaction technology using TaqMan allelic discrimination assay. The required reagents for the TaqMan assay including TaqMan genotyping assay and TaqMan genotyping master mix were brought from Applied Biosystems (Foster City, CA, USA). The assay ID for rs5743708 is C_27860663_10 and for rs4986790 is C_11722238_20. PCR was run with a total reaction volume of 25 μL reaction volume. The components of PCR reaction were 12.5 μL TaqMan genotyping master mix; No AmpErase UNG (2×), genomic DNA (20 ng) diluted to 11.25 μL with DNase-RNase free water, and 1.25 μL TaqMan SNP genotyping assay mix (Cat. No. 4351379, Applied Biosystems, Foster City, USA). Nuclease-free water was used as a negative control.

The PCR amplification was carried out in a StepOne™ real-time PCR system (Applied Biosystems, Foster City, CA, USA) according to the following conditions: a hold cycle (95 °C for 10 min) followed by a 40-cycle PCR consisting of 95 °C for 15 s and 60 °C for one minute. SDS software version 1.3.1 (Applied Biosystems) was used for allelic discrimination. Genotyping was performed with blindness to sepsis/non-sepsis status.

4.6. Statistical Analysis

Statistical analysis was carried out using Microsoft® Excel 2010 and the "Statistical Package for the Social Sciences (SPSS) for windows" software, version 24. Odds ratios (OR) with a 95% confidence interval (CI) were calculated. Descriptive statistics were expressed as percentages for qualitative variables and mean ± standard deviation (SD) for quantitative variables. Testing differences between septic patients and no septic patients were performed using Student's *t*-test, Chi-square (χ^2) test, or Fisher's exact tests. *p*-value was considered statistically significant below 0.05. The Hardy–Weinberg equilibrium (HWE) was calculated by the Online Encyclopedia for Genetic Epidemiology (OEGE) software (http://www.oege.org/software/hwe-mrcalc.shtml (accessed on 10 March 2019)). The relationship between the risk factors including our polymorphisms and the development of sepsis was further determined using logistic regression after adjustment of factors. Survival analysis was performed as well. Log-rank, Breslow, and Tarone–Ware tests were used to find Kaplan–Meier estimates for survival. Cox regression analysis was applied to the data to determine if any of the variables were independently associated with the duration of survival.

5. Conclusions

Rs5743708 was predicted by nearly all used bioinformatics tools to possess a damaging impact on TLR2 which was not the case with rs4986790 of TLR4. Meanwhile, the conducted pilot study concluded that the *TLR2* genotype may be a risk factor for sepsis in adult patients, Moreover, our study showed that Asp299Gly polymorphism in TLR4 may be associated with an increased risk of *Acinetobacter baumannii* infection. In addition, a significant association was found between the *TLR4* polymorphism and the post-surgical category of patients admitted to intensive care units. Identification of the role of *TLR2* and *TLR4* polymorphisms in developing infection and sepsis could allow early prediction, prevention, and management of these serious diseases.

Author Contributions: M.Y.B. (Investigation, Methodology, Conceptualization, Formal Analysis, Writing–Original Draft Preparation); M.M.A. (Supervision, Conceptualization, Formal Analysis, Methodology, Writing–Original Draft Preparation); E.E.-D.A.I. (Supervision, Methodology, Writing–Review and Editing); E.A.T. (Supervision, Methodology, Formal Analysis, Writing–Review and Editing); A.A.A. (Supervision, Methodology, Conceptualization, Writing–Original Draft Preparation); H.R.H. and A.A.S. (Supervision, Formal Analysis, Writing–Review and Editing). All authors have read and agreed to the published version of the manuscript.

Funding: This research received no external funding.

Institutional Review Board Statement: The study protocol was approved by Scientific Research Ethics Commission at Suez Canal University (reference No. 201709MH1).

Informed Consent Statement: All subjects or their next of kin gave informed consent prior to inclusion in the study.

Data Availability Statement: All supporting data of the study are available from the corresponding authors upon request.

Conflicts of Interest: The authors declare no conflict of interest.

References

1. Black, R.E.; Morris, S.S.; Bryce, J. Where and why are 10 million children dying every year? *Lancet* **2003**, *361*, 2226–2234. [CrossRef]
2. Vincent, J.-L.; Rello, J.; Marshall, J.; Silva, E.; Anzueto, A.; Martin, C.D.; Moreno, R.; Lipman, J.; Gomersall, C.; Sakr, Y.; et al. International study of the prevalence and outcomes of infection in intensive care units. *JAMA* **2009**, *302*, 2323–2329. [CrossRef] [PubMed]
3. Nachtigall, I.; Tamarkin, A.; Tafelski, S.; Weimann, A.; Rothbart, A.; Heim, S.; Wernecke, K.D.; Spies, C. Polymorphisms of the toll-like receptor 2 and 4 genes are associated with faster progression and a more severe course of sepsis in critically ill patients. *J. Int. Med. Res.* **2014**, *42*, 93–110. [CrossRef] [PubMed]
4. Singer, M.; Deutschman, C.S.; Seymour, C.W.; Shankar-Hari, M.; Annane, D.; Bauer, M.; Bellomo, R.; Bernard, G.R.; Chiche, J.-D.; Coopersmith, C.M.; et al. The Third International Consensus Definitions for Sepsis and Septic Shock (Sepsis-3). *JAMA* **2016**, *315*, 801–810. [CrossRef]
5. Angus, D.C.; Linde-Zwirble, W.T.; Lidicker, J.; Clermont, G.; Carcillo, J.; Pinsky, M.R. Epidemiology of severe sepsis in the United States: Analysis of incidence, outcome, and associated costs of care. *Crit. Care Med.* **2001**, *29*, 1303–1310. [CrossRef]
6. Frodsham, A.J.; Hill, A.V.S. Genetics of infectious diseases. *Hum. Mol. Genet.* **2004**, *13*, R187–R194. [CrossRef]
7. Feng, B.; Mao, Z.; Pang, K.; Zhang, S.; Li, L. Association of tumor necrosis factor α -308G/A and interleukin-6 -174G/C gene polymorphism with pneumonia-induced sepsis. *J. Crit. Care.* **2015**, *30*, 920–923. [CrossRef]
8. Yomade, O.; Spies-Weisshart, B.; Glaser, A.; Schnetzke, U.; Hochhaus, A.; Scholl, S. Impact of NOD2 polymorphisms on infectious complications following chemotherapy in patients with acute myeloid leukaemia. *Ann. Hematol.* **2013**, *92*, 1071–1077. [CrossRef]
9. Schenten, D.; Medzhitov, R. The control of adaptive immune responses by the innate immune system. *Adv. Immunol.* **2011**, *109*, 87–124.
10. Noreen, M.; Arshad, M. Association of TLR1, TLR2, TLR4, TLR6, and TIRAP polymorphisms with disease susceptibility. *Immunol. Res.* **2015**, *62*, 234–252. [CrossRef]
11. Sameer, A.S.; Nissar, S. Toll-Like Receptors (TLRs): Structure, Functions, Signaling, and Role of Their Polymorphisms in Colorectal Cancer Susceptibility. *Biomed. Res. Int.* **2021**, *2021*, 1157023. [CrossRef]
12. Mukherjee, S.; Karmakar, S.; Babu, S.P.S. TLR2 and TLR4 mediated host immune responses in major infectious diseases: A review. *Braz. J. Infect. Dis.* **2016**, *20*, 193–204. [CrossRef]
13. Mukherjee, S.; Huda, S.; Sinha Babu, S.P. Toll-like receptor polymorphism in host immune response to infectious diseases: A review. *Scand. J. Immunol.* **2019**, *90*, e12771. [CrossRef]
14. Zhu, G.; Li, C.; Cao, Z.; Corbet, E.F.; Jin, L. Toll-like receptors 2 and 4 gene polymorphisms in a Chinese population with periodontitis. *Quintessence Int.* **2008**, *39*, 217–226.
15. Ohto, U.; Yamakawa, N.; Akashi-Takamura, S.; Miyake, K.; Shimizu, T. Structural analyses of human Toll-like receptor 4 polymorphisms D299G and T399I. *J. Biol. Chem.* **2012**, *287*, 40611–40617. [CrossRef]
16. Xiong, Y.; Song, C.; Snyder, G.A.; Sundberg, E.J.; Medvedev, A.E. R753Q polymorphism inhibits Toll-like receptor (TLR) 2 tyrosine phosphorylation, dimerization with TLR6, and recruitment of myeloid differentiation primary response protein 88. *J. Biol. Chem.* **2012**, *287*, 38327–38337. [CrossRef]
17. Ferwerda, B.; McCall, M.B.B.; Alonso, S.; Giamarellos-Bourboulis, E.J.; Mouktaroudi, M.; Izagirre, N.; Syafruddin, D.; Kibiki, G.; Cristea, T.; Hijmans, A.; et al. TLR4 polymorphisms, infectious diseases, and evolutionary pressure during migration of modern humans. *Proc. Natl. Acad. Sci. USA* **2007**, *104*, 16645–16650. [CrossRef]
18. Ioana, M.; Ferwerda, B.; Plantinga, T.S.; Stappers, M.; Oosting, M.; McCall, M.; Cimpoeru, A.; Burada, F.; Panduru, N.; Sauerwein, R.; et al. Different patterns of Toll-like receptor 2 polymorphisms in populations of various ethnic and geographic origins. *Infect. Immun.* **2012**, *80*, 1917–1922. [CrossRef]
19. Lorenz, E.; Mira, J.P.; Frees, K.L.; Schwartz, D.A. Relevance of mutations in the TLR4 receptor in patients with gram-negative septic shock. *Arch. Intern. Med.* **2002**, *162*, 1028–1032. [CrossRef]
20. Kumpf, O.; Giamarellos-Bourboulis, E.J.; Koch, A.; Hamann, L.; Mouktaroudi, M.; Oh, D.-Y.; Latz, E.; Lorenz, E.; A Schwartz, D.; Ferwerda, B.; et al. Influence of genetic variations in TLR4 and TIRAP/Mal on the course of sepsis and pneumonia and cytokine release: An observational study in three cohorts. *Crit. Care* **2010**, *14*, R103. [CrossRef]
21. Gao, J.; Zhang, A.; Wang, X.; Li, Z.; Yang, J.; Zeng, L.; Gu, W.; Jiang, J.-X. Association between the TLR2 Arg753Gln polymorphism and the risk of sepsis: A meta-analysis. *Crit. Care* **2015**, *19*, 416. [CrossRef]

22. Hossain, M.S.; Roy, A.S.; Islam, M.S. In silico analysis predicting effects of deleterious SNPs of human RASSF5 gene on its structure and functions. *Sci. Rep.* **2020**, *10*, 14542. [CrossRef] [PubMed]
23. Behairy, M.Y.; Abdelrahman, A.l.A.; Abdallah, H.Y.; Ibrahim, E.E.-D.A.; Sayed, A.A.; Azab, M.M. In silico analysis of missense variants of the *C1qA* gene related to infection and autoimmune diseases. *J. Taibah. Univ. Med. Sci.* **2022**. Available online: https://www.sciencedirect.com/science/article/pii/S1658361222000890 (accessed on 4 August 2022).
24. Behairy, M.Y.; Soltan, M.A.; Adam, M.S.; Refaat, A.M.; Ezz, E.M.; Albogami, S.; Fayad, E.; Althobaiti, F.; Gouda, A.M.; Sileem, A.E.; et al. Computational Analysis of Deleterious SNPs in NRAS to Assess Their Potential Correlation with Carcinogenesis. *Front. Genet.* **2022**, *13*, 872845. Available online: https://www.frontiersin.org/articles/10.3389/fgene.2022.872845 (accessed on 19 August 2022).
25. Behairy, M.Y.; Abdelrahman, A.A.; Abdallah, H.Y.; Ibrahim, E.E.-D.A.; Hashem, H.R.; Sayed, A.A.; Azab, M.M. Role of MBL2 Polymorphisms in Sepsis and Survival: A Pilot Study and In Silico Analysis. *Diagnostics* **2022**, *12*, 460. [CrossRef] [PubMed]
26. Takeda, K.; Akira, S. TLR signaling pathways. *Semin. Immunol.* **2004**, *16*, 3–9. [CrossRef] [PubMed]
27. Deller, M.C.; Kong, L.; Rupp, B. Protein stability: A crystallographer's perspective. *Acta Cryst. Sect. F Struct. Biol. Commun.* **2016**, *72 Pt 2*, 72–95. [CrossRef]
28. Berezin, C.; Glaser, F.; Rosenberg, J.; Paz, I.; Pupko, T.; Fariselli, P.; Casadio, R.; Ben-Tal, N. ConSeq: The identification of functionally and structurally important residues in protein sequences. *Bioinformatics* **2004**, *20*, 1322–1324. [CrossRef]
29. Saleh, M.A.; Ramadan, M.M.; Arram, E.O. Toll-like receptor-2 Arg753Gln and Arg677Trp polymorphisms and susceptibility to pulmonary and peritoneal tuberculosis. *APMIS* **2017**, *125*, 558–564. [CrossRef]
30. Liu, F.; Lu, W.; Qian, Q.; Qi, W.; Hu, J.; Feng, B. Frequency of TLR 2, 4, and 9 gene polymorphisms in Chinese population and their susceptibility to type 2 diabetes and coronary artery disease. *J. Biomed. Biotechnol.* **2012**, *2012*, 373945. [CrossRef]
31. Shan, X.; Wu, Y.; Ye, J.; Ding, Z.; Qian, C.; Zhou, A. Gene polymorphisms of Toll-like receptors in Chinese Han children with sepsis in Wenzhou. *Zhonghua Er Ke Za Zhi = Chin. J. Pediatr.* **2010**, *48*, 15–18.
32. Fawzy, M.S.; Hussein, M.H.; Abdelaziz, E.Z.; Yamany, H.A.; Ismail, H.M.; Toraih, E.A. Association of MicroRNA-196a2 Variant with Response to Short-Acting β2-Agonist in COPD: An Egyptian Pilot Study. *PLoS ONE* **2016**, *11*, e0152834. [CrossRef]
33. Tumangger, H.; Jamil, K.F. Contribution of genes polymorphism to susceptibility and outcome of sepsis. *Egypt. J. Med. Hum. Genet.* **2010**, *11*, 97–103. [CrossRef]
34. David, V.L.; Ercisli, M.F.; Rogobete, A.F.; Boia, E.S.; Horhat, R.; Nitu, R.; Diaconu, M.M.; Pirtea, L.; Ciuca, I.; Horhat, D.I.; et al. Early Prediction of Sepsis Incidence in Critically Ill Patients Using Specific Genetic Polymorphisms. *Biochem. Genet.* **2017**, *55*, 193–203. [CrossRef]
35. Chatzi, M.; Papanikolaou, J.; Makris, D.; Papathanasiou, I.; Tsezou, A.; Karvouniaris, M.; Zakynthinos, E. Toll-like receptor 2, 4 and 9 polymorphisms and their association with ICU-acquired infections in Central Greece. *J. Crit. Care.* **2018**, *47*, 1–8. [CrossRef]
36. March, C.; Regueiro, V.; Llobet, E.; Moranta, D.; Morey, P.; Garmendia, J.; Bengoechea, J.A. Dissection of host cell signal transduction during Acinetobacter baumannii-triggered inflammatory response. *PLoS ONE* **2010**, *5*, e10033. [CrossRef]
37. Knapp, S.; Wieland, C.W.; Florquin, S.; Pantophlet, R.; Dijkshoorn, L.; Tshimbalanga, N.; Akira, S.; van der Poll, T. Differential roles of CD14 and toll-like receptors 4 and 2 in murine Acinetobacter pneumonia. *Am. J. Respir. Crit. Care Med.* **2006**, *173*, 122–129. [CrossRef]
38. Kim, C.-H.; Jeong, Y.-J.; Lee, J.; Jeon, S.-J.; Park, S.-R.; Kang, M.-J.; Park, J.-H.; Park, J.-H. Essential role of toll-like receptor 4 in Acinetobacter baumannii-induced immune responses in immune cells. *Microb. Pathog.* **2013**, *54*, 20–25. [CrossRef]
39. Erridge, C.; Moncayo-Nieto, O.L.; Morgan, R.; Young, M.; Poxton, I.R. Acinetobacter baumannii lipopolysaccharides are potent stimulators of human monocyte activation via Toll-like receptor 4 signalling. *J. Med. Microbiol.* **2007**, *56 Pt 2*, 165–171. [CrossRef]
40. Koch, A.; Hamann, L.; Schott, M.; Boehm, O.; Grotemeyer, D.; Kurt, M.; Schwenke, C.; Schumann, R.R.; Bornstein, S.R.; Zacharowski, K. Genetic variation of TLR4 influences immunoendocrine stress response: An observational study in cardiac surgical patients. *Crit. Care.* **2011**, *15*, R109. [CrossRef]
41. Mayr, F.B.; Yende, S.; Angus, D.C. Epidemiology of severe sepsis. *Virulence* **2014**, *5*, 4–11. [CrossRef]
42. Wong, D.T.; Gomez, M.; McGuire, G.P.; Kavanagh, B. Utilization of intensive care unit days in a Canadian medical-surgical intensive care unit. *Crit. Care Med.* **1999**, *27*, 1319–1324. [CrossRef]
43. Abelha, F.J.; Castro, M.A.; Landeiro, N.M.; Neves, A.M.; Santos, C.C. Mortality and length of stay in a surgical intensive care unit. *Rev. Bras. Anestesiol.* **2006**, *56*, 34–45. [CrossRef] [PubMed]
44. Sim, N.L.; Kumar, P.; Hu, J.; Henikoff, S.; Schneider, G.; Ng, P.C. SIFT web server: Predicting effects of amino acid substitutions on proteins. *Nucleic Acids Res.* **2012**, *40*, W452–W457. [CrossRef] [PubMed]
45. Adzhubei, I.A.; Schmidt, S.; Peshkin, L.; Ramensky, V.E.; Gerasimova, A.; Bork, P.; Kondrashov, A.S.; Sunyaev, S.R. A method and server for predicting damaging missense mutations. *Nat. Methods* **2010**, *7*, 248–249. [CrossRef] [PubMed]
46. Tang, H.; Thomas, P.D. PANTHER-PSEP: Predicting disease-causing genetic variants using position-specific evolutionary preservation. *Bioinformatics* **2016**, *32*, 2230–2232. [CrossRef]
47. Choi, Y.; Chan, A.P. PROVEAN web server: A tool to predict the functional effect of amino acid substitutions and indels. *Bioinformatics* **2015**, *31*, 2745–2747. [CrossRef]
48. Capriotti, E.; Calabrese, R.; Fariselli, P.; Martelli, P.L.; Altman, R.B.; Casadio, R. WS-SNPs&GO: A web server for predicting the deleterious effect of human protein variants using functional annotation. *BMC Genom.* **2013**, *14* (Suppl. S3), S6.

49. Blum, M.; Chang, H.Y.; Chuguransky, S.; Grego, T.; Kandasaamy, S.; Mitchell, A.; Nuka, G.; Paysan-Lafosse, T.; Qureshi, M.; Raj, S.; et al. The InterPro protein families and domains database: 20 years on. *Nucleic Acids Res.* **2021**, *49*, D344–D354. [CrossRef]
50. Capriotti, E.; Fariselli, P.; Casadio, R. I-Mutant2.0: Predicting stability changes upon mutation from the protein sequence or structure. *Nucleic Acids Res.* **2005**, *33*, W306–W310. [CrossRef]
51. Bava, K.A.; Gromiha, M.M.; Uedaira, H.; Kitajima, K.; Sarai, A. ProTherm, version 4.0: Thermodynamic database for proteins and mutants. *Nucleic Acids Res.* **2004**, *32*, D120–D121. [CrossRef]
52. Ashkenazy, H.; Abadi, S.; Martz, E.; Chay, O.; Mayrose, I.; Pupko, T.; Ben-Tal, N. ConSurf 2016: An improved methodology to estimate and visualize evolutionary conservation in macromolecules. *Nucleic Acids Res.* **2016**, *44*, W344–W350. [CrossRef]
53. Venselaar, H.; Te Beek, T.A.; Kuipers, R.K.; Hekkelman, M.L.; Vriend, G. Protein structure analysis of mutations causing inheritable diseases. An e-Science approach with life scientist friendly interfaces. *BMC Bioinform.* **2010**, *11*, 548. [CrossRef]

International Journal of *Molecular Sciences*

MDPI

Article

RAGE–TLR4 Crosstalk Is the Key Mechanism by Which High Glucose Enhances the Lipopolysaccharide-Induced Inflammatory Response in Primary Bovine Alveolar Macrophages

Longfei Yan [1,†], Yanran Li [1,†], Tianyu Tan [1], Jiancheng Qi [1], Jing Fang [1], Hongrui Guo [1], Zhihua Ren [1], Liping Gou [1], Yi Geng [1], Hengmin Cui [1], Liuhong Shen [1], Shumin Yu [1], Zhisheng Wang [2] and Zhicai Zuo [1,*]

1 College of Veterinary Medicine, Sichuan Agricultural University, Chengdu 611134, China
2 Institute of Animal Nutrition, Sichuan Agricultural University, Chengdu 611134, China
* Correspondence: 11966@sicau.edu.cn
† These authors contributed equally to this work.

Abstract: The receptor of advanced glycation end products (RAGE) and Toll-like receptor 4 (TLR4) are important receptors for inflammatory responses induced by high glucose (HG) and lipopolysaccharide (LPS) and show crosstalk phenomena in inflammatory responses. However, it is unknown whether RAGE and TLR4 can influence each other's expression through a crosstalk mechanism and whether the RAGE–TLR4 crosstalk related to the molecular mechanism of HG enhances the LPS-induced inflammatory response. In this study, the implications of LPS with multiple concentrations (0, 1, 5, and 10 μg/mL) at various treatment times (0, 3, 6, 12, and 24 h) in primary bovine alveolar macrophages (BAMs) were explored. The results showed that a 5 μg/mL LPS treatment at 12 h had the most significant increment on the pro-inflammatory cytokine interleukin 1β (IL-1β), IL-6, and tumor necrosis factor (TNF)-α levels in BAMs ($p < 0.05$) and that the levels of TLR4, RAGE, MyD88, and NF-κB p65 mRNA and protein expression were upregulated ($p < 0.05$). Then, the effect of LPS (5 μg/mL) and HG (25.5 mM) co-treatment in BAMs was explored. The results further showed that HG significantly enhanced the release of IL-1β, IL-6, and TNF-α caused by LPS in the supernatant ($p < 0.01$) and significantly increased the levels of RAGE, TLR4, MyD88, and NF-κB p65 mRNA and protein expression ($p < 0.01$). Pretreatment with FPS-ZM1 and TAK-242, the inhibitors of RAGE and TLR4, significantly alleviated the HG + LPS-induced increment of RAGE, TLR4, MyD88, and NF-κB p65 mRNA and protein expression in the presence of HG and LPS ($p < 0.01$). This study showed that RAGE and TLR4 affect each other's expression through crosstalk during the combined usage of HG and LPS and synergistically activate the MyD88/NF-κB signaling pathway to promote the release of pro-inflammatory cytokines in BAMs.

Keywords: TLR4–RAGE crosstalk; glucose; lipopolysaccharide (LPS); inflammatory; alveolar macrophages

Citation: Yan, L.; Li, Y.; Tan, T.; Qi, J.; Fang, J.; Guo, H.; Ren, Z.; Gou, L.; Geng, Y.; Cui, H.; et al. RAGE–TLR4 Crosstalk Is the Key Mechanism by Which High Glucose Enhances the Lipopolysaccharide-Induced Inflammatory Response in Primary Bovine Alveolar Macrophages. *Int. J. Mol. Sci.* **2023**, *24*, 7007. https://doi.org/10.3390/ijms24087007

Academic Editors: Oliver Planz and Ralf Kircheis

Received: 28 February 2023
Revised: 4 April 2023
Accepted: 6 April 2023
Published: 10 April 2023

1. Introduction

Long-distance transportation is an essential part of the beef cattle industry, but transportation stress caused by long-distance transportation can weaken cattle's immune systems causing the occurrence of bovine respiratory disease complex (BRDC) [1]. During the occurrence of BRDC, gram-negative bacterial infections are a major cause of pneumonia [2,3]. Lipopolysaccharide (LPS) and hyperglycemia were considered the most common contributors to the pulmonary inflammatory response [4]. LPS, the primary element of gram-negative bacteria's outer membrane, can activate alveolar macrophages (AMs) and promote the release of pro-inflammatory cytokines such as IL-1β, IL-6, and TNF-α [5,6]. Furthermore, transportation stress elevates blood glucose in cattle, leading to hyperglycemia [7,8]. Hyperglycemia is not only related to the release of pro-inflammatory cytokines in AMs, such as IL-1β, IL-6, and TNF-α [9–11], but is also associated with the production of a cytokine storm [12]. Several studies have reported that HG and LPS promote AM activation and that HG can exacerbate LPS-induced inflammatory response [4,13,14], but the molecular mechanism needs to be further investigated.

TLR4 is an intrinsic immune receptor that is closely associated with inflammatory responses and cytokine storm formation [15,16]. It is the receptor for LPS in triggering myeloid differentiation factor 88 (MyD88) and nuclear factor kappa-B (NF-κB) inflammatory signaling pathways and promoting the production and release of pro-inflammatory cytokines IL-1β, IL-6, and TNF-α [17–20]. Dasu et al. [21] found that HG-induced upregulation of Toll-like receptor 2 (TLR2) and TLR4 expression in human monocytes further promoted the activation of NF-κB. Mudaliar et al. [22] reported that inhibition of TLR2 or TLR4 significantly inhibited HG-induced NF-κB activation and the synthesis and release of pro-inflammatory cytokines and chemokines in human microvascular endothelial cells. These studies suggested that TLR4 is also closely associated with the HG-induced inflammatory response.

RAGE, another important receptor involved in the inflammatory response, can bind with various ligands, including advanced glycosylation end products (AGEs) [23], high mobility group box 1 (HMGB1) [24], and the S100A8/A9 heterodimer [25]. The ligands mentioned above are increased in the HG-induced inflammatory response [26,27]. HG induces significant upregulation of RAGE and its ligands, which then bind to RAGE to enhance RAGE oligomerization and further activate the cellular inflammatory signaling pathways [27]. Wang et al. [28] found that the use of RAGE-blocking antibodies effectively inhibited LPS-induced NF-κB activation in endothelial cells, suggesting that RAGE plays an important role in the LPS-induced inflammatory response.

Previous studies found that RAGE and TLRs have a similar mechanism of action and similar ligands and inflammatory signaling pathways are shared by TLR4 and RAGE, which have a synergistic activation effect on downstream signaling pathways when both receptors are activated. The above phenomenon is called "cross-talk" or crosstalk [29,30]. Specifically, upon activation of RAGE and TLR4, downstream signaling pathways of the two receptors converge and activate synergistically at multiple levels. The use of TLR4-related signaling pathways to transduce activation signals from RAGE can activate a wider range of signaling pathways as well as cause more intense signaling after RAGE binds to ligands [29,31]. Both HG and LPS can activate RAGE and TLR4 and share downstream inflammatory signaling pathways. However, it is unclear whether RAGE–TLR4 crosstalk is involved in the combination treatment of HG and LPS-induced inflammation response in BAMs. In addition, whether RAGE interacts with TLR4 to influence each other's expression at the receptor level and synergistic activation of the MyD88/NF-κB pathway in the inflammatory response of BAMs needs to be further investigated.

In this experiment, we aimed to investigate the role of RAGE–TLR4 crosstalk in the activation of the MyD88/NF-κB signaling pathway and release of pro-inflammatory cytokines TNF-α, IL-1β, and IL-6 in BAMs induced by the combination of HG and LPS. We used FPS-ZM1 (a RAGE inhibitor) and TAK-242 (a TLR4 inhibitor) to confirm the involvement of the RAGE–TLR4 crosstalk in this inflammatory response. Our results suggest that the crosstalk between RAGE and TLR4 plays a critical role in the inflammatory response induced by the combination of HG and LPS in BAMs. These findings highlight the potential therapeutic significance of targeting the RAGE–TLR4 crosstalk in the treatment of inflammation-related diseases.

2. Results

2.1. LPS Increased Pro-Inflammatory Cytokine Release in Primary BAMs in a Dose- and Time-Dependent Manner

BAMs were treated with LPS at a range of various doses (0, 1, 5, and 10 μg/mL) for 12 h. The results are depicted in Figure 1A–C. Contrasted with the control group

(0 μg/mL), all three LPS concentrations significantly raised the levels of IL-1β, IL-6, and TNF-α ($p < 0.05$). In addition, 5 μg/mL of LPS showed the strongest increment in the release of IL-6 and TNF-α ($p < 0.01$) compared to 1 μg/mL and 10 μg/mL LPS treatment groups. In comparison to the control group, the 10 μg/mL LPS treatment significantly raised the levels of IL-1β, IL-6, and TNF-α ($p < 0.05$), but there was no significant difference with the 1 μg/mL LPS treatment ($p > 0.05$). Therefore, we selected 5 μg/mL as the LPS concentration for subsequent experiments.

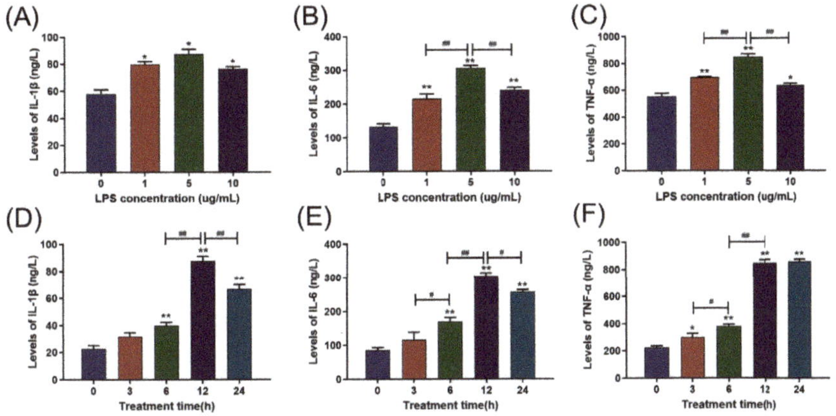

Figure 1. Effects of different LPS treatment times/doses on pro-inflammatory cytokine release in BAMs. (**A–C**) Effects of various LPS concentrations (0, 1, 5, and 10 μg/mL) on the release of pro-inflammatory cytokines IL-1β, IL-6, and TNF-α in BAMs. (**D–F**) Effects of different LPS treatment times (0, 3, 6, 12, and 24 h) on the release of pro-inflammatory cytokines IL-1β, IL-6, and TNF-α in BAMs. Each experiment was carried out at least three times, and the data were displayed using mean ± SD. *, $p < 0.05$; **, $p < 0.01$ vs. control group; $^{\#}$, $p < 0.05$; $^{\#\#}$, $p < 0.01$ vs. other treatment groups; One-way ANOVA.

Similarly, the BAMs were exposed to 5 μg/mL LPS at a series of times (0, 3, 6, 12, and 24 h), and the results are displayed in Figure 1D–F. The levels of IL-1β and IL-6 significantly raised from the 6th h ($p < 0.05$) and reached their highest values at the 12th h and then reduced ($p < 0.05$) (Figure 1D,E). In addition, the TNF-α level significantly increased from the 3rd h ($p < 0.05$), reached its peak value at the 12th h, and then maintained. Therefore, we selected 12 h as the treatment duration for the following trials.

2.2. TLR4, RAGE, and Their Interaction Were Involved in the Inflammatory Response Caused by LPS

The levels of TLR4 and RAGE mRNA expression in the BAMs were tested, and the results were shown in Figure 2A,B,D,E. The levels of TLR4 and RAGE mRNA expression were dramatically elevated in groups with LPS treatment ($p < 0.05$), and the 5 μg/mL LPS treatment group showed the most significant effects ($p < 0.01$). Intriguingly, there was no effect on the levels of TLR4 mRNA expression from 10 μg/mL LPS treatment (Figure 2A). In addition, the levels of RAGE and TLR4 mRNA expression significantly raised from the 6th h ($p < 0.05$) and reached the peak value at the 24th h after LPS treatment ($p < 0.01$) (Figure 2B,E).

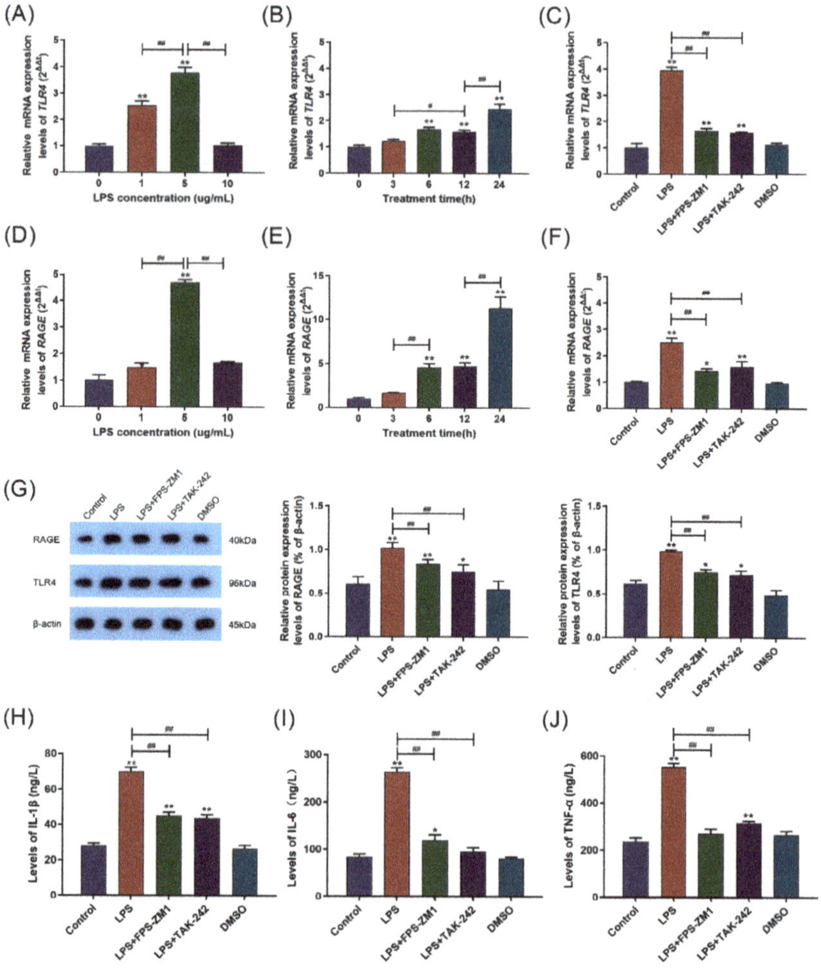

Figure 2. TLR4, RAGE and their interaction were involved in LPS-induced inflammatory response in BAMs. (**A**,**D**) Effects of different LPS concentrations (0, 1, 5, and 10 μg/mL) on the levels of TLR4 and RAGE mRNA expression. (**B**,**E**) Effects of different LPS treatment time (0, 3, 6, 12, and 24 h) on the levels of TLR4 and RAGE mRNA expression. (**C**,**F**) Effects of pretreatment with FPS-ZM1 and TAK-242 on the levels of TLR4 and RAGE mRNA expression. (**G**) Comparison of gel images and grayscale for RAGE and TLR4 immunoblot detection. (**H**–**J**) The levels of IL-1β, IL-6, and TNF- α in the supernatants of each group. Each experiment was carried out at least three times, and the data were displayed using mean ± SD. *, $p < 0.05$; **, $p < 0.01$ vs. control group; #, $p < 0.05$; ##, $p < 0.01$ vs. other treatment groups; One-way ANOVA.

The BAMs were pretreated with 10 μM TLR4 inhibitor TAK-242 and 1.0 μM RAGE inhibitor FPS-ZM1 for 1 h, respectively, and dimethyl sulfoxide (DMSO) treatment performed as the solvent control group. According to Figure 2C,F, in the FPS-ZM1 pretreatment group, not only did the level of RAGE mRNA expression significantly reduce but the level of TLR4 mRNA expression also significantly decreased ($p < 0.01$). Similarly, in the TAK-242 pretreatment group, the levels of TLR4 and RAGE mRNA expression also significantly decreased ($p < 0.01$). The results of the levels of RAGE and TLR4 protein expression are displayed in Figure 2G. The levels of TLR4 and RAGE protein expression were considerably raised in the LPS treatment group ($p < 0.01$). As a result of the FPS-ZM1 pretreatment, the levels of RAGE and TLR4 protein expression in the LPS treatment group were significantly reduced ($p < 0.01$), and the TAK-242 pretreatment showed the same results ($p < 0.01$). The mRNA and protein levels of RAGE decreased when TLR4 expression was inhibited with TAK-242, and the mRNA and protein levels of TLR4 decreased when RAGE expression was inhibited with FPS-ZM1, suggesting crosstalk between RAGE and TLR4. Then, the levels of IL-1β, IL-6, and TNF-α were detected, and the results are displayed in Figure 3H–J. Pretreatment with 1.0 μM FPS-ZM1 and 10 μM TAK-242 both significantly decreased the levels of IL-1β, IL-6, and TNF-α in BAMs ($p < 0.01$).

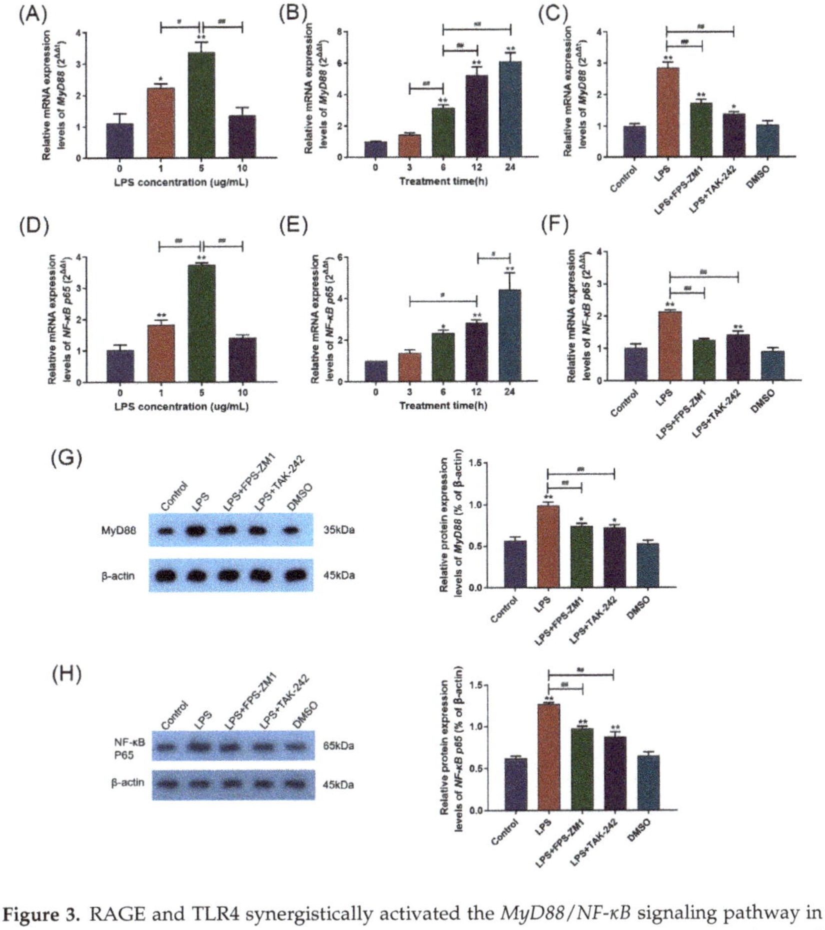

Figure 3. RAGE and TLR4 synergistically activated the *MyD88/NF-κB* signaling pathway in the inflammation caused by LPS. (**A**,**D**) Effects of different LPS concentrations (0, 1, 5, and 10 μg/mL) on the levels of MyD88 and NF-κB p65 mRNA expression. (**B**,**E**) Effects of different LPS treatment times (0, 3, 6, 12, and 24 h) on the levels of MyD88 and NF-κB p65 mRNA expression. (**C**,**F**) Effects of FPS-ZM1 and TAK-242 pretreatment on the levels of MyD88 and NF-κB p65 mRNA expression. (**G**,**H**) Comparison of gel images and grayscale for MyD88 and NF-κB p65 immunoblot detection. Each experiment was carried out at least three times, and the data were displayed using mean ± SD. *, $p < 0.05$; **, $p < 0.01$ vs. control group; #, $p < 0.05$; ##, $p < 0.01$ vs. other treatment groups; One-way ANOVA.

2.3. RAGE and TLR4 Synergistically Activate the MyD88/NF-κB Signaling Pathway in the Inflammation Response Caused by LPS

The levels of MyD88 and NF-κB p65 mRNA and protein expression were detected. Figure 3A,B,D,E showed that the 5 μg/mL and 24 h of LPS treatment group showed the most significantly increased levels of MyD88 and NF-κB p65 mRNA expression compared with the control group ($p < 0.01$). Similarly, in the inhibitor experiments, pretreatment with FPS-ZM1 and TAK-242 blocked the LPS-induced increase in the levels of MyD88 and NF-κB P65 mRNA expression ($p < 0.01$) (Figure 4C,F). The levels of MyD88 and NF-κB P65 protein expression were similar in the DMSO and control groups ($p > 0.05$), but, compared to the control group, they are significantly increased in the LPS treatment group ($p < 0.01$) (Figure 3G). However, in contrast to the LPS treatment group, pretreatment with FPS-ZM1 and TAK-242 significantly blocked the LPS-induced increases in MyD88 and NF-κB P65 protein expression levels ($p < 0.01$) (Figure 3G,H).

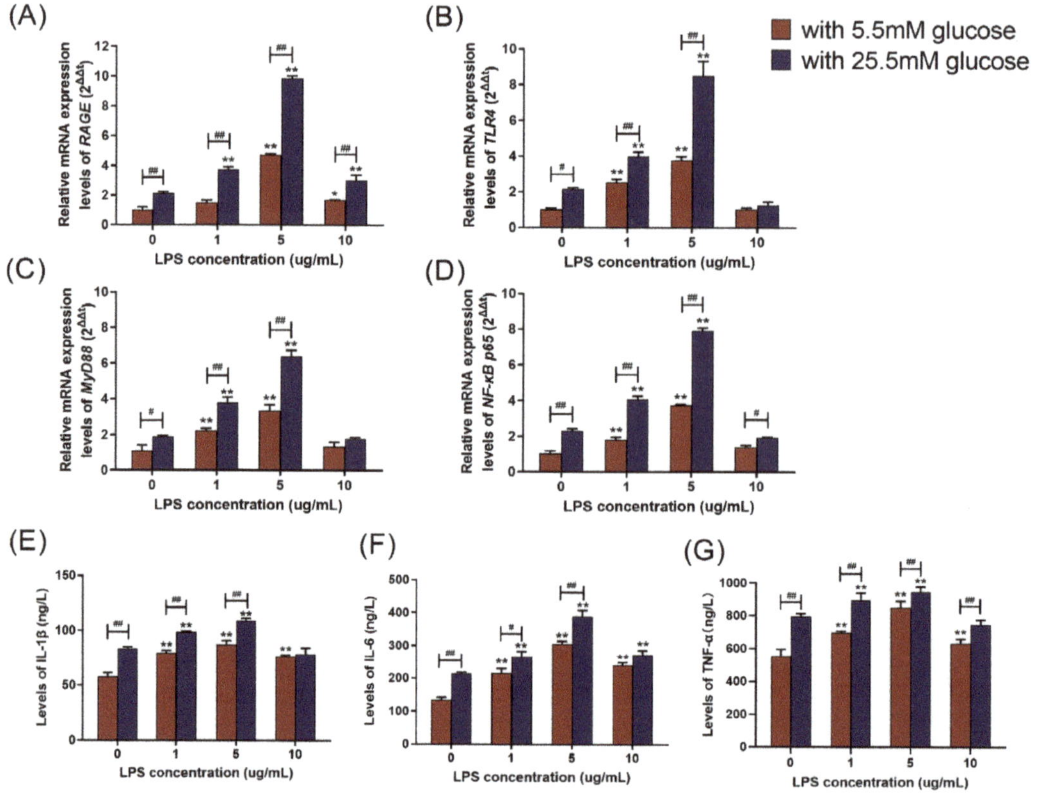

Figure 4. Effects of different glucose concentrations on LPS-induced pro-inflammatory cytokine secretion and the *RAGE/TLR4/MyD88/NF-κB p65* pathway in BAMs. (**A–D**) The impact of NG and HG on the levels of RAGE, TLR4, MyD88, and NF-κB p65 mRNA expression in the presence of different concentrations of LPS (0, 1, 5, and 10 μg/mL). (**E–G**) The impact of NG and HG on the levels of IL-1β, IL-6, and TNF-α in the presence of various concentrations of LPS (0, 1, 5, and 10 μg/mL). Each experiment was carried out at least three times, and the data were displayed using mean ± SD.*, $p < 0.05$; **, $p < 0.01$ vs. control group; #, $p < 0.05$; ##, $p < 0.01$ vs. other treatment groups; One-way ANOVA.

2.4. HG Enhanced LPS-Induced Pro-Inflammatory Cytokine Secretion and Upregulated the RAGE/TLR4/MyD88/NF-κB p65 Pathway in BAMs

The levels of RAGE, TLR4, MyD88, and NF-κB p65 mRNA expression in the HG group and normal glucose (NG, 5.5 mM) group were assessed, and Figure 4A–D displays the result. At 1 μg/mL and 5 μg/mL LPS stimulus, compared to the NG group, the levels of RAGE, TLR4, MyD88, and NF-κB p65 mRNA expression in the HG group were significantly increased ($p < 0.01$). Under 10 μg/mL LPS treatment, the levels of *NF-κB p65* ($p < 0.05$) and *RAGE* ($p < 0.01$) mRNA expression in the HG group were significantly increased compared with the NG group, but the levels of TLR4 and MyD88 mRNA expression were comparable to those of the NG group ($p > 0.05$). Additionally, the levels of IL-1β, IL-6, and TNF-α in the HG and NG groups were tested, and Figure 4E–G showed the results. After 1 and 5 μg/mL LPS treatment, the levels of IL-1β, IL-6, and TNF-α in the HG group were significantly higher than those in the NG group ($p < 0.05$). Under 10 μg/mL LPS treatment, the level of TNF-α in the HG group was significantly raised ($p < 0.01$) compared with the NG group, although the levels of IL-1β and IL-6 were comparable to those of the NG group ($p > 0.05$).

2.5. RAGE–TLR4 Crosstalk Regulated the Synergism between HG and LPS on the Inflammatory Response in BAMs

The levels of RAGE, TLR4, MyD88, and NF-κB p65 mRNA expression in each group were detected, and the results are displayed in Figure 5A–D. In the HG group or LPS group, the levels of RAGE, TLR4, MyD88, and NF-κB p65 mRNA expression were significantly raised compared with the control group ($p < 0.05$) and were significantly lower than in the HG + LPS group ($p < 0.01$). Similarly, pretreatment with FPS-ZM1 and TAK-242 significantly blocked the HG + LPS-induced increase in RAGE, TLR4, MyD88, and NF-κB p65 mRNA expression in the HG + LPS group ($p < 0.01$). The levels of RAGE, TLR4, MyD88, and NF-κB p65 protein expression in each group were also detected, and Figure 5E–J displays the results. The levels of RAGE, TLR4, MyD88, and NF-κB p65 protein expression in the HG or LPS group were significantly increased compared to the control group ($p < 0.05$) and were significantly reduced than those in the HG + LPS group ($p < 0.01$). In the HG + LPS group, pretreatment with FPS-ZM1 or TAK-242 significantly inhibited the HG + LPS-induced increase in RAGE, TLR4, MyD88, and NF-κB p65 protein expression ($p < 0.01$). Figure 5K–M displayed the results of IL-1β, IL-6, and TNF-α levels in each group. The levels of IL-1β, IL-6, and TNF-α in the DMSO group were comparable to the control group ($p > 0.05$). The levels of IL-1β, IL-6, and TNF-α levels in the HG group and LPS group were significantly increased ($p < 0.01$) compared to the control group but considerably reduced in contrast to the HG + LPS group ($p < 0.01$). Pretreatment with FPS-ZM1 and TAK-242 significantly ameliorated the HG + LPS-induced increase in IL-1β, IL-6, and TNF-α levels in the HG + LPS group ($p < 0.01$).

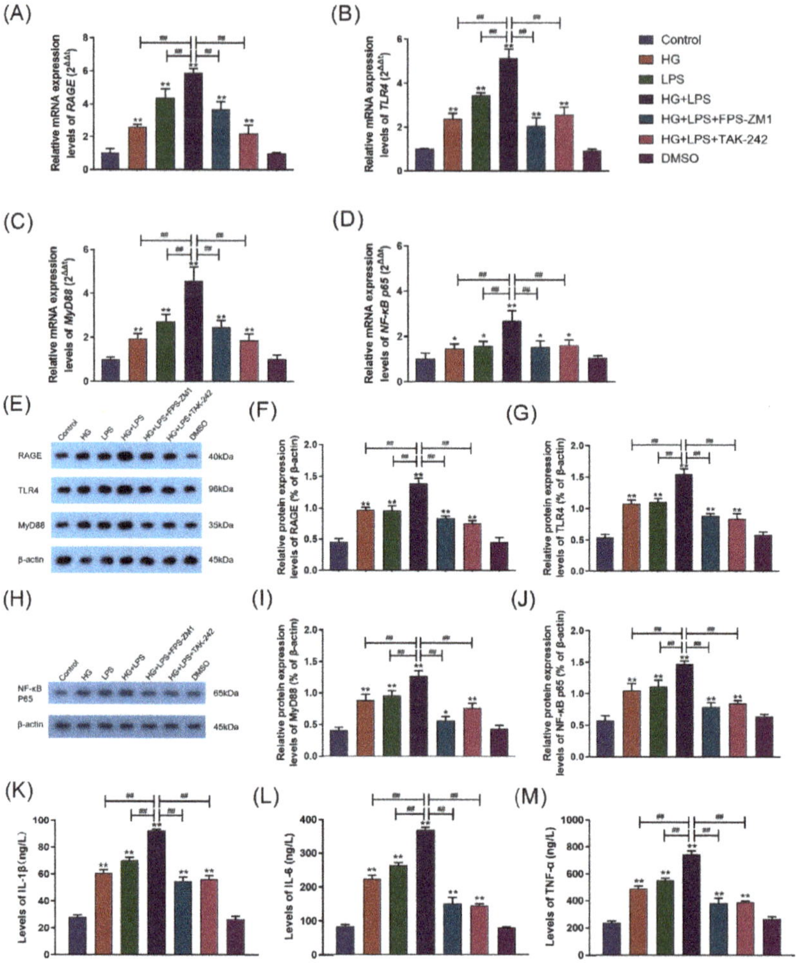

Figure 5. RAGE–TLR4 crosstalk regulated the synergism between high glucose and LPS on the inflammatory response in BAMs. (**A–D**) The levels of RAGE, TLR4, MyD88 and NF-κB p65 mRNA expression in each group. (**E–J**) The levels of RAGE, TLR4, MyD88 and NF-κB p65 protein expression in each group. (**K–M**) The levels of IL-1β, IL-6, and TNF-α in each group. Each experiment was carried out at least three times, and the data were displayed using mean ± SD. *, $p < 0.05$; **, $p < 0.01$ vs. control group; ##, $p < 0.01$ vs. other treatment groups; One-way ANOVA.

3. Discussions

Stress-induced hyperglycemia and gram-negative bacterial infections are two of the primary factors contributing to severe pulmonary inflammation of bovine during long-distance transportation [1,32]. RAGE and TLR4 are the main receptors for LPS and HG leading to the development of cellular inflammatory responses, and RAGE–TLR4 crosstalk occurs when RAGE and TLR4 are activated, thereby synergistically activating shared downstream inflammatory signaling pathways [31]. However, it is not clear whether RAGE–TLR4 crosstalk occurs during HG and LPS co-treatment-induced inflammation in BAMs, and the effects of RAGE–TLR4 crosstalk on downstream-related signaling pathways and specific mechanisms need to be further elucidated. Previous studies have demonstrated that LPS can promote the synthesis and release of pro-inflammatory cytokines, such as TNF-α, IL-1β, and IL-6, from a variety of cell types, including placental cells, endothelial cells, and macrophages [33–35]. In our study, the levels of pro-inflammatory cytokines IL-1β, IL-6, and TNF-α in the supernatant considerably elevated with the increased concentration of LPS and the duration of treatment, indicating that LPS induces BAMs to release pro-inflammatory cytokines in a dose- and time-dependent manner.

LPS promotes the release of pro-inflammatory cytokines from cells associated with various transmembrane receptors on the cell surface, mainly TLR4 [18] and RAGE [28]. We assumed that the LPS-induced inflammatory response in BAMs via receptor crosstalk occurred when RAGE and TLR4 were simultaneously activated. In our work, the levels of TLR4 and RAGE expression were increased after LPS treatment. However, TAK-242 (TLR4 inhibitor) and FPS-ZM1 (RAGE inhibitor) inhibited TLR4 and RAGE expression as well as the release of IL-1β, IL-6, and TNF-α in BAMs. These results indicated that both RAGE and TLR4 are involved in LPS-induced inflammatory response in BAMs. Intriguingly, we found that the inhibition of RAGE could down-regulate the level of TLR4 expression, and inhibition of TLR4 could also down-regulate the level of RAGE expression, indicating that RAGE and TLR4 mutually affect each other's expression at the receptor level in the inflammatory response of BAMs caused by LPS. Prior research has demonstrated that LPS could trigger RAGE and TLR4 and cause RAGE–TLR4 crosstalk [36].

MyD88, a classical downstream signaling adaptor molecule of TLR4, is important for triggering TLR4-related inflammatory signaling pathways [37]. However, some studies have found that MyD88 could bind to RAGE and transduce a signal to downstream molecules, blocking the function of MyD88 abrogated intracellular signaling from HMGB1-activated RAGE [38], but the study had shortcomings. The HMGB1 used in the study had been shown to bind not only to RAGE but also to TLR4; so, blocking MyD88 may cause signal attenuation through the TLR4 pathway rather than the RAGE pathway. It is still unclear whether MyD88 can be coordinately regulated by RAGE and TLR4. The binding of LPS to TLR4 transmits inflammatory signals intracellularly through the MyD88-independent (TRAM) and MyD88-dependent pathways, activating NF-κB and promoting the synthesis and release of multiple pro-inflammatory cytokines [39,40]. Gąsiorowski et al. [31] found that RAGE and TLR4 exert synergistic activation on common downstream signaling pathways, such as NF-κB and AP-1. In the present study, the levels of MyD88 and NF-κB p65 expression were significantly upregulated after LPS treatment. However, TAK-242 and FPS-ZM1 significantly decreased the levels of MyD88 and NF-κB p65 expression. These results indicated that RAGE and TLR4 synergistically activate the MyD88/NF-κB signaling pathway in LPS-induced inflammatory responses of BAMs. With similar results to ours, Byrd et al. [41] found that LPS induced upregulation of TLR4, MyD88, NF-κB expression and pro-inflammatory cytokine TNF-α release in primary mouse macrophages. Wang et al. [28] also found that LPS caused the upregulation of RAGE protein expression and activated NF-κB, while inhibition of RAGE inhibited LPS-induced NF-κB activation in venous endothelial cells.

Glucose is an important source of energy for mammals, and NG (mostly 5.5 mM in in vitro experiments) ensures the energy demand and utilization of cells [29]. However, when glucose concentration is elevated (mostly 25.5 mM in in vitro experiments), it has a pro-inflammatory response to cells [42,43]. In our study, the co-treatment of LPS and HG caused a significant increase in the levels of RAGE, TLR4, MyD88, and NF-κB p65 mRNA expression and the release of IL-1β, IL-6, and TNF-α. The results indicated that HG enhanced the LPS-induced inflammatory response through further activation of the RAGE/TLR4/MyD88/NF-κB p65 signaling pathway in BAMs. Consistent with our results, Nielsen et al. [44] and Kong et al. [45] have reported that the co-treatment of HG and LPS might cause a more intense inflammatory response in cells and organisms compared to HG or LPS treatment alone, suggesting that HG and LPS had a synergistic pro-inflammatory effect on the inflammatory response.

Crosstalk is an important direction to study the correlation between different signaling pathways. Relevant studies reported that when RAGE and TLR4 were activated, a crosstalk phenomenon occurred between RAGE and TLR4, and the RAGE–TLR4 crosstalk synergistically maintained and amplified the inflammatory response [31,46]. In previous studies, Ayala et al. [47] found that HG can upregulate the protein level of TLR4 on the cell membrane surface, which increases the sensitivity of cells to LPS and enhances the intensity of the initial inflammatory signal after LPS binds to TLR4, thus intensifying the LPS-induced inflammatory response. Nareika et al. [48] found that HG promoted CD14 expression, thereby amplifying the intensity of the initial inflammatory signal stimulated by LPS. The mechanism was that HG significantly upregulated NF-κB and AP-1 activity, thus promoting LPS-induced CD14 expression and inflammatory response. In the present study, LPS + HG treatment further increased the levels of RAGE, TLR4, MyD88, and NF-κB p65 expression and the release of IL-1β, IL-6, and TNF-α. Meanwhile, TAK-242 and FPS-ZM1 blocked the HG + LPS-induced increase in RAGE, TLR4, MyD88, and NF-κB p65 expression levels and the levels of IL-1β, IL-6, and TNF-α. These results suggest that HG and LPS exert synergistic pro-inflammatory effects in BAMs through the RAGE/TLR4/MyD88/NF-κB p65 pathway. Interestingly, the inhibition of RAGE could down-regulate TLR4 expression levels, and the inhibition of TLR4 could down-regulate RAGE expression levels in the presence of HG and LPS co-treatment in BAMs. These results indicate that RAGE–TLR4 crosstalk plays an important role in the activation of the RAGE/TLR4/MyD88/NF-κB inflammatory signaling pathway and the release of pro-inflammatory cytokines IL-1β, IL-6, and TNF-α in the inflammatory response of BAMs caused by HG and LPS co-treatment. Consistent with our results, previous studies have demonstrated that different ligands, such as S100A8/A9, LPS, and HMGB1 can cause RAGE–TLR4 crosstalk, which can further upregulate NF-κB expression as well as pro-inflammatory cytokine release to cause more severe inflammatory responses [46,49,50]. Collectively, when HG and LPS co-treatment-induced inflammation occurred, the TLR4/RAGE levels on the surface of BAMs increased and elevated levels of TLR4/RAGE via RAGE–TLR4 crosstalk. This further increased the possibility of binding with the gram-negative bacteria derived LPS, HG-derived HGBM1, and S100A8/A9 in the synergistic activation of downstream MyD88/NF-κB inflammatory signaling pathways, thus promoting the inflammation response in BAMs. However, there are still many gaps in the research on the mechanisms by which HG exacerbates the inflammatory response of cells induced by LPS. Although HG does not bind to RAGE or TLR4 directly, it can induce the release of pro-inflammatory mediators such as HMGB1 and AGEs and upregulate RAGE and TLR4 expression through the oxidative stress pathway. This, in turn, leads to the binding of mediators to RAGE or TLR4 and the subsequent activation of downstream inflammatory signaling pathways [23,31]. However, existing studies tend to ignore the process of RAGE or TLR4 activation by HG and focus on the changes in downstream signaling pathways after RAGE or TLR4 activation by HG. The crosstalk mechanism involved in this study is also included in this list. Subdividing the pro-inflammatory process of HG to investigate the involvement of additional HG-related pro-inflammatory mediators or pathways in RAGE and TLR4 activation, as well as examining the impact of HG-related pro-inflammatory mediators such as HMGB1 and AGEs on the promotion of inflammatory response by LPS through RAGE/TLR4, instead of treating the HG pro-inflammatory process as a whole, could provide further insight into the molecular mechanisms by which HG exacerbates the inflammatory response of cells induced by LPS. In addition, whether pro-inflammatory mediators produced by HG through oxidative stress such as HMGB1, AGEs, and S100A8/A9 can compete with LPS to bind RAGE and TLR4, and whether they have antagonistic effects with LPS needs to be further investigated. The possible mechanisms of RAGE–TLR4 crosstalk in the inflammatory response of BAMs caused by LPS and HG co-treatment are illustrated in Figure 6.

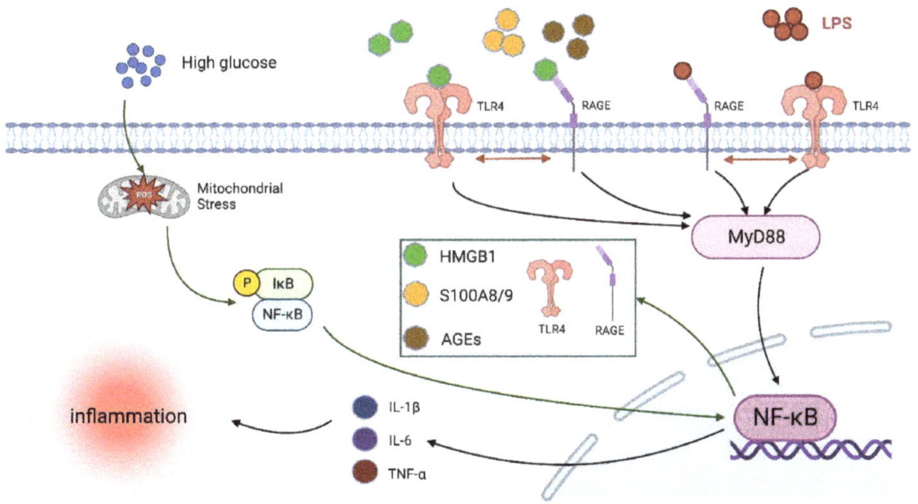

Figure 6. Schematic representation of the TLR4–RAGE crosstalk participates in the HG-enhanced LPS-induced inflammation in BAMs (created with BioRender.com). HG firstly upregulated the levels of RAGE and TLR4 genes and protein expression in the BAM membrane and the secretion of pro-inflammatory-related mediators (such as HMGB1, S100A8/9, and AGEs) through the ROS/NF-κB pathway. Then, pro-inflammatory-related mediators bind to RAGE and TLR4, causing RAGE–TLR4 crosstalk (increasing the levels of RAGE and TLR4) to synergistically activate the downstream MyD88/NF-κB signaling pathway and promote the release of pro-inflammatory cytokines IL-1β, IL-6, and TNF-α. When HG acts together with LPS, HG upregulates the levels of RAGE and TLR4 genes and protein expression, which can provide more receptor sites for LPS binding, and the LPS also upregulates the levels of RAGE and TLR4 genes and protein expression that can offer more receptor sites for pro-inflammatory-related mediators. The combination of HG and LPS through RAGE–TLR4 crosstalk further activates the downstream MyD88/NF-κB signaling pathway and exacerbates pro-inflammatory cytokine release in BAMs.

4. Materials and Methods

4.1. Isolation and Treatment of BAMs

BAMs were obtained from the intact bovine lungs of five healthy Chinese Simmental cattle (400–500 kg b.w., male, the lung showed no signs of bacterial infection) from a local slaughterhouse, and the isolation protocol was performed following the previous description [51]. Briefly, the lungs were lavaged with approximately 2 L of high-glucose Dulbecco's modified Eagle medium (DMEM) (Solarbio, Beijing, China) 3 times, and the irrigation fluid was filtered by 200 mesh sterile gauze to remove tissues. Cells were washed with PBS three times and then were cultured in DMEM supplemented with 10% fetal bovine serum (FBS) (ThermoFisher, Shanghai, China), 1% penicillin (100 U/mL), and streptomycin (100 mg/mL) (Gibco, Shanghai, China). Cells were cultured with 5% carbon dioxide (CO_2) at 37 °C. The non-adherent cells were removed 3 h later by PBS washing. Then, the remaining cells were digested, resuspended, and adjusted to 1×10^6 per mL. The cells were dispensed into T25 cell culture flasks and incubated for 3 h to allow re-adherence, pending subsequent processing.

In the LPS treatment experiments, BAMs were treated with (0, 1, 5, 10 μg/mL) LPS (from *Escherichia coli* 055: B5, SIGMA-L6529, St. Louis, MO, USA) for 12 h or with 5 μg/mL LPS for 0, 3, 6, 12, and 24 h. In the combined use of LPS and HG experiments, BAMs were pretreated with inhibitors for 1 h before adding LPS (5 μg/mL), and inhibitors were dissolved in 100% DMSO (Sigma-Aldrich, St. Louis, MO, USA) and then blended in a culture medium. After adding HG or LPS, the same concentration of inhibitor was re-added to the culture medium. Table 1 displays the grouping as well as the processing. FPS-ZM1 and TAK-242 were purchased from MCE, Shanghai, China.

Table 1. Experimental grouping of the combined effects of LPS and HG.

Group	Treatment
Control	0 μg/mL LPS + 5.5 mM glucose
HG	0 μg/mL LPS + 25.5 mM glucose
LPS	5 μg/mL LPS + 5.5 mM glucose
HG +LPS	5 μg/mL LPS + 25.5 mM glucose
HG + LPS + FPS-ZM1	5 μg/mL LPS + 25.5 mM glucose +1 μM FPS-ZM1
HG + LPS + TAK-242	5 μg/mL LPS + 25.5 mM glucose +10 μM TAK-242
DMSO	0 μg/mL LPS + 5.5 mM glucose + DMSO

4.2. Real-Time Quantitative Polymerase Chain Reaction (RT-qPCR)

A Trizol reagent was used to extract the total ribonucleic acid (RNA) from each treatment in cultivated BAMs (ThermoFisher, Shanghai, China). Reverse transcription was performed on the isolated total RNA using a kit to create complementary deoxyribonucleic acid (cDNA) (Takara, Tokyo, Japan). The 96-well plate used for the RT-qPCR contained 4.0 μL diluted cDNA, 5.0 μL of TaKaRa SYBR Green PCR MIX, and 0.5 μL of upstream and downstream primers. For RT-qPCR, the following amplification procedures were used: 95 °C pre-denaturation for 3 min, 95 °C for 10 s; 58 °C for 30 s, and 40 cycles. Using a Real-Time PCR Kit (Takara, Tokyo, Japan), the CFX96 Touch Real-Time PCR Detection System was used to perform the RT-qPCR experiments (Bio-Rad, CA, USA). Table 2 displays the primer sequences that were utilized. The relative gene expression level was estimated using the $2^{-\Delta\Delta Ct}$ method with β-actin as a standard to normalize the RT-qPCR data.

Table 2. The primer sequence of genes for Real-Time PCR.

Gene Name	Forward Primer Sequence (5′–3′)	Reverse Primer Sequence (3′–5′)
β-actin	GCCCATCTATGAGGGGTACG	TCACGGACGATTTCCGCT
RAGE	GACAGTCGCCCTGCTCATT	CCTCTGGCTGGTTCAGTTCC
TLR4	TGCCTTCACTACAGGGACTTT	TGGGACACCACGACAATAAC
NF-κB p65	GAGATCATCGAGCAGCCCAA	ATAGTGGGGTGGGTCTTGGT
MyD88	AGAAGAGGTGCCGTCGGATGG	TTGGTGTAGTCACAGACAGTGATGAAG

4.3. Immunoblot Assay

The cultured BAMs from each treatment were collected, and the total proteins were extracted using the RIPA lysis buffer. After centrifuging the lysates, the supernatants were analyzed using the BCA Protein Assay Kit (Beyotime, Shanghai, China). Each sodium dodecyl sulfate-polyacrylamide gel (10%) lane was filled with identical quantities of protein (20 μL), and the gels were electrophoresed to separate them. Using the Bio-Rad Trans-Blot device, the separated proteins were subsequently transferred through an electron transfer procedure to polyvinylidene fluoride membranes (Bio-Rad, Hercules, CA, USA). Next, 5% skim milk was used to soak the polyvinylidene fluoride (PVDF) membranes before blocking. They were then incubated with RAGE antibody (ab37647, Abcam, 1:1000), TLR4 antibody (19811-1-AP, Proteintech 1:1000), MyD88 antibody (70R-50098, Fitzgerald 1:1000), NF-κB p65 (C22B4, CST, 1:1000), and β-actin antibody (4970, CST, 1:5000) overnight at 4 °C. The membranes were then incubated with HRP anti-rabbit secondary antibodies (SE134, Solarbio, 1:5000) for 1 h with moderate shaking. With the use of an enhanced chemiluminescence (ECL) kit, blots were seen (Thermo Scientific, Rockford, UK). With the help of the Tanon-5200 automated chemiluminescence imaging analysis system, the band intensity of the pictures was assessed (Tanon, Shanghai, China). Image-Pro Plus 6.0 software was used to examine the integrated optical density (IOD) of each protein band (Media Cybernetics, Inc., Rockville, MD, USA).

4.4. Enzyme-Linked Immunosorbent Assay (ELISA)

Isolated BAMs were seeded in a 96-well plate at 1×10^5 cells per well and allowed to settle for 3 h at 37 °C in a 5% CO_2 incubator. Then, the levels of pro-inflammatory cytokines IL-1β, IL-6, and TNF-α in the supernatants of cultured BAMs in different groups (grouping processing steps as stated in Section 4.1) were measured using ELISA kits (double-antibody sandwich method). All measurements were performed following the manufacturer's instructions (Jingmei Biotechnology, Jiangsu, China). All experiments were performed three times independently, once in triplicate.

4.5. Statistical Analysis

All experiments were performed in triplicate, and the test results are expressed as mean ± standard deviation (SD). The experimental data were sorted and unified in Excel, and experimental data visualization was performed using GraphPad Prism 9.0 (GraphPad Software, CA, USA). The SPSS 26.0 software (IBM, New York, NY, USA) was used to carry out a One-way analysis of variance (ANOVA) test, followed by a least significant difference (LSD) post hoc test. $p < 0.05$ and $p < 0.01$ indicate statistical significance.

5. Conclusions

In conclusion, this study demonstrated that RAGE–TLR4 crosstalk was critical in the HG-enhanced LPS-induced inflammation response in BAMs. LPS and HG enhance the inflammatory response of BAMs not only through activation of RAGE and TLR4 but also through the crosstalk between RAGE and TLR4. The crosstalk between RAGE and TLR4 can regulate the genes' transcription and translation at the receptor level and synergistically activate the downstream MyD88/NF-κB inflammatory signaling pathway that enhances the synthesis and release of pro-inflammatory cytokines TNF-α, IL-1β, and IL-6 in BAMs. These results highlighted the potential of disrupting the crosstalk between RAGE and TLR4 to mitigate the cellular inflammatory response triggered by HG and LPS, emphasizing that simultaneous induction of inflammation responses by multiple pro-inflammatory factors should be avoided.

Author Contributions: L.Y., conceptualization, methodology, software, formal analysis, resources, data curation, writing—original draft, writing—review and editing, and visualization; Y.L., conceptualization, methodology, validation, investigation, resources, data curation, writing—original draft, and supervision; T.T., methodology, validation, data curation, writing—original draft, and supervision; J.Q., validation, investigation, and resources; J.F., resources and supervision; H.G., resources and data curation; Z.R., investigation and resources; L.G., resources; Y.G., resources; H.C., resources; L.S., software and resources; S.Y., resources; Z.W., resources and supervision; Z.Z., conceptualization, validation, investigation, resources, supervision, project administration, and funding acquisition. All authors have read and agreed to the published version of the manuscript.

Funding: This research was funded by the National Key Research and Development project (No. 2021YFD1600200) and China Agriculture Research System of MOF and MARA (Beef Cattle/Yak, CARS-37).

Institutional Review Board Statement: Not applicable.

Informed Consent Statement: Not applicable.

Data Availability Statement: The data used to support the findings of this study are available from the corresponding author upon request.

Conflicts of Interest: There are no conflicts of interest according to the authors.

References

1. Earley, B.; Buckham Sporer, K.; Gupta, S. Invited review: Relationship between cattle transport, immunity and respiratory disease. *Animal* **2017**, *11*, 486–492. [CrossRef] [PubMed]
2. Harada, N.; Takizawa, K.; Matsuura, T.; Yokosawa, N.; Tosaki, K.; Katsuda, K.; Tanimura, N.; Shibahara, T. Bovine peritonitis associated with *Mannheimia haemolytica* serotype 2 in a three-day-old Japanese Black calf. *J. Vet. Med. Sci.* **2019**, *81*, 143–146. [CrossRef] [PubMed]
3. Holman, D.B.; Timsit, E.; Amat, S.; Abbott, D.W.; Buret, A.G.; Alexander, T.W. The nasopharyngeal microbiota of beef cattle before and after transport to a feedlot. *BMC Microbiol.* **2017**, *17*, 70.
4. Baker, E.H.; Archer, J.R.H.; Srivastava, S.A. Hyperglycemia, Lung Infection, and Inflammation. *Clin. Pulm. Med.* **2009**, *16*, 258–264. [CrossRef]
5. Yu, J.; Shi, J.; Wang, D.; Dong, S.; Zhang, Y.; Wang, M.; Gong, L.; Fu, Q.; Liu, D. Heme Oxygenase-1/Carbon Monoxide-regulated Mitochondrial Dynamic Equilibrium Contributes to the Attenuation of Endotoxin-induced Acute Lung Injury in Rats and in Lipopolysaccharide-activated Macrophages. *Anesthesiology* **2016**, *125*, 1190–1201. [CrossRef]
6. He, S.; Shi, J.; Liu, W.; Du, S.; Zhang, Y.; Gong, L.; Dong, S.; Li, X.; Gao, Q.; Yang, J.; et al. Heme oxygenase-1 protects against endotoxin-induced acute lung injury depends on NAD+-mediated mitonuclear communication through PGC1α/PPARγ signaling pathway. *Inflamm. Res.* **2022**, *71*, 1095–1108. [CrossRef]
7. Mitchell, G.; Hattingh, J.; Ganhao, M. Stress in cattle assessed after handling, after transport and after slaughter. *Vet. Rec.* **1988**, *123*, 201–205. [CrossRef]
8. Hagenmaier, J.A.; Reinhardt, C.D.; Bartle, S.J.; Henningson, J.N.; Ritter, M.J.; Calvo-Lorenzo, M.S.; Vogel, G.J.; Guthrie, C.A.; Siemens, M.G.; Thomson, D.U. Effect of handling intensity at the time of transport for slaughter on physiological response and carcass characteristics in beef cattle fed ractopamine hydrochloride. *J. Anim. Sci.* **2017**, *95*, 1963–1976.
9. Wu, C.-P.; Huang, K.-L.; Peng, C.-K.; Lan, C.-C. Acute Hyperglycemia Aggravates Lung Injury via Activation of the SGK1-NKCC1 Pathway. *Int. J. Mol. Sci.* **2020**, *21*, 4803. [CrossRef]
10. Aljada, A.; Friedman, J.; Ghanim, H.; Mohanty, P.; Hofmeyer, D.; Chaudhuri, A.; Dandona, P. Glucose ingestion induces an increase in intranuclear nuclear factor kappaB, a fall in cellular inhibitor kappaB, and an increase in tumor necrosis factor alpha messenger RNA by mononuclear cells in healthy human subjects. *Metabolism* **2006**, *55*, 1177–1185. [CrossRef]
11. Lapar, D.J.; Hajzus, V.A.; Zhao, Y.; Lau, C.L.; French, B.A.; Kron, I.L.; Sharma, A.K.; Laubach, V.E. Acute hyperglycemic exacerbation of lung ischemia-reperfusion injury is mediated by receptor for advanced glycation end-products signaling. *Am. J. Respir. Cell Mol. Biol.* **2012**, *46*, 299–305. [CrossRef] [PubMed]
12. Viurcos-Sanabria, R.; Escobedo, G. Immunometabolic bases of type 2 diabetes in the severity of COVID-19. *World J. Diabetes* **2021**, *12*, 1026–1041. [CrossRef] [PubMed]
13. Liao, H.; Li, Y.; Zhang, X.; Zhao, X.; Li, R. Protective effects of thalidomide on high-glucose-induced podocyte injury through in vitro modulation of macrophage M1/M2 differentiation. *J. Immunol. Res.* **2020**, 8263598. [CrossRef] [PubMed]
14. Xiao, J.; Tang, J.; Chen, Q.; Tang, D.; Liu, M.; Luo, M.; Wang, Y.; Wang, J.; Zhao, Z.; Tang, C.; et al. miR-429 regulates alveolar macrophage inflammatory cytokine production and is involved in LPS-induced acute lung injury. *Biochem. J.* **2015**, *471*, 281–291. [CrossRef] [PubMed]
15. Wang, L.; Wang, J.; Fang, J.; Zhou, H.; Liu, X.; Su, S.B. High glucose induces and activates Toll-like receptor 4 in endothelial cells of diabetic retinopathy. *Diabetol. Metab. Syndr.* **2015**, *7*, 89. [CrossRef]
16. Karki, R.; Kanneganti, T.-D. The 'cytokine storm': Molecular mechanisms and therapeutic prospects. *Trends Immunol.* **2021**, *42*, 681–705. [CrossRef]
17. Miller, S.I.; Ernst, R.K.; Bader, M.W. LPS, TLR4 and infectious disease diversity. *Nat. Rev. Microbiol.* **2005**, *3*, 36–46. [CrossRef]
18. Lu, Y.-C.; Yeh, W.-C.; Ohashi, P.S. LPS/TLR4 signal transduction pathway. *Cytokine* **2008**, *42*, 145–151. [CrossRef]
19. Fitzgerald, K.A.; Rowe, D.C.; Barnes, B.J.; Caffrey, D.R.; Visintin, A.; Latz, E.; Monks, B.; Pitha, P.M.; Golenbock, D.T. LPS-TLR4 signaling to IRF-3/7 and NF-κB involves the toll adapters TRAM and TRIF. *J. Exp. Med.* **2003**, *198*, 1043–1055. [CrossRef]
20. Cao, C.; Yin, C.; Shou, S.; Wang, J.; Yu, L.; Li, X.; Chai, Y. Ulinastatin protects against LPS-induced acute lung injury by attenuating TLR4/NF-κB pathway activation and reducing inflammatory mediators. *Shock* **2018**, *50*, 595–605. [CrossRef]
21. Dasu, M.R.; Devaraj, S.; Ling, Z.; Hwang, D.H.; Jialal, I. High Glucose Induces Toll-Like Receptor Expression in Human Monocytes: Mechanism of Activation. *Diabetes* **2008**, *57*, 3090–3098. [CrossRef]
22. Mudaliar, H.; Pollock, C.; Jin, M.; Wu, H.; Chadban, S.; Panchapakesan, U. The Role of TLR2 and 4-Mediated Inflammatory Pathways in Endothelial Cells Exposed to High Glucose. *PLoS ONE* **2014**, *9*, e108844. [CrossRef] [PubMed]
23. Ohtsu, A.; Shibutani, Y.; Seno, K.; Iwata, H.; Kuwayama, T.; Shirasuna, K. Advanced glycation end products and lipopolysaccharides stimulate interleukin-6 secretion via the RAGE/TLR4-NF-κB-ROS pathways and resveratrol attenuates these inflammatory responses in mouse macrophages. *Exp. Ther. Med.* **2017**, *14*, 4363–4370. [PubMed]
24. Luan, Z.-G.; Zhang, H.; Yang, P.-T.; Ma, X.-C.; Zhang, C.; Guo, R.-X. HMGB1 activates nuclear factor-κB signaling by RAGE and increases the production of TNF-α in human umbilical vein endothelial cells. *Immunobiology* **2010**, *215*, 956–962. [CrossRef] [PubMed]
25. Ma, L.; Sun, P.; Zhang, J.-C.; Zhang, Q.; Yao, S.-L. Proinflammatory effects of S100A8/A9 via TLR4 and RAGE signaling pathways in BV-2 microglial cells. *Int. J. Mol. Med.* **2017**, *40*, 31–38. [CrossRef] [PubMed]

26. Allam VS, R.R.; Faiz, A.; Lam, M.; Rathnayake, S.N.H.; Ditz, B.; Pouwels, S.D.; Brandsma, C.-A.; Timens, W.; Hiemstra, P.S.; Tew, G.W.; et al. RAGE and TLR4 differentially regulate airway hyperresponsiveness: Implications for COPD. *Allergy* **2021**, *76*, 1123–1135. [CrossRef] [PubMed]
27. Yao, D.; Brownlee, M. Hyperglycemia-Induced Reactive Oxygen Species Increase Expression of the Receptor for Advanced Glycation End Products (RAGE) and RAGE Ligands. *Diabetes* **2010**, *59*, 249–255. [CrossRef] [PubMed]
28. Wang, L.; Wu, J.; Guo, X.; Huang, X.; Huang, Q. RAGE Plays a Role in LPS-Induced NF-κB Activation and Endothelial Hyperpermeability. *Sensors* **2017**, *17*, 722. [CrossRef] [PubMed]
29. Ibrahim, Z.A.; Armour, C.L.; Phipps, S.; Sukkar, M.B. RAGE and TLRs: Relatives, friends or neighbours? *Mol. Immunol.* **2013**, *56*, 739–744. [CrossRef]
30. Lin, L. RAGE on the Toll Road? *Cell Mol. Immunol.* **2006**, *3*, 351–358.
31. Gąsiorowski, K.; Brokos, B.; Echeverria, V.; Barreto, G.E.; Leszek, J. RAGE-TLR Crosstalk Sustains Chronic Inflammation in Neurodegeneration. *Mol. Neurobiol.* **2018**, *55*, 1463–1476. [CrossRef]
32. Van Engen, N.K.; Coetzee, J.F. Effects of transportation on cattle health and production: A review. *Anim. Health Res. Rev.* **2018**, *19*, 142–154. [CrossRef]
33. Talwar, H.; Bauerfeld, C.; Bouhamdan, M.; Farshi, P.; Liu, Y.; Samavati, L. MKP-1 negatively regulates LPS-mediated IL-1β production through p38 activation and HIF-1α expression. *Cell Signal.* **2017**, *34*, 1–10. [CrossRef] [PubMed]
34. Lee, W.-S.; Shin, J.-S.; Jang, D.S.; Lee, K.-T. Cnidilide, an alkylphthalide isolated from the roots of Cnidium officinale, suppresses LPS-induced NO, PGE2, IL-1β, IL-6 and TNF-α production by AP-1 and NF-κB inactivation in RAW 264.7 macrophages. *Int. Immunopharmacol.* **2016**, *40*, 146–155. [CrossRef]
35. Li, X.; Zhao, E.; Li, L.; Ma, X. Unfractionated Heparin Modulates Lipopolysaccharide-Induced Cytokine Production by Different Signaling Pathways in THP-1 Cells. *J. Interferon Cytokine Res.* **2018**, *38*, 283–289. [CrossRef] [PubMed]
36. Kim, S.-J.; Lee, S.-M. Necrostatin-1 protects against D-Galactosamine and lipopolysaccharide-induced hepatic injury by preventing TLR4 and RAGE signaling. *Inflammation* **2017**, *40*, 1912–1923. [CrossRef] [PubMed]
37. Kayama, H.; Ramirez-Carrozzi, V.R.; Yamamoto, M.; Mizutani, T.; Kuwata, H.; Iba, H.; Matsumoto, M.; Honda, K.; Smale, S.T.; Takeda, K. Class-specific regulation of pro-inflammatory genes by MyD88 pathways and IkappaBzeta. *J. Biol. Chem.* **2008**, *283*, 12468–12477. [CrossRef]
38. Sakaguchi, M.; Murata, H.; Yamamoto, K.-I.; Ono, T.; Sakaguchi, Y.; Motoyama, A.; Hibino, T.; Kataoka, K.; Huh, N.-h. TIRAP, an adaptor protein for TLR2/4, transduces a signal from RAGE phosphorylated upon ligand binding. *PLoS ONE* **2011**, *6*, e23132. [CrossRef]
39. Wang, Y.; Yang, Y.; Liu, X.; Wang, N.; Cao, H.; Lu, Y.; Zhou, H.; Zheng, J. Inhibition of clathrin/dynamin-dependent internalization interferes with LPS-mediated TRAM-TRIF-dependent signaling pathway. *Cell Immunol.* **2012**, *274*, 121–129. [CrossRef]
40. Niederberger, E.; Geisslinger, G. Proteomics and NF-κB: An update. *Expert Rev. Proteom.* **2013**, *10*, 189–204. [CrossRef]
41. Byrd-Leifer, C.A.; Block, E.F.; Takeda, K.; Akira, S.; Ding, A. The role of MyD88 and TLR4 in the LPS-mimetic activity of Taxol. *Eur. J. Immunol.* **2001**, *31*, 2448–2457. [CrossRef] [PubMed]
42. Gonzalez, Y.; Herrera, M.T.; Soldevila, G.; Garcia-Garcia, L.; Fabián, G.; Pérez-Armendariz, E.M.; Bobadilla, K.; Guzmán-Beltrán, S.; Sada, E.; Torres, M. High glucose concentrations induce TNF-α production through the down-regulation of CD33 in primary human monocytes. *BMC Immunol.* **2012**, *13*, 19. [CrossRef] [PubMed]
43. Chen, F.; Zhang, N.; Ma, D.; Ma, X.; Wang, Q. Effect of the Rho kinase inhibitor Y-27632 and fasudil on inflammation and fibrosis in human mesangial cells (HMCs) under high glucose via the Rho/ROCK signaling pathway. *Int. J. Clin. Exp. Med.* **2017**, *10*, 13224–13234.
44. Nielsen, T.B.; Pantapalangkoor, P.; Yan, J.; Luna, B.M.; Dekitani, K.; Bruhn, K.; Tan, B.; Junus, J.; Bonomo, R.A.; Schmidt, A.M.; et al. Diabetes Exacerbates Infection via Hyperinflammation by Signaling through TLR4 and RAGE. *mBio* **2017**, *8*, e00818-17. [CrossRef] [PubMed]
45. Kong, H.; Zhao, H.; Chen, T.; Song, Y.; Cui, Y. Targeted P2X7/NLRP3 signaling pathway against inflammation, apoptosis, and pyroptosis of retinal endothelial cells in diabetic retinopathy. *Cell Death Dis.* **2022**, *13*, 336. [CrossRef]
46. Paudel, Y.N.; Angelopoulou, E.; Piperi, C.; Othman, I.; Aamir, K.; Shaikh, M.F. Impact of HMGB1, RAGE, and TLR4 in Alzheimer's Disease (AD): From Risk Factors to Therapeutic Targeting. *Cells* **2020**, *9*, 383. [CrossRef]
47. Ayala, T.S.; Tessaro, F.H.G.; Jannuzzi, G.P.; Bella, L.M.; Ferreira, K.S.; Martins, J.O. High Glucose Environments Interfere with Bone Marrow-Derived Macrophage Inflammatory Mediator Release, the TLR4 Pathway and Glucose Metabolism. *Sci. Rep.* **2019**, *9*, 11447. [CrossRef]
48. Nareika, A.; Im, Y.-B.; Game, B.A.; Slate, E.H.; Sanders, J.J.; London, S.D.; Lopes-Virella, M.F.; Huang, Y. High glucose enhances lipopolysaccharide-stimulated CD14 expression in U937 mononuclear cells by increasing nuclear factor kappaB and AP-1 activities. *J. Endocrinol.* **2008**, *196*, 45–55. [CrossRef]
49. Zhong, H.; Li, X.; Zhou, S.; Jiang, P.; Liu, X.; Ouyang, M.; Nie, Y.; Chen, X.; Zhang, L.; Liu, Y.; et al. Interplay between RAGE and TLR4 Regulates HMGB1-Induced Inflammation by Promoting Cell Surface Expression of RAGE and TLR4. *J. Immunol.* **2020**, *205*, 767–775. [CrossRef]

50. Prantner, D.; Nallar, S.; Vogel, S.N. The role of RAGE in host pathology and crosstalk between RAGE and TLR4 in innate immune signal transduction pathways. *FASEB J.* **2020**, *34*, 15659–15674. [CrossRef]
51. Xu, G.; Zhang, Y.; Jia, H.; Li, J.; Liu, X.; Engelhardt, J.F.; Wang, Y. Cloning and identification of microRNAs in bovine alveolar macrophages. *Mol. Cell Biochem.* **2009**, *332*, 9–16. [CrossRef] [PubMed]

International Journal of
Molecular Sciences

MDPI

Article

A Xanthohumol-Rich Hop Extract Diminishes Endotoxin-Induced Activation of TLR4 Signaling in Human Peripheral Blood Mononuclear Cells: A Study in Healthy Women

Finn Jung [1], Raphaela Staltner [1], Anja Baumann [1], Katharina Burger [1], Emina Halilbasic [2], Claus Hellerbrand [3] and Ina Bergheim [1,*]

[1] Department of Nutritional Sciences, Molecular Nutritional Science, University of Vienna, Josef-Holaubek Platz 2, 1090 Vienna, Austria
[2] Department of Medicine III, Medical University of Vienna, Währinger Gürtel 18-20, 1090 Vienna, Austria
[3] Institute of Biochemistry, Friedrich-Alexander University Erlangen, 91054 Erlangen, Germany
* Correspondence: ina.bergheim@univie.ac.at; Tel.: +43-1-4277-54981

Citation: Jung, F.; Staltner, R.; Baumann, A.; Burger, K.; Halilbasic, E.; Hellerbrand, C.; Bergheim, I. A Xanthohumol-Rich Hop Extract Diminishes Endotoxin-Induced Activation of TLR4 Signaling in Human Peripheral Blood Mononuclear Cells: A Study in Healthy Women. *Int. J. Mol. Sci.* **2022**, *23*, 12702. https://doi.org/10.3390/ijms232012702

Academic Editors: Oliver Planz and Ralf Kircheis

Received: 31 August 2022
Accepted: 17 October 2022
Published: 21 October 2022

Abstract: Infections with Gram-negative bacteria are still among the leading causes of infection-related deaths. Several studies suggest that the chalcone xanthohumol (XN) found in hop (Humulus lupulus) possesses anti-inflammatory effects. In a single-blinded, placebo controlled randomized cross-over design study we assessed if the oral intake of a single low dose of 0.125 mg of a XN derived through a XN-rich hop extract (75% XN) affects lipopolysaccharide (LPS)-induced immune responses in peripheral blood mononuclear cells (PBMCs) ex vivo in normal weight healthy women (n = 9) (clinicaltrials.gov: NCT04847193) and determined associated molecular mechanisms. LPS-stimulation of PBMCs isolated from participants 1 h after the intake of the placebo for 2 h resulted in a significant induction of pro-inflammatory cytokine release which was significantly attenuated when participants had consumed XN. The XN-dependent attenuation of proinflammatory cytokine release was less pronounced 6 h after the LPS stimulation while the release of sCD14 was significantly reduced at this timepoint. The LPS-dependent activation of hTLR4 transfected HEK293 cells was significantly and dose-dependently suppressed by the XN-rich hop extract which was attenuated when cells were co-challenged with sCD14. Taken together, our results suggest even a one-time intake of low doses of XN consumed in a XN-rich hop extract can suppress LPS-dependent stimulation of PBMCs and that this is related to the interaction of the hop compound with the CD14/TLR4 signaling cascade.

Keywords: CD14; LPS; hop; TLR4; inflammation

1. Introduction

Despite a large selection of antibiotics, infections with Gram-negative bacteria are still among the leading causes of morbidity and mortality in the world, also as antibacterial multidrug resistance (AMR) is constantly increasing. Indeed, it has been estimated that in 2019 4.95 million deaths worldwide were associated with AMR. From those, 2.2 million deaths were related with AMR against the four most frequent Gram-negative bacteria *Eschericia coli*, *Klebsiella pneunomoniae*, *Acinetobacter baumannii* and *Pseudomonas aeruginosa* [1]. Furthermore, results of studies in patients with metabolic diseases, e.g., type 2 diabetes and non-alcoholic fatty liver disease (NAFLD) but also in patients with alcohol-related liver disease suggest that Gram-negative bacteria and even more so components of their outer cell wall, e.g., lipopolysaccharides (LPS), may also be an important trigger in the development of these diseases [2]. For instance, it has been shown that the loss or blockage of toll-like receptor 4 (TLR4), attenuates the development of NAFLD and alcohol-related liver diseases and improves insulin signaling in settings of type 2 diabetes [3–5]. TLR4 has been shown to bind bacterial LPS through a complex interplay of cluster of differentiation 14 (CD14) and myeloid differentiation factor 2 (MD-2) [6]. Furthermore, in macrophages and monocytes the LPS-dependent activation of TLR4 signaling has been shown to result in a marked induction of cytokines such as interleukin-1β (IL-1β), interleukin-6 (IL-6) and tumor necrosis factor α (TNF-α) [7,8].

Results of in vivo and in vitro studies suggest that secondary plant compounds found in hop-like α-acids and β-acids as well as xanthohumol (XN) may possess anti-inflammatory effects in various disease settings [9]. For example, despite having been suggested to be poorly absorbed [10], XN has been shown to possess anti-inflammatory effects in the development of NAFLD and insulin resistance but upon stimulation with lipoteichoic acid derived from Gram positive bacteria [11,12]. Furthermore, XN has been suggested to interfere with the TLR4 signaling cascades [13–15]. Specifically, in in vitro cell and molecular docking studies it has been suggested that XN may suppress endotoxin-induced TLR4 activation through interfering with endotoxin binding to MD-2 [13,16]. However, molecular mechanisms underlying the beneficial effects of XN on inflammatory processes induced by LPS are still not fully understood. In addition, the doses of XN used in most studies were within the pharmacological range, i.e., 12 mg/d per person or higher [17,18]. If XN, when ingested at low doses (as they could be found in approximately 250 mL beer (XN concentrations vary between 0.002 and 0.69 mg/L, depending on the type of beer [19,20]), also exerts beneficial effects in humans, e.g., attenuates the inflammatory responses triggered by Gram-negative bacteria or LPS, has not yet been assessed.

In the present study, we therefore aimed to determine if, in healthy, normal weight women, the intake of a low dose of XN (0.125 mg XN derived through a XN-rich hop extract (75% XN), dose based on concentrations found in 250 mL beer [19,20] contained in a beverage consumed along with a light breakfast affects LPS-dependent immune response of peripheral blood mononuclear cells (PBMCs) isolated after the intake of the hop compound. Furthermore, molecular mechanisms underlying the effects of XN on TLR4 signaling cascade were assessed.

2. Results

2.1. Co-Culture Cell Assay

To determine if XN passes the intestinal barrier and binds or is taken up by blood immune cells, a co-culture model, consisting of differentiated Caco-2 cells and isolated human PBMCs obtained from buffy coats, was employed (for experimental design see Figure 1a). PBMCs were spotted on glass slides and counterstained with DAPI (480 nm). One hour after the exposure to 0.125 (Figure 1b), 0.375 and 0.750 mg/mL XN (data not shown), respectively, autofluorescence of XN at 530 nm was detected in PBMCs seeded in the basolateral compartment of the transwell system (indicated with white arrow) whereas no fluorescence was detected at 530 nm in untreated cells. Merged pictures indicate that XN is either bound or taken up by PBMCs. Based on these results, it was concluded that that XN is rapidly taken up/passes through enterocytes and an exposure time of 1 h was selected for the human intervention.

2.2. Effect of the Oral Intake of XN on LPS-Induced Inflammation

Of the 12 normal weight, healthy women enrolled in the study, 9 were analyzed. The study design is summarized in Figure 2a,b. Three women had to be excluded from the analysis as, for some time points, the numbers of cells obtained were not sufficient for the stimulation experiments. Characteristics of the analyzed participants and routine laboratory parameters are shown in Table 1.

As expected, after a 2 h challenge with LPS, protein levels of IL-1β, IL-6, and TNF-α in cell supernatant of PBMCs isolated before the intake of the beverages were significantly higher than in cells only treated with plain cell culture media. In line with these findings, in cell supernatant of LPS-treated PBMCs isolated after the intake of the placebo, protein levels of the three pro-inflammatory cytokines were also significantly higher than in cells without the

LPS challenge. In contrast, when PBMCs isolated 1 h after the intake of XN-rich hop extract were challenged with LPS, protein levels of IL-1β IL-6 and TNF-α in cell supernatant were not significantly different from those without the LPS challenge (Figure 3a–c). Furthermore, when being challenged for 6 h IL-1β, IL-6 and TNF-α protein levels in cell culture supernatant were still higher in LPS-stimulated cells isolated after the ingestion of the placebo than when subjects had consumed the XN enriched beverages; however, differences were less pronounced (Supplementary Figure S1). As protein levels of the three cytokines determined in untreated cells after the ingestion of the placebo and the XN-rich hop extract, respectively, were similar, only those measured after the intake of the placebo are shown.

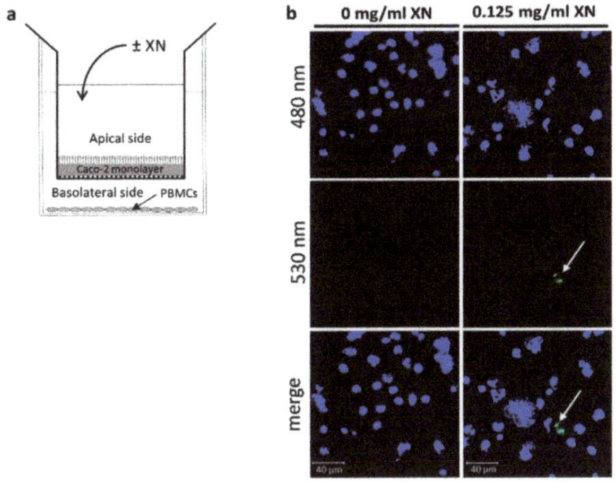

Figure 1. Fluorescence imaging and XN determination of PBMCs isolated from buffy coat of healthy donors co-cultured with Caco-2 cells incubated with XN. (**a**) Graphical illustration of the experimental co-culture setup using Caco-2 cells and PBMCs. (**b**) Representative pictures of fluorescence of XN in PBMCs cells after incubation of Caco-2 cells ± XN (0.125 mg/mL derived through a XN-rich hop extract) for 1 h (magnification 400×) in a co-culture model. White arrows indicate autofluorescence of XN. PBMC, peripheral blood mononuclear cell, XN, xanthohumol. Data are expressed as means ± SEM.

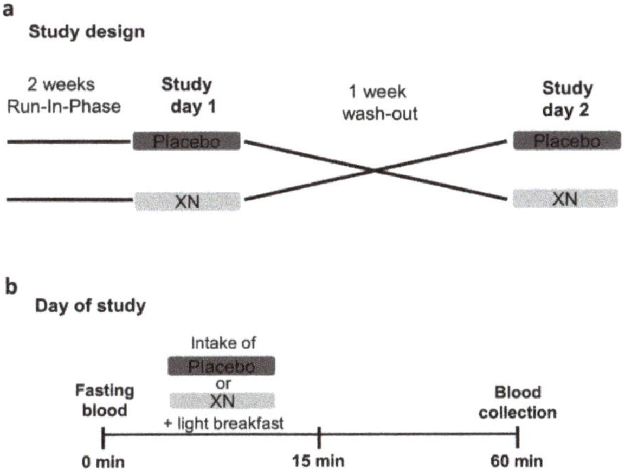

Figure 2. Graphical visualization of the study design. (**a**) Study design and (**b**) the procedure performed on each day of the study. XN, xanthohumol derived though a xanthohumol-rich hop extract.

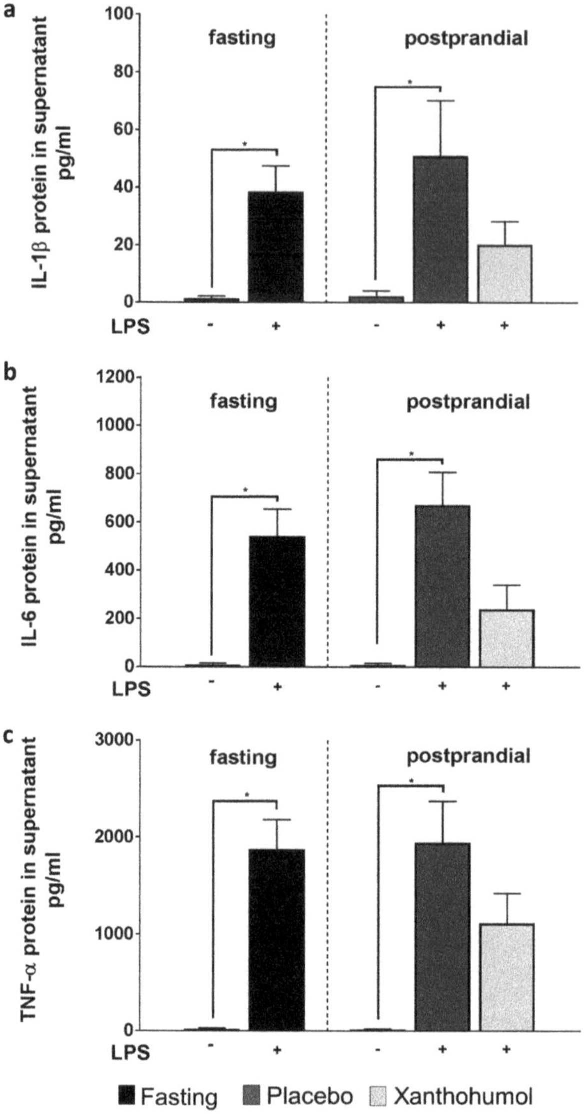

Figure 3. Cytokine concentrations in supernatant of LPS−stimulated PBMCs obtained from healthy study participants. Protein concentrations of IL−−1β (**a**), IL−−6 (**b**) and TNF−−α (**c**) in cell culture supernatant of PBMCs stimulated with 0 or 100 ng/mL LPS for 2 h isolated from healthy study participants receiving either a placebo or the study drink containing XN derived through a XN−rich hop extract. IL, interleukin; LPS, lipopolysaccharide; PBMC, peripheral blood mononuclear cell; XN, xanthohumol. Data are expressed as means ± SEM. * = $p < 0.005$.

Table 1. Anthropometric and health characteristics of study participants. Data are expressed as means ± SEM. AST, aspartate amino transferase; ALT, alanine amino transferase; CRP, c reactive protein; γ-GT, γ-glutamyltransferase; HDL, high density lipoprotein; and LDL, low density lipoprotein.

Parameter	Healthy Participants
Sex (m/f)	0/12
Age (years)	26.1 ± 1.1
Body weight (kg)	60.9 ± 2.1
Height (m)	1.66 ± 0.02
BMI (kg/m^2)	22.1 ± 0.5
Blood pressure	
Systolic (mmHg)	126.6 ± 7.57
Diastolic (mmHg)	78.3 ± 3.6
Fasting glucose (mg/dL)	90.1 ± 2.2
Uric acid (mg/dL)	4.1 ± 0.2
AST (U/L)	29.8 ± 8.7
ALT (U/L)	33.7 ± 16.6
γ-GT (U/L)	12.2 ± 1.2
Cholesterol (mg/dL)	171.8 ± 8.4
HDL cholesterol (mg/dL)	66.4 ± 4.8
LDL cholesterol (mg/dL)	91.6 ± 6.1
Triglycerides (mg/dL)	77.6 ± 9.2
CRP (mg/dL)	0.03 ± 0.03

2.3. Effect of the Oral Intake of XN on MD-2, TLR4 and sCD14 Protein in PBMCs

As it has been reported before that XN may dampen the LPS-dependent TLR4-response of immune cells through MD-2-dependent mechanisms [13,17], we next determined MD−2 and TLR4 protein levels in cells and sCD14 in cell supernatant. Neither protein levels of MD-2 nor of TLR4 were altered by the LPS-challenge of cells isolated before or after ingestion of the placebo and the XN enriched beverage, respectively (Figure 4a–c). In contrast, protein levels of sCD14 were significantly higher in LPS challenged cells isolated before the consumption of the beverages and those isolated 1 h after the intake of the placebo. In contrast, concentrations of sCD14 protein were unchanged in cell supernatant of LPS-challenged cells isolated after the intake of XN (Figure 4d).

2.4. Effect of XN on the LPS-Dependent Activation of TLR4 and the Effect of sCD14 Herein

To further delineate mechanisms underlying the suppressive effects of XN on the LPS-dependent activation of PBMCs, we next determined if XN derived through a XN-rich hop extract alters the LPS-dependent activation of a commercially available HEK blue cells assay, in which cells are transfected with human TLR4, MD-2 and CD14. Results are shown in Figure 5. The XN-rich hop extract attenuated the activation of cells in an almost dose-dependent manner, with 4 µg/mL of XN diminishing the LPS-dependent activation of cells by ~50%. To further determine if the suppressive effects on TLR4 signaling were related to an interaction of the hop compound with MD-2, CD14 or TLR4, we adapted the in vitro assay of Chen et al. [21]. None of the doses of XN (0–8 µg/mL) used affected the binding of biotinylated LPS to MD-2 (Figure 6a) or TLR4 (Figure 6b). In contrast, the binding of biotinylated LPS to CD14 was dose-dependently inhibited by XN (Figure 6c). Furthermore, when stimulating the TLR4, MD-2 and CD14 transfected HEK blue cells with increasing sCD14 concentrations in the presences of 0 or 4 µg/mL XN and 0 or 100 ng/mL LPS, the ~50% suppression of the LPS-dependent activation of cells was attenuated when 1000 ng/mL sCD14 were added to the cell media (Figure 6d).

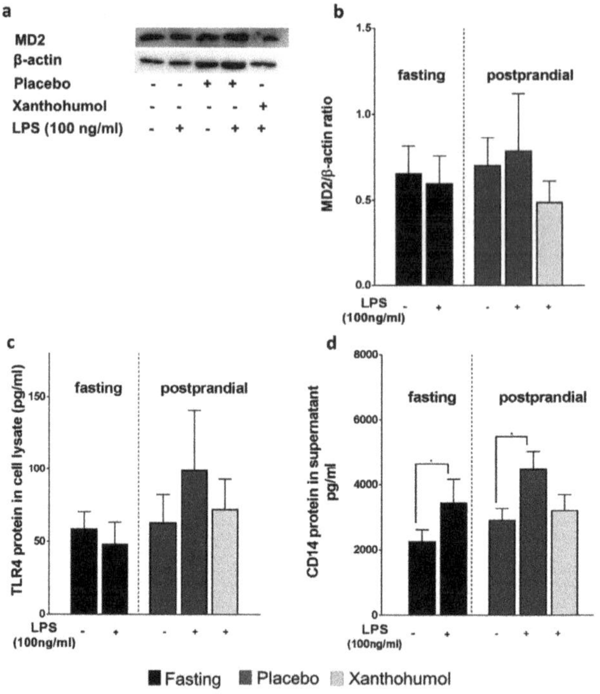

Figure 4. Protein concentration of MD−2, TLR4 and CD14 in LPS−stimulated PBMCs obtained from healthy study participants. Representative blots (**a**) and densitometric analysis of MD−2 western blot (**b**), TLR4 protein concentration in total protein lysate (**c**) and sCD14 protein concentration in cell culture supernatant (**d**) of PBMCs obtained from study participants either receiving a placebo or XN stimulated with LPS (100 ng/mL) for 6 h. LPS, lipopolysaccharide; MD−2, myeloid differentiation factor 2; PBMC, peripheral blood mononuclear cell; sCD14, soluble cluster of differentiation 14; TLR, toll−like receptor; XN, xanthohumol derived through a XN−rich hop extract. Data are expressed as means ± SEM. * = $p < 0.05$.

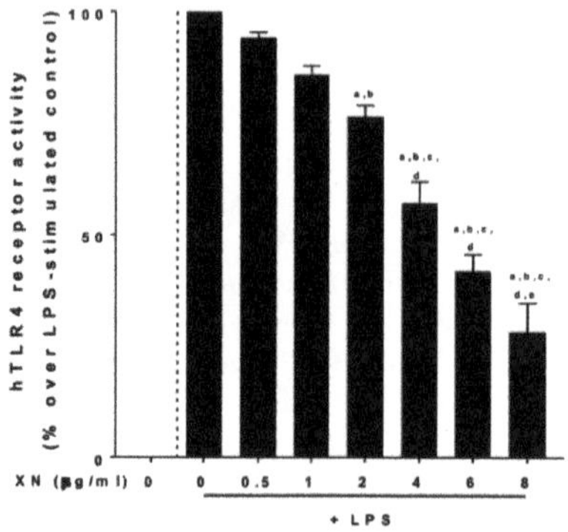

Figure 5. Receptor activities of HEK293 cells co-stimulated with LPS and XN for 12h. HEK293 cells were stimulated with LPS (100 ng/mL) and increasing concentrations of XN (0–8 µg/mL) for 12 h. LPS, lipopolysaccharide; XN, xanthohumol derived through a XN-rich hop extract. Data are expressed as means ± SEM. $^a = p < 0.05$ compared to 0 XN + LPS, $^b = p < 0.05$ compared to 0.5 XN + LPS, $^c = p < 0.05$ compared to 1 XN + LPS, $^d = p < 0.05$ compared to 2 XN, $^e = p < 0.05$ compared to 4 XN + LPS.

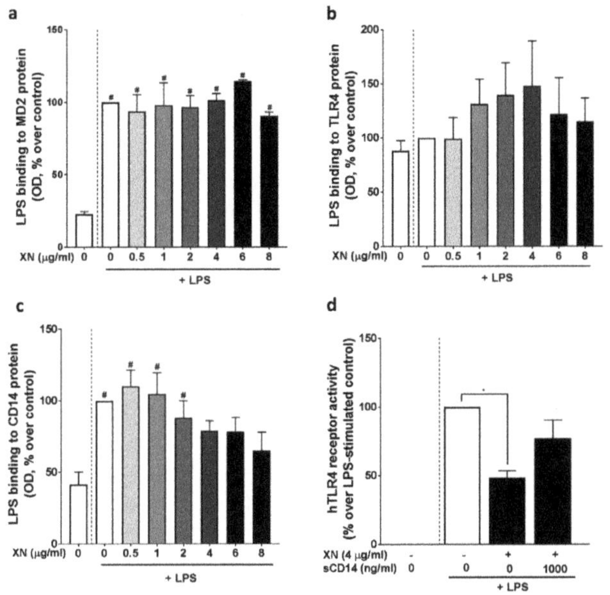

Figure 6. Inhibitory effect of XN on LPS−binding to MD−2, TLR4 and CD14. Effect of increasing concentrations of XN (0–8 µg/mL) on LPS−binding to MD−2 (**a**), TLR4 (**b**) and CD14 (**c**) as well as receptor activity of hTLR4 HEK293 cells co−stimulated with LPS (100 ng/mL), XN (4 µg/mL) and sCD14 (1000 ng/mL) for 12 h (**d**). LPS, lipopolysaccharide; MD−2, myeloid differentiation factor 2; sCD14, soluble cluster of differentiation 14; TLR, toll−like receptor; XN, xanthohumol derived through a XN−rich hop extract. Data are expressed as means ± SEM. # = $p < 0.05$ compared to unstimulated control (0 ng/mL LPS, 0 µg/mL XN). * = $p < 0.05$ compared to LPS−stimulated cells.

3. Discussion

Infections caused by bacteria are common and can lead to severe and even life-threatening health condition frequently demanding extended and cost-intensive treatments. Antibiotics are still the treatment option of choice for most bacterial disease. However, antibiotic resistances are increasingly often limiting therapeutic options [22]. Here, we determined the effect of an acute ingestion of low doses of a XN-derived through a XN-rich hop extract on the LPS-dependent immune responses in isolated PBMCs of healthy young women. The dose of XN used in our study was based on the average XN amount in brewing of ~250 mL beer [19,20] and on studies employing a co-cultural model to mimic the gut/blood barrier, where a clear permeation of XN through a Caco-2 cell monolayer and binding to immune cells was shown when cells were exposed to 0.125 mg of XN. Somewhat in line with the findings of others [10,23], we found that XN rapidly crosses the intestinal barrier here mimicked by Caco-2 cells and binds or is taken up by some cells in the PBMC fraction. Recently, we showed, that XN predominantly binds to the monocyte/ macrophage fraction of PBMCs, while no binding to T- and B-cells were shown in these experiments [14]. Furthermore, while using lower doses than other groups when showing anti-inflammatory effects in mice and rats [11,24,25], in the present study, LPS-dependent activation of PBMCs was dampened by XN derived through a XN-rich hop extract when compared to the placebo. These results are in line with recent findings of our own group showing that and oral intake of XN in doses alike can diminish the LTA-dependent immune responses of PBMCs in humans [14]. Still, the immune response was not completely diminished to that of unstimulated cells but rather, it was dampened by ~43–64% after 2 h and ~17–20% after 6 h, depending on the cytokine measured. Others have suggested before that a total suppression of immune responses to viral or bacterial challenges may be deleterious and may even worsen the severity of the disease and extend recovery [26]. Therefore, a dampening of the immune response may even be more wishful than a total abolishment of the immune response. Taken together, our results suggest that XN derived through a XN-rich hop extract when consumed at low doses may dampen the LPS-dependent immune response of PBMCs in healthy and normal weight women. However, as in the present study PBMCs were exposed to LPS only ex vivo after one acute ingestion of the hop compound, further studies are needed to determine if (1) effects alike are also present in humans suffering from an infection with Gram-negative bacteria and (2) when the compound is consumed repeatedly and (3) when 100% pure XN is used.

Through which mechanisms does XN dampen LPS-dependent activation of PBMCs?

The mechanisms involved in the uptake of XN, and if blood cells bind or take up the hop compound, are still not fully understood. Results of other groups have suggested before that XN may bind to MD-2 and may thereby interfere and dampen endotoxin-dependent activation of TLR4 in immune cells such as monocytes or macrophages [16]. Contrasting these findings, in the present study, while blocking LPS-dependent activation of HEK blue cells transfected with TLR4, MD-2 and CD14, neither LPS binding to MD-2 nor protein levels of MD-2 in LPS stimulated PBMCs were altered by the presence of XN. In contrast, we found that XN, almost in a dose-dependent manner, inhibited the bindings of biotinylated LPS to CD14 in a cell-free in vitro assay and diminished the release of sCD14 from LPS-stimulated PBMCs. Furthermore, sCD14 attenuated the inhibitory effects of XN derived through a XN-rich hop extract on the LPS-dependent activation of HEK blue cells transfected with TLR4. Interestingly, similarly to MD-2, XN derived through the XN-rich hop extract had no effect on TLR4 protein or LPS binding to TLR4. Somewhat in line with these findings, we recently reported that XN also attenuated the LTA-dependent stimulation of TLR2-signaling and that the protective effects of XN on the LTA-dependent activation of TLR2 seemed also to depend upon its effects on CD14 [14]. Furthermore, results of several studies have shown that LPS is bound to CD14 and then transferred to TLR4/MD-2 [27]. It further has been shown that CD14 is critical for the recognition of LPS by TLR4 and MD-2 [28,29]. In addition, other results suggest that CD14 is "shedded" from the cell surface [30] and that this may be critical in the inflammatory response to bacterial toxins [31]. In summary, our data suggest that XN derived through a XN-rich hop extract dampens the LPS-dependent activation of the TLR4 signaling cascade through CD14-dependent mechanisms. Still, our results by no means preclude that XN may also affect other (intra) cellular signaling cascades, especially, when consumed at higher doses and/ or over an extended period of time. Rather, our results suggest, that when consumed at non-pharmacological concentrations, XN derived through a XN-rich hop extract may at least temporally attenuate the LPS/CD14-dependent activation of the TLR4-signaling cascade in blood immune cells.

Limitations

When interpreting the data of the present study, some limitations need to be considered. For one, all cell stimulation experiments were carried out ex vivo with cells stemming only from healthy, normal weight women. Therefore, additional studies need to assess if these effects are also found in patients infected with Gram-negative bacteria. Furthermore, here, XN was only eaten once with PBMCs being isolated 1 h after the intake. Accordingly, from the presented data, no estimations regarding persistence and long-term effects of XN on bacterial toxin-triggered immune response can be made. Furthermore, in the present study, for ethical and safety reasons, XN was presented to study participants in form of a commercially available hop extract containing 75% XN but also other hop compounds (e.g., prenylflavonoids and geranyl flavonoids). It has been shown before by others that these by-products are so low in concentration, that physiological effects are considered unlikely [32]. To be consistent within the study, the same extract used in the human intervention was also employed in the cell culture experiments. In addition, in the present study, we only assessed effects in healthy, normal weight women. If effects alike are also found in overweight and/or older individuals as well as male subjects needs to be determined in future studies. Another limitation that needs to be taken in consideration is the use of a commercially available transfected HEK blue cell system to assess the effects of XN derived through the XN-rich hop extract on CD14/MD−2/TLR4 signaling. Indeed, while these cells are transfected with CD14, MD−2 and TLR4 it cannot be ruled out that this cell culture system does not resemble all molecular interactions found in isolated immune cells and in vivo, respectively. Moreover, in the present study the blockage of LPS-binding to CD14 by XN derived through the XN-rich hop extract was only shown in a cell-free in vitro assay bearing some limitations when extrapolating results to the in vivo situation.

4. Materials and Methods

4.1. Co-Culture Cell Assay

To determine the uptake and binding of XN to PBMCs, a co-culture model of Caco-2 cells and human PBMCs, isolated from buffy coat was employed (also see Figure 1a for study design). In brief, Caco-2 cells were grown according to the instructions of the manufacturer in a humidified, 5% carbon dioxide atmosphere to 100% confluence in semipermeable transwells by using DMEM medium containing 10% fetal bovine serum (FBS) (Pan-Biotech GmbH, Aidenbach, Germany), 100 µg/mL streptomycin and 100 U/mL penicillin (Pan-Biotech GmbH, Aidenbach, Germany). Integrity of Caco-2 monolayer was checked daily for 9 days by transepithelial electrical resistance (TEER). Once the Caco-2 cell monolayer had reached confluency and TEER was stable (day 9), PBMCs isolated from buffy coats of healthy donors were isolated by gradient centrifugation as detailed before by others [33] and were seeded in the basolateral compartment below the transwell containing the differentiated Caco-2 cells. Caco-2 cells were then incubated with XN (0–0.750 mg/mL) derived from a XN-rich hop extract for 1 and 2 h. Cell culture medium was collected in the basolateral compartment and PBMCs were washed, fixed on glass slides via cytospin and stained with DAPI (Sigma-Aldrich Corp., St. Louis, MO, USA). DAPI staining was detected using DAPI filter system included in the microscope (Leica Camera AG, Wetzlar, Germany). Binding of XN to cells was determined as detailed before by others [34,35]. In brief, spotted cells were excited at 480/40 nm and emission was detected at 530 nm to detect autofluorescence of XN.

4.2. Study Participants

Based on sample size calculations assuming 1–2 drop-outs, a total of 12 normal weight healthy female subjects were enrolled in the study. As the yield of PBMCs for the ex vivo stimulation experiments was insufficient at some time points, three of the subjects had to be excluded from the final analysis. Previous studies of others [36] assessing the reduction in pro-inflammatory cytokine release in human PBMCs were used to estimate the sample size via power analysis (a priori) (Gpower, Version 3.1.9.2). Only participants not reporting food intolerances or food allergies that would require a particular dietary intervention were enrolled. All study participants confirmed the absence of metabolic diseases, chronic inflammatory diseases or viral and bacterial infections within the last 3 weeks before the study. Furthermore, the use of anti-inflammatory medication was defined as exclusion criteria for the study. The study, which is part of a larger project assessing the acute and chronic effect of the intake of hop compounds such as XN and iso-α-acids on the immune response of bacterial-toxin-activated PBMCs (ClinicalTrials.gov: NCT04847193), was approved by the Ethics Committee of the University of Vienna, Vienna, Austria (00367) and was carried out in accordance with the ethical standards laid down in the Declaration of Helsinki of 1975 as revised in 1983.

4.3. Intervention Study

Before the intervention the participants underwent a wash-out phase for 14 days during which they refrained from all hops containing products such as beer. After the wash-out, participants were then randomly and single-blinded assigned to receive a study drink containing 10 mL water, thickener (Nestlé S.A., Vevey, Switzerland), 70 mg skim milk powder, and lemon flavor (Pepsico Inc., Purchase, NY, USA) enriched with 0.125 mg XN (XanthoFlav, generous gift from Hopsteiner GmbH, Au an der Hallertau, Germany) or a placebo. XanthoFlavTM consists of XN (75%) and other prenylated flavonoids (<25%) occurring naturally in hops (for further details, also see: https://www.hopsteiner.com/wp-content/uploads/2021/12/26_21_ls_xanthoflav.pdf, accessed on 10 August 2022). Among these prenylated flavonoids kaempferol-3-O-β-D-glucopyranosid, cis-/trans-p-coumaric acid methylester and n-multifidol-di-C-glucopyranoside are the quantitively most common flavonoids next to XN. However, concentrations are very low and only reliably verifiable by LC-MS [33]. The placebo was similar to the study drink but lacked the addition of XN. The study drink was consumed within 15 min in combination with a standardized breakfast containing 2 medium sized pretzels and 30 g butter. Before, in fasted state, and 60 min after the intake of the study drink, blood was collected, and PBMCs were isolated as detailed below. After a second wash-out phase lasting at least 7 days, in which the participants were again asked to refrain from all hops containing foods and beverages, the intervention was repeated in a cross-over design.

4.4. Isolation and Culture of PBMCs

Following the manufacturer instructions, a Vacutainer® CPTTM System (Becton Dickinson GmbH, Heidelberg, Germany) was used to isolate PBMCs from whole blood samples acquired from each participant at fasted state and after the consumption of the combination of study drink and the standardized breakfast. Isolated PBMCs were then cultivated in RPMI-1640 medium (Sigma-Aldrich Corp., St. Louis, MO, USA) with 10% fetal bovine serum (Pan-Biotech GmbH, Aidenbach Germany), 100 µg/mL streptomycin and 100 U/mL penicillin (Pan-Biotech GmbH, Aidenbach, Germany) for 1 h. Subsequently, cells were either challenged with 0 or 100 ng/mL LPS for 2 h and 6 h, respectively. Cell culture supernatant was collected and PBMCs were lysed using RIPA buffer to obtain total protein. Supernatant and protein were stored at −80 °C until further use.

4.5. Cell Culture Experiments with hTLR4 Transfected HEK Cells Response

To assess the effects of XN on CD14/TLR4 signaling, a commercially available reporter gene assay with HEK-BlueTM TLR4 cells, co-transfected with human TLR4, human MD-2 and human CD14 as well as an inducible secreted embryonic alkaline phosphatase (SEAP) fused to nuclear factor 'kappa-light-chain-enhancer' of B-cells (NF-κB) and activator protein 1 (AP-1) was used (InvivoGen, CA, USA, Cat.Number: hTLR4 = hkb-htlr4). Following the instructions of the manufacturer, cells were grown in a humidified, 5% carbon dioxide atmosphere and were grown up to 80% confluence using DMEM media (Pan-Biotech, GmbH, Aidenbach, Germany). In a first set of experiments, cells were challenged with 100 ng/mL LPS for 12 h in the presence of 0–8 µg/mL XN derived through the XN-rich hop extract. Activity of TLR4 was indirectly determined by measuring SEAP induced color change of cell culture medium at 655 nm. In a second set of experiments, cells were again grown to 80% confluence and challenged with 0 or 100 ng/mL LPS in the presence of 0 or 4 µg/mL XN and 0–1000 ng/mL sCD14. The XN dose used in this experiment was determined in the first set of experiments. After 12 h, color changes of medium were determined at 655 nm.

4.6. Western Blot

RIPA buffer was used to extract total protein from cells as detailed before [37]. Protein extracts and cell culture supernatant, respectively, were separated in a 10% polyacrylamide gel and were transferred on a polyvinylidene difluoride membrane (Bio-Rad Laboratories, Hercules, CA, USA). Membranes were blocked with either 3% skim milk powder or 5% BSA for 1 h prior to the incubation with a specific monoclonal antibody for MD-2 (#NB100-56655; Novus Bio, Centennial, CO, USA) followed by a 1 h incubation with an appropriate secondary antibody (#7074S for MD-2, Cell Signaling Technology Inc., MA, USA). Bands were detected using Super Signal West Dura kit (Thermo Fisher Scientific, Waltham, MA, USA). Densitometric analysis of bands was performed using Image Lab 6.0 Software (Bio-Rad Laboratories, Hercules, CA, USA).

4.7. Enzyme-Linked Immunosorbent Assays (ELISA's)

IL-1β, IL-6, TNF-α, TLR4 and sCD14 protein concentrations were analyzed in cell culture supernatant and extract from total protein using commercially available ELISA kits (IL-1β: Sigma-Aldrich Corp., St. Louis, MO, USA; IL-6: Bio-Techne Corp., Minneapolis, MN, USA; TNF-α: BioVendor R&D®, Czech Republic; CD14: Sigma-Aldrich Corp., St. Louis, MO, USA).

4.8. LPS Binding Assays

To determine if XN binds to MD-2, CD14 or TLR4, a LPS binding assay adapted from Zhang et al. was used [38]. In brief, polystyrene 96 well plates (R&D Systems Inc., Minneapolis, MN, USA) were coated with MD-2 (#NB100-56655; Novus Bio, Centennial, CO, USA), CD14 (#56082S, Cell Signaling Technology Inc, Danvers, MA, USA) or TLR4 (#NB100-56723SS; Novus Bio, Centennial, CO, USA) antibodies over night at 4 °C. Plates were blocked with 3% BSA for 2 h followed by an incubation with their respective recombinant protein (MD-2: 1787-MD, Bio-Techne Corp., Minneapolis, MN, USA; CD14: 110-01, PeproTech, Cranbury, NJ, USA; TLR4: 160-06, PeproTech, Cranbury, NJ, USA) (1 μg/mL) for 1.5 h. XN in concentrations ranging from 0–8 μg/mL derived through the XN-rich hop extract was added to the plate in presence or absence of LPS-biotin (50 ng/mL, InvivoGen, CA, USA,) for 1 h. After a 1 h incubation with HRP-Streptavidin (Bio-Techne Corp., Minneapolis, MN, USA) plates were incubated with TMB (Thermo Fisher Scientific, Waltham, MA, USA) for another 15 min. Excess antibody and streptavidin-HRP solution was removed by washing with 0.05% PBST. To stop the reaction, 2 N H_2SO_4 was added to the plate and absorbance was measured at 450 nm.

4.9. Statistical Analyses

Data are presented as means ± standard error of the means (SEMs). The Friedman test with Dunn's multiple comparison test and the Wilcoxon test was used to determine statistically significant differences between interventions and one-way ANOVA was used for all other comparisons (GraphPad Prism Software, CA, USA). $p \leq 0.05$ was selected as the level of significance.

5. Conclusions

Taken together, the results of our study suggest that the acute consumption of low doses of XN derived through a XN-rich hop extract may dampen the LPS-dependent immune response of PBMCs in healthy women. Our results further suggest that the beneficial effects of the hop compound are related to an inhibition of the binding of LPS to CD14. While in the present study it was shown that the immunosuppressed effects of XN are found within 1 h after the oral intake of XN derived through a XN-rich hop extract, further study is needed to determine dose- and time-responses as well as the exact mechanisms underlying the inhibitory effects of XN.

Supplementary Materials: The following supporting information can be downloaded at: https://www.mdpi.com/article/10.3390/ijms232012702/s1.

Author Contributions: Conceptualization, C.H. and I.B.; formal analysis: F.J., R.S., K.B. and A.B.; investigation: F.J., R.S., K.B. and A.B.; writing—original draft preparation: F.J. and I.B.; writing—review and editing: F.J., R.S., K.B., A.B., E.H. and I.B.; supervision: I.B.; project administration: F.J. and I.B.; funding acquisition: I.B. All authors have read and agreed to the published version of the manuscript.

Funding: This research received was funded by the Joint Grant of the European Foundation for Alcohol Research (ERAB) and European Brewery Convention (EBC) (EEP 18 07 to IB and CH), the Wissenschaftsförderung der Deutschen Brauwirtschaft e.V. (WiFö) (B115 to IB and CH) and in part by the Austrian Science Fund (FW, P-35165 to IB). Open Access Funding by the Austrian Science Fund (FWF).

Institutional Review Board Statement: The present study was approved by the Ethics Committee of the University of Vienna, Vienna, Austria (00367) and was carried out in accordance with the ethical standards laid down in the Declaration of Helsinki of 1975 as revised in 1983. The study is registered at ClinicalTrials.gov (NCT04847193).

Informed Consent Statement: Informed consent was obtained from all subjects involved in the study.

Conflicts of Interest: The authors declare no conflict of interest.

References

1. Antimicrobial Resistance Collaborators. Global burden of bacterial antimicrobial resistance in 2019: A systematic analysis. *Lancet* **2022**, *399*, 629–655. [CrossRef]
2. Massier, L.; Blüher, M.; Kovacs, P.; Chakaroun, R.M. Impaired Intestinal Barrier and Tissue Bacteria: Pathomechanisms for Metabolic Diseases. *Front. Endocrinol.* **2021**, *12*, 616506. [CrossRef] [PubMed]
3. Hritz, I.; Mandrekar, P.; Velayudham, A.; Catalano, D.; Dolganiuc, A.; Kodys, K.; Kurt-Jones, E.; Szabo, G. The critical role of toll-like receptor (TLR) 4 in alcoholic liver disease is independent of the common TLR adapter MyD88. *Hepatology* **2008**, *48*, 1224–1231. [CrossRef] [PubMed]
4. Lu, Z.; Zhang, X.; Li, Y.; Lopes-Virella, M.F.; Huanga, Y. TLR4 antagonist attenuates atherogenesis in LDL receptor-deficient mice with diet-induced type 2 diabetes. *Immunobiology* **2015**, *220*, 1246–1254. [CrossRef]
5. Spruss, A.; Kanuri, G.; Wagnerberger, S.; Haub, S.; Bischoff, S.C.; Bergheim, I. Toll-like receptor 4 is involved in the development of fructose-induced hepatic steatosis in mice. *Hepatology* **2009**, *50*, 1094–1104. [CrossRef]
6. Lu, Y.C.; Yeh, W.C.; Ohashi, P.S. LPS/TLR4 signal transduction pathway. *Cytokine* **2008**, *42*, 145–151. [CrossRef]
7. Hege, M.; Jung, F.; Sellmann, C.; Jin, C.; Ziegenhardt, D.; Hellerbrand, C.; Bergheim, I. An iso-alpha-acid-rich extract from hops (Humulus lupulus) attenuates acute alcohol-induced liver steatosis in mice. *Nutrition* **2018**, *45*, 68–75. [CrossRef]
8. Hsu, H.-Y.; Hua, K.-F.; Lin, C.-C.; Lin, C.-H.; Hsu, J.; Wong, C.-H. Extract of Reishi polysaccharides induces cytokine expression via TLR4-modulated protein kinase signaling pathways. *J. Immunol.* **2004**, *173*, 5989–5999. [CrossRef]
9. Iniguez, A.B.; Zhu, M.J. Hop bioactive compounds in prevention of nutrition-related noncommunicable diseases. *Crit. Rev. Food Sci. Nutr.* **2020**, *61*, 1900–1913. [CrossRef]
10. Legette, L.; Karnpracha, C.; Reed, R.L.; Choi, J.; Bobe, G.; Christensen, J.M.; Rodriguez-Proteau, R.; Purnell, J.Q.; Stevens, J.F. Human pharmacokinetics of xanthohumol, an antihyperglycemic flavonoid from hops. *Mol. Nutr. Food Res.* **2014**, *58*, 248–255. [CrossRef]
11. Dorn, C.; Kraus, B.; Motyl, M.; Weiss, T.S.; Gehrig, M.; Schölmerich, J.; Heilmann, J.; Hellerbrand, C. Xanthohumol, a chalcon derived from hops, inhibits hepatic inflammation and fibrosis. *Mol. Nutr. Food Res.* **2010**, *54* (Suppl. S2), S205–S213. [CrossRef]
12. Miranda, C.L.; Elias, V.D.; Hay, J.J.; Choi, J.; Reed, R.L.; Stevens, J. Xanthohumol improves dysfunctional glucose and lipid metabolism in diet-induced obese C57BL/6J mice. *Arch. Biochem. Biophys.* **2016**, *599*, 22–30. [CrossRef]
13. Cho, Y.-C.; Kim, H.J.; Kim, Y.-J.; Lee, K.Y.; Choi, H.J.; Lee, I.-S.; Kang, B.Y. Differential anti-inflammatory pathway by xanthohumol in IFN-gamma and LPS-activated macrophages. *Int. Immunopharmacol.* **2008**, *8*, 567–573. [CrossRef]
14. Jung, F.; Staltner, R.; Tahir, A.; Baumann, A.; Burger, K.; Halilbasic, E.; Hellerbrand, C.; Bergheim, I. Oral intake of xanthohumol attenuates lipoteichoic acid-induced inflammatory response in human PBMCs. *Eur. J. Nutr.* **2022**, 61. [CrossRef]
15. Peluso, M.R.; Miranda, C.L.; Hobbs, D.J.; Proteau, R.R.; Stevens, J.F. Xanthohumol and related prenylated flavonoids inhibit inflammatory cytokine production in LPS-activated THP-1 monocytes: Structure-activity relationships and in silico binding to myeloid differentiation protein-2 (MD-2). *Planta Med.* **2010**, *76*, 1536–1543. [CrossRef]
16. Chen, G.; Xiao, B.; Chen, L.; Bai, B.; Zhang, Y.; Xu, Z.; Fu, L.; Liu, Z.; Li, X.; Zhao, Y.; et al. Discovery of new MD2-targeted anti-inflammatory compounds for the treatment of sepsis and acute lung injury. *Eur. J. Med. Chem.* **2017**, *139*, 726–740. [CrossRef]
17. Langley, B.O.; Ryan, J.J.; Hanes, D.; Phipps, J.; Stack, E.; Metz, T.O.; Stevens, J.F.; Bradley, R. Xanthohumol Microbiome and Signature in Healthy Adults (the XMaS Trial): Safety and Tolerability Results of a Phase I Triple-Masked, Placebo-Controlled Clinical Trial. *Mol. Nutr. Food Res.* **2021**, *65*, e2001170. [CrossRef]
18. Pichler, C.; Ferk, F.; Al-Serori, H.; Huber, W.; Jäger, W.; Waldherr, M.; Mišík, M.; Kundi, M.; Nersesyan, A.; Herbacek, I.; et al. Xanthohumol Prevents DNA Damage by Dietary Carcinogens: Results of a Human Intervention Trial. *Cancer Prev. Res.* **2017**, *10*, 153–160. [CrossRef]
19. Stevens, J.F.; Page, J.E. Xanthohumol and related prenylflavonoids from hops and beer: To your good health! *Phytochemistry* **2004**, *65*, 1317–1330. [CrossRef]
20. Stevens, J.F.; Taylor, A.W.; Clawson, J.E.; Deinzer, M.L. Fate of xanthohumol and related prenylflavonoids from hops to beer. *J. Agric. Food Chem.* **1999**, *4*, 2421–2428. [CrossRef]

21. Chen, H.; Zhang, Y.; Zhang, W.; Liu, H.; Sun, C.; Zhang, B.; Bai, B.; Wu, D.; Xiao, Z.; Lum, H.; et al. Inhibition of myeloid differentiation factor 2 by baicalein protects against acute lung injury. *Phytomedicine* **2019**, *63*, 152997. [CrossRef]
22. Ventola, C.L. The antibiotic resistance crisis: Part 1: Causes and threats. *Pharmacol. Ther.* **2015**, *40*, 277–283.
23. Ferk, F.; Mišík, M.; Nersesyan, A.; Pichler, C.; Jäger, W.; Szekeres, T.; Marculescu, R.; Poulsen, H.E.; Henriksen, T.; Bono, R.; et al. Impact of xanthohumol (a prenylated flavonoid from hops) on DNA stability and other health-related biochemical parameters: Results of human intervention trials. *Mol. Nutr. Food Res.* **2016**, *60*, 773–786. [CrossRef]
24. Chen, X.; Li, Z.; Hong, H.; Wang, N.; Chen, J.; Lu, S.; Zhang, H.; Zhang, X.; Bei, C. Xanthohumol suppresses inflammation in chondrocytes and ameliorates osteoarthritis in mice. *Biomed. Pharmacother.* **2021**, *137*, 111238. [CrossRef] [PubMed]
25. Zhang, M.; Zhang, R.; Zheng, T.; Chen, Z.; Ji, G.; Peng, F.; Wang, W. Xanthohumol Attenuated Inflammation and ECM Degradation by Mediating HO-1/C/EBPbeta Pathway in Osteoarthritis Chondrocytes. *Front. Pharmacol.* **2021**, *12*, 680585. [CrossRef] [PubMed]
26. Bouchard-Boivin, F.; Désy, O.; Béland, S.; Houde, I.; De Serres, S.A. TNF-alpha Production by Monocytes Stimulated with Epstein-Barr Virus-Peptides as a Marker of Immunosuppression-Related Adverse Events in Kidney Transplant Recipients. *Kidney Int. Rep.* **2019**, *4*, 1446–1453. [CrossRef]
27. Ciesielska, A.; Matyjek, M.; Kwiatkowska, K. TLR4 and CD14 trafficking and its influence on LPS-induced pro-inflammatory signaling. *Cell Mol. Life Sci.* **2021**, *78*, 1233–1261. [CrossRef] [PubMed]
28. Da Silva Correia, J.; Soldau, K.; Christen, U.; Tobias, P.S.; Ulevitch, R.J. Lipopolysaccharide is in close proximity to each of the proteins in its membrane receptor complex. transfer from CD14 to TLR4 and MD-2. *J. Biol. Chem.* **2001**, *276*, 21129–21135. [CrossRef]
29. Da Silva Correia, J.; Ulevitch, R.J. MD-2 and TLR4 N-linked glycosylations are important for a functional lipopolysaccharide receptor. *J. Biol. Chem.* **2002**, *277*, 1845–1854. [CrossRef]
30. Bazil, V.; Strominger, J.L. Shedding as a mechanism of down-modulation of CD14 on stimulated human monocytes. *J. Immunol.* **1991**, *147*, 1567–1574.
31. Rokita, E.; Menzel, E.J. Characteristics of CD14 shedding from human monocytes. Evidence for the competition of soluble CD14 (sCD14) with CD14 receptors for lipopolysaccharide (LPS) binding. *APMIS* **1997**, *105*, 510–518. [CrossRef] [PubMed]
32. Dresel, M.; Dunkel, A.; Hofmann, T. Sensomics analysis of key bitter compounds in the hard resin of hops (*Humulus lupulus* L.) and their contribution to the bitter profile of Pilsner-type beer. *J. Agric. Food Chem.* **2015**, *63*, 3402–3418. [CrossRef] [PubMed]
33. Menck, K.; Behme, D.; Pantke, M.; Reiling, N.; Binder, C.; Pukrop, T.; Klemm, F. Isolation of human monocytes by double gradient centrifugation and their differentiation to macrophages in teflon-coated cell culture bags. *J. Vis. Exp.* **2014**, *91*, e51554. [CrossRef] [PubMed]
34. Motyl, M.; Kraus, B.; Heilmann, J. Pitfalls in cell culture work with xanthohumol. *Pharmazie* **2012**, *67*, 91–94.
35. Wolff, H.; Motyl, M.; Hellerbrand, C.; Heilmann, J.; Kraus, B. Xanthohumol uptake and intracellular kinetics in hepatocytes, hepatic stellate cells, and intestinal cells. *J. Agric. Food Chem.* **2011**, *59*, 12893–12901. [CrossRef]
36. Capó, X.; Martorell, M.; Sureda, A.; Batle, J.M.; Tur, J.A.; Pons, A. Docosahexaenoic diet supplementation, exercise and temperature affect cytokine production by lipopolysaccharide-stimulated mononuclear cells. *J. Physiol. Biochem.* **2016**, *72*, 421–434. [CrossRef]
37. Brandt, A.; Jin, C.J.; Nolte, K.; Sellmann, C.; Engstler, A.J.; Bergheim, I. Short-Term Intake of a Fructose-, Fat- and Cholesterol-Rich Diet Causes Hepatic Steatosis in Mice: Effect of Antibiotic Treatment. *Nutrients* **2017**, *9*, 1013. [CrossRef]
38. Zhang, Y.; Wu, J.; Ying, S.; Chen, G.; Wu, B.; Xu, T.; Liang, G. Discovery of new MD2 inhibitor from chalcone derivatives with anti-inflammatory effects in LPS-induced acute lung injury. *Sci. Rep.* **2016**, *6*, 25130. [CrossRef]

International Journal of
Molecular Sciences

MDPI

Article

Advanced Glycation End Products and Activation of Toll-like Receptor-2 and -4 Induced Changes in Aquaporin-3 Expression in Mouse Keratinocytes

Yonghong Luo [1], Rawipan Uaratanawong [1,2], Vivek Choudhary [1,3], Mary Hardin [1], Catherine Zhang [1], Samuel Melnyk [1], Xunsheng Chen [1] and Wendy B. Bollag [1,3,4,5,*]

1 Department of Physiology, Medical College of Georgia at Augusta University, Augusta, GA 30912, USA
2 Department of Medicine (Dermatology), Faculty of Medicine, Vajira Hospital, Navamindradhiraj University, Bangkok 10300, Thailand
3 Charlie Norwood VA Medical Center, Augusta, GA 30904, USA
4 Department of Dermatology, Medical College of Georgia, Augusta University, Augusta, GA 30912, USA
5 Department of Medicine, Medical College of Georgia, Augusta University, Augusta, GA 30912, USA
* Correspondence: wbollag@augusta.edu or wendy.bollag@va.gov; Tel.: +1-706-721-0698

Abstract: Prolonged inflammation and impaired re-epithelization are major contributing factors to chronic non-healing diabetic wounds; diabetes is also characterized by xerosis. Advanced glycation end products (AGEs), and the activation of toll-like receptors (TLRs), can trigger inflammatory responses. Aquaporin-3 (AQP3) plays essential roles in keratinocyte function and skin wound re-epithelialization/re-generation and hydration. Suberanilohydroxamic acid (SAHA), a histone deacetylase inhibitor, mimics the increased acetylation observed in diabetes. We investigated the effects of TLR2/TLR4 activators and AGEs on keratinocyte AQP3 expression in the presence and absence of SAHA. Primary mouse keratinocytes were treated with or without TLR2 agonist Pam_3Cys-Ser-$(Lys)_4$ (PAM), TLR4 agonist lipopolysaccharide (LPS), or AGEs, with or without SAHA. We found that (1) PAM and LPS significantly upregulated AQP3 protein basally (without SAHA) and PAM downregulated AQP3 protein with SAHA; and (2) AGEs (100 µg/mL) increased AQP3 protein expression basally and decreased AQP3 levels with SAHA. PAM and AGEs produced similar changes in AQP3 expression, suggesting a common pathway or potential crosstalk between TLR2 and AGEs signaling. Our findings suggest that TLR2 activation and AGEs may be beneficial for wound healing and skin hydration under normal conditions via AQP3 upregulation, but that these pathways are likely deleterious in diabetes chronically through decreased AQP3 expression.

Keywords: toll-like receptor-2 (TLR2); TLR4; advanced glycation end products (AGEs); aquaporin-3 (AQP3); histone deacetylase inhibitor; diabetes; inflammation; keratinocytes; skin

Citation: Luo, Y.; Uaratanawong, R.; Choudhary, V.; Hardin, M.; Zhang, C.; Melnyk, S.; Chen, X.; Bollag, W.B. Advanced Glycation End Products and Activation of Toll-like Receptor-2 and -4 Induced Changes in Aquaporin-3 Expression in Mouse Keratinocytes. *Int. J. Mol. Sci.* **2023**, *24*, 1376. https://doi.org/10.3390/ijms24021376

Academic Editors: Oliver Planz and Ralf Kircheis

Received: 1 July 2022
Revised: 3 January 2023
Accepted: 4 January 2023
Published: 10 January 2023

1. Introduction

Impaired wound healing is one of the major complications of diabetes mellitus. This impairment creates an enormous financial burden on and stress to patients as well as therapeutic challenges for physicians. Unlike acute wounds, wounds with impaired healing do not progress through the four dynamic and overlapping phases necessary for proper healing: hemostasis, inflammation, proliferation, and remodeling. Acute wound healing shows a quick inflammatory response and rapid resolution. In contrast, chronic wound healing is characterized by excessive inflammation and impaired re-epithelization [1,2]. Another characteristic of diabetes is xerosis, i.e., skin dryness resulting from reduced hydration [3]. Aquaporin-3 (AQP3), a water channel that can also transport glycerol and hydrogen peroxide [4–9], has been shown to play an important role in regulating proliferation, differentiation and migration of skin epidermal keratinocytes [10–20], as well as skin function, including the water permeability barrier and skin hydration in vivo [20,21]. For example, Verkman and colleagues demonstrated that global AQP3 knockout mice exhibit a skin phenotype of decreased water-holding capacity (hydration) and delayed water permeability barrier recovery, as well as impaired skin wound healing [20,21]. Indeed, in vivo studies have indicated that a diabetes-associated reduction in AQP3 levels may contribute to the skin xerosis observed in diabetes [10,22]. The importance of changes in AQP3 in wound healing in diabetes has also been demonstrated, using a streptozotocin (STZ)-induced diabetic rat model in which impaired re-epithelialization correlated with reduced AQP3 expression during the wound healing process [23]. In addition, siRNA-mediated knockdown of AQP3 in normal human keratinocytes reduces proliferation and migration in wound healing in vitro [10]. Finally, it was recently shown that down-regulation of AQP3 in STZ-induced diabetic mice was not the result of hyperglycemia per se, since one week after STZ injection, serum glucose levels were increased without an accompanying reduction in AQP3 expression. However, a week later (or two weeks after STZ injection), although serum glucose was elevated to a similar level as at one week, AQP3 mRNA and protein levels were decreased [24], suggesting that a diabetic product might need to accumulate to induce AQP3 reduction. We hypothesized that this product might be advanced glycation end products (AGEs).

AGEs are formed through a complicated biochemical process involving non-enzymatic reactions between reducing sugars and free amino groups on proteins, lipids, or nucleic acids. This process is accelerated under chronic hyperglycemic and oxidative stress conditions, which results in AGEs accumulation in high amounts in diabetes [25,26]. In turn, AGEs, through interaction with the receptor for AGEs (RAGE), can activate downstream intracellular signaling pathways that lead to further oxidative stress and production of pro-inflammatory mediators [25,27]. In a diabetic mouse model AGEs have been shown to impair wound healing via delayed infiltration of inflammatory cells, sustained expression of inflammatory mediators, and diminished re-epithelization, and these adverse effects can be inhibited via the use of blockers of RAGE [28,29].

On the other hand, diabetes not only results in increased serum levels of AGEs but has also been found to enhance protein acetylation in various cell types both in vitro and in vivo [30–33]. This effect is presumed to be mediated by diabetes-related hyperglycemia-induced increases in the generation of the acetyl-CoA required for lysine acetylation. Protein acetylation can also be enhanced by treating cells with the pan-histone deacetylase inhibitor, suberanilohydroxamic acid (SAHA). Therefore, it is possible that SAHA treatment of mouse keratinocytes may mimic some of the effects of diabetes on protein acetylation. Diabetes is also characterized by inflammation, as well as increased serum levels of high-mobility group box 1 (HMGB1) protein [34], an endogenous protein reported to activate toll-like receptors (TLRs), such as TLR2 and TLR4 [34]. TLR2 and TLR4 activation has been demonstrated to both enhance and impair skin wound healing [35]. TLR2 or TLR4 activation during the early healing process improves wound closure, and long-term deficiency of these TLRs delays normal wound healing [1,36,37]. In contrast, in diabetic wounds, the observed extensive expression of TLR2 and TLR4 seems to contribute to increased inflammation and impaired wound closure, while their knockout improves healing [38,39].

The link between AGEs, as well as TLR activation, and AQP3 expression has not been explored. Therefore, we activated TLR2, TLR4, or RAGE with the triacylated synthetic lipopeptide, Pam_3CSK_4 (PAM), lipopolysaccharide (LPS) or AGEs, respectively, in the presence or absence of SAHA, to examine AQP3 expression in primary mouse keratinocytes basally and in the presence of SAHA. Treatment with SAHA served to increase the low-level basal expression of AQP3 [40], such that inhibitory effects might be more readily detected. SAHA also enhances protein acetylation, which has been observed in the diabetic setting [30–33].

2. Results

2.1. AGE-BSA Used Initially Was Contaminated with Endotoxin

Our initial experiments using AGEs obtained from Sigma showed that this reagent reduced the levels of the glycosylated form of AQP3 in mouse and human keratinocytes either in the absence (human) or presence (mouse) of SAHA (Supplementary Figure S1). In human keratinocytes, this reduction was accompanied by a decrease in its function, as measured by glycerol uptake (Supplementary Figure S2). The question then arose as to the mechanism by which the AGEs were affecting AQP3 levels. Reports in the literature suggested that AGEs might serve as endogenous damage-associated molecular patterns (DAMPs) to activate TLR4 [41,42]. To determine whether AGEs activated TLR4, we used a TLR4 reporter cell line and incubated with different concentrations of AGEs or BSA (as a control) obtained from Sigma (St. Louis, MO, USA). As shown in Figure 1a, Sigma AGEs did, in fact, activate TLR4; however, Sigma BSA itself caused as much activation of TLR4 as did the AGEs, suggesting that the Sigma BSA-AGEs (and the BSA from which the AGEs are generated) might be contaminated with a TLR4 activator. In fact, when we measured the amount of endotoxin in the Sigma AGEs using a Pierce™ Chromogenic Endotoxin Quant Kit, the AGEs (at both 25 and 100 µg/mL) showed endotoxin levels greater than the highest concentration tested in the standard curve (4 EU/mL) (Figure 1b). These results confirmed that the Sigma AGEs were contaminated with endotoxin, and it became unclear whether the changes in AQP3 protein expression observed initially were due to the AGEs themselves, to the contaminating endotoxin, or to both.

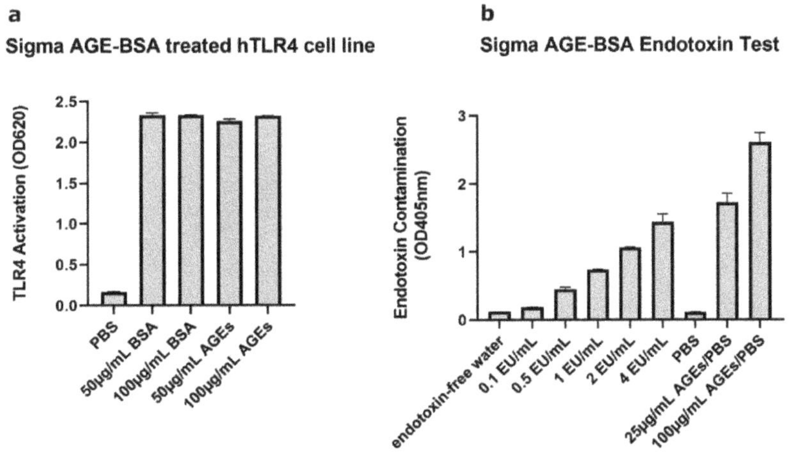

Figure 1. Sigma AGE-BSA Showed Contamination with TLR-Activating Endotoxin. (**a**) The TLR4 reporter cell line HEK-Blue hTLR4 was incubated with Sigma BSA or Sigma AGE-BSA at the indicated concentrations and TLR4 activation measured as a change in absorbance at 620 nm after 24 h. (**b**) Endotoxin contaminating the Sigma AGE-BSA was determined using a Pierce Chromogenic Endotoxin Quant Kit, according to the manufacturer's directions. Shown are representative experiments performed in duplicate.

2.2. TLR2 Activation Downregulated AQP3 mRNA Expression; TLR2/TLR4 Activation Upregulated AQP3 Protein in the Absence of SAHA, and TLR2 Activation Downregulated AQP3 Protein in the Presence of SAHA in Primary Mouse Epidermal Keratinocytes

To test the relative importance of TLR activation in regulating AQP3 expression, keratinocytes were treated with and without a TLR2 or a TLR4 activator in the presence and absence of SAHA. Pam_3CSK_4 has been previously shown to increase inflammatory mediator production and NF-κB activation in keratinocytes downstream of TLR2 [43,44]; LPS stimulates keratinocyte inflammatory mediator production downstream of TLR4. TLR2, but not TLR4, activation significantly inhibited AQP3 mRNA expression either in the presence or absence of SAHA (Figure 2). Similarly, TLR2 activation decreased SAHA-induced AQP3 protein expression (Figure 3b, right panel). In contrast, TLR2 and TLR4 activation significantly increased AQP3 protein level basally (Figure 3b, left panel). AQP3 presents as two bands upon Western analysis: an approximately 28 kDa non-glycosylated form and an about 40 kDa glycosylated form [45] (Figure 3a). We, therefore, further analyzed glycosylated and non-glycosylated AQP3 separately. We found that (1) TLR4 activation significantly increased non-glycosylated AQP3 protein expression in the absence of SAHA (Figure 3c, left panel); and (2) TLR2 activation significantly increased both non-glycosylated and glycosylated AQP3 protein expression without SAHA (Figure 3c,d, left panels) but decreased non-glycosylated AQP3 levels in the presence of SAHA (Figure 3c, right panel). Interestingly, we observed increased AQP3 protein levels, despite an inhibitory effect of TLR2 activation on mRNA expression. Since reductions in protein levels tend to lag behind decreases in mRNA expression, particularly if the protein is stable, it seems possible that later time points might be required to observe a decrease in AQP3 levels with PAM treatment.

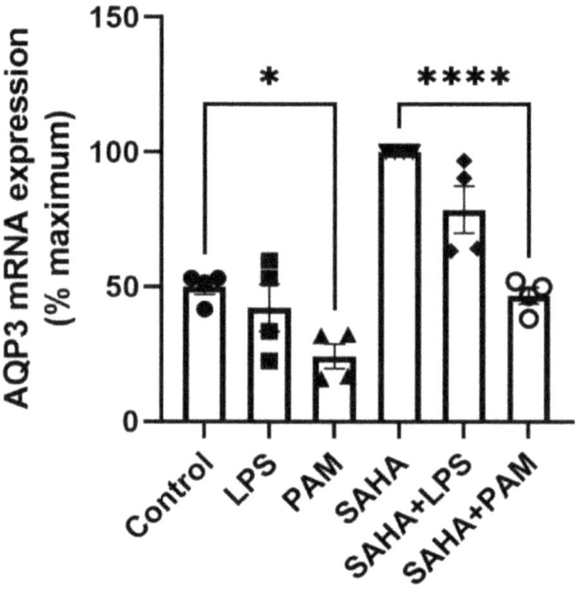

Figure 2. TLR2 Activation Decreased AQP3 mRNA Expression. Mouse keratinocytes were treated with 0 and 2.0 μg/mL LPS or 2.5 μg/mL PAM, in the presence or absence of 1 μM SAHA, for 24 h. mRNA expression of AQP3 was monitored by quantitative RT-PCR. The data are shown as % maximal response and represent the means ± SEM from at least four independent experiments; symbols represent individual experiments. * $p < 0.05$ and **** $p < 0.0001$ versus the control value.

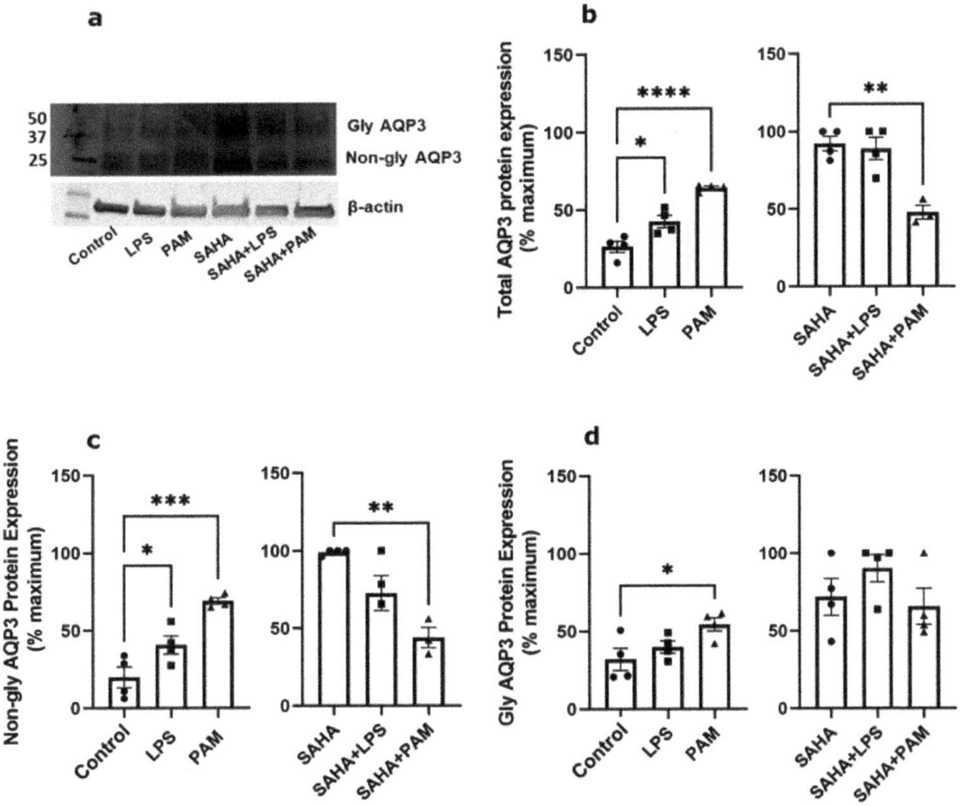

Figure 3. TLR2/TLR4 Activation Increased AQP3 Protein Expression Basally and TLR2 Reduced SAHA-induced AQP3 Levels. Mouse keratinocytes were treated with 0 and 2.0 µg/mL LPS or 2.5 µg/mL PAM, in the presence or absence of 1 µM SAHA, for 24 h. Cells were lysed in 3% boiling SDS lysis buffer and processed for Western analysis. (**a**) Representative Western blot showing APQ3 protein expression. (**b**) Left panel: total AQP3 protein expression in the absence of SAHA. Right panel: total AQP3 protein expression in the presence of SAHA. (**c**) Left panel: non-glycosylated AQP3 protein expression in the absence of SAHA. Right panel: non-glycosylated AQP3 protein expression in the presence of SAHA. (**d**) Left panel: glycosylated AQP3 protein expression in the absence of SAHA. Right panel: glycosylated AQP3 protein expression in the presence of SAHA. The data are expressed as % maximal response among all six groups, with the values in the absence and presence of SAHA analyzed separately and representing the means ± SEM from 3 to 4 independent experiments; symbols represent individual experiments. Note that one data point corresponding to the PAM treatment in the presence of SAHA from the total AQP3 protein expression ((**b**), right panel) and one from the non-glycosylated AQP3 protein expression ((**c**), right panel) were identified as significant outliers ($p < 0.05$) by the GraphPad Outlier Calculator and were removed from the statistical analysis. * $p < 0.05$, ** $p < 0.01$, *** $p < 0.001$, **** $p < 0.0001$ versus the control value.

2.3. AGE-BSA from BioVision, Inc. Was Not Contaminated with TLR-Activating Endotoxin

The datasheet provided by BioVision, Inc., (Waltham, MA, USA) for their AGE-BSA indicates endotoxin contamination of less than 0.1 IU/mg protein. To determine whether uncontaminated AGEs can serve as TLR-activating DAMPs, the HEK-Blue TLR4 reporter cell line was incubated with various doses of AGE-BSA from BioVision, Inc. Concentrations of AGEs ranging from 10 to 50 µg/mL showed no activation of TLR4 compared to the phosphate-buffered saline (PBS) control. Even at a concentration of 100 µg/mL, AGEs produced minimal TLR4 activation (Figure 4). Therefore, the AGEs from BioVision, Inc. did not serve as DAMPs to activate TLR4 and did not appear to be contaminated with endotoxin.

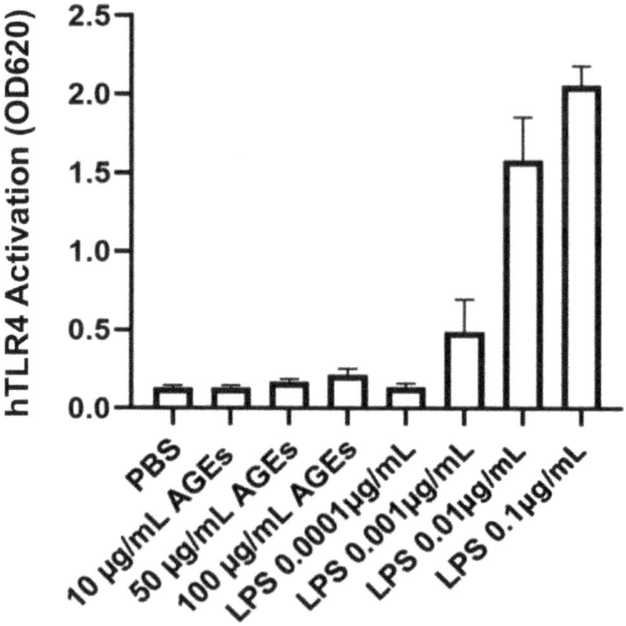

Figure 4. BioVision AGE-BSA Exhibited Little TLR4-Stimulating Activity. HEK-Blue hTLR4 reporter cells were incubated with the indicated concentrations of BioVision AGE-BSA or LPS (as a positive control), and TLR4 activation was measured as a change in absorbance at 620 nm after 24 h. Results represent the means ± SEM of four independent experiments performed in duplicate.

2.4. AGEs Increased AQP3 Protein in the Absence of SAHA but Reduced AQP3 Protein in Its Presence in Primary Mouse Epidermal Keratinocytes

We next investigated whether uncontaminated AGEs from BioVision, Inc. would affect AQP3 expression in primary mouse keratinocytes at 24 h (Figure 5), 48 h (Figure 6) and 72 h (Figure 7). Our results revealed that a 24- and 72-h treatment with AGEs at a concentration of 100 µg/mL (AGEs100) tended to increase AQP3 protein levels in the absence of SAHA (Figures 5b and 7b, left panels), although the results did not attain statistical significance. For the 48-h treatment, AGEs100 significantly increased AQP3 protein levels without SAHA (Figure 6b, left panel). In contrast, AGEs100 significantly reduced AQP3 protein expression in the presence of SAHA for each of the 24-, 48-, and 72-h treatments (Figures 5b, 6b and 7b, right panels). We further analyzed glycosylated and non-glycosylated AQP3 separately with or without SAHA. We found that (1) in the absence of SAHA, for the 48-h treatment, AGEs100 significantly increased both non-glycosylated and glycosylated AQP3 protein levels (Figure 6c,d, left panels). For the 24- and 72-h AGEs100 treatment, AQP3 protein expression showed the same tendency as the 48-h AGEs100 treatment even though statistical significance was not achieved (Figures 5c,d and 7c,d, left panels); (2) in the presence of SAHA, AGEs100 significantly decreased both non-glycosylated and glycosylated AQP3 protein (Figure 5c,d, right panels) for the 24-h treatment and significantly reduced glycosylated AQP3 for the 48- and 72-h treatment (Figures 6d and 7d, right panels). Our results suggest that AGEs affected AQP3 protein expression in a dose-dependent manner, with different effects basally versus under conditions of enhanced protein acetylation as has been observed in diabetes [30–33].

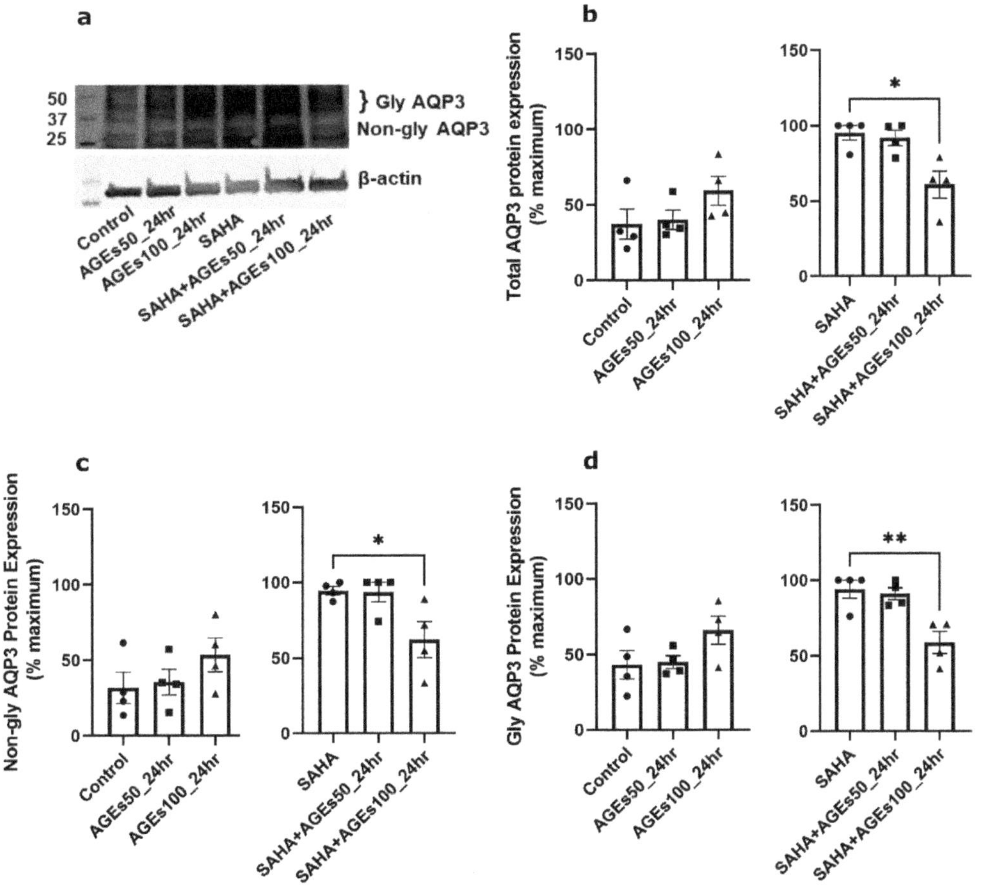

Figure 5. A 24-h Treatment with BioVision AGEs Altered Keratinocyte AQP3 Protein Levels. Western blotting of mouse keratinocytes treated for 24 h with 50 or 100 μg/mL BioVision AGEs. (**a**) Representative Western blot showing AQP3 protein expression. (**b**) Left panel: total AQP3 protein expression in the absence of SAHA. Right panel: total AQP3 protein expression in the presence of SAHA. (**c**) Left panel: non-glycosylated AQP3 protein expression in the absence of SAHA. Right panel: non-glycosylated AQP3 protein expression in the presence of SAHA. (**d**) Left panel: glycosylated AQP3 protein expression in the absence of SAHA. Right panel: glycosylated AQP3 protein expression in the presence of SAHA. The data are analyzed as in Figure 3 and expressed as % maximal response; values represent the means ± SEM from four independent experiments, with symbols representing individual experiments. * $p < 0.05$ and ** $p < 0.01$ versus the control value.

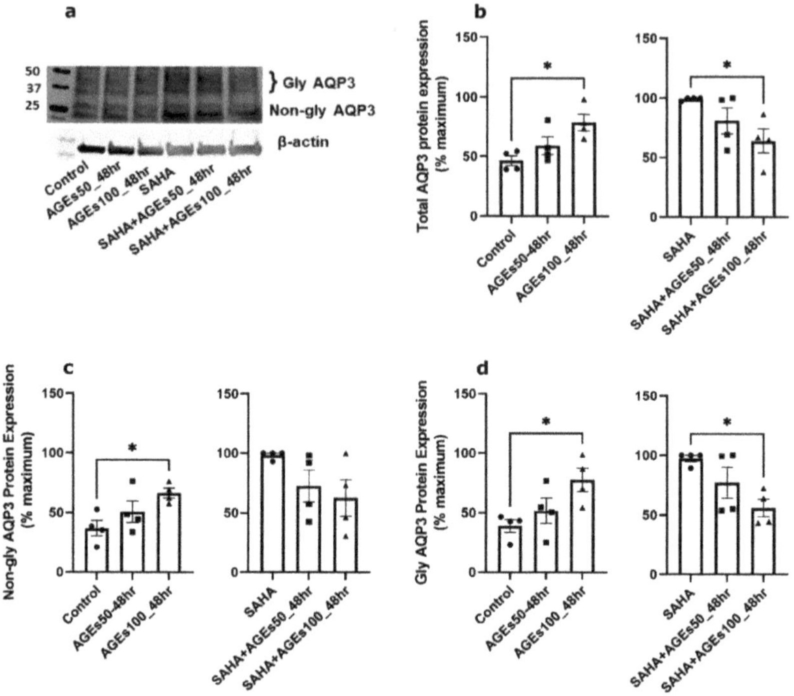

Figure 6. A 48-h Treatment with BioVision AGEs Altered Keratinocyte AQP3 Protein Levels. Western blotting of mouse keratinocytes treated for 48 h with 50 or 100 µg/mL AGEs. (**a**) Representative Western blot showing AQP3 protein expression. (**b**) Left panel: total AQP3 protein expression in the absence of SAHA. Right panel: total AQP3 protein expression in the presence of SAHA. (**c**) Left panel: non-glycosylated AQP3 protein expression in the absence of SAHA. Right panel: non-glycosylated AQP3 protein expression in the presence of SAHA. (**d**) Left panel: glycosylated AQP3 protein expression in the absence of SAHA. Right panel: glycosylated AQP3 protein expression in the presence of SAHA. The data are analyzed as in Figure 3 and expressed as % maximal response; values represent the means ± SEM from four independent experiments, with symbols representing individual experiments. * $p < 0.05$ versus the control value.

2.5. In Primary Mouse Epidermal Keratinocytes, AGEs, but Not SAHA, Increased Cell Number, a Measure of Proliferation

We also determined the effect of a 48-h treatment with BioVision AGEs (50 and 100 µg/mL) on cell numbers in primary mouse keratinocytes treated with or without 1 µM SAHA, as measured using 3-(4,5-dimethylthiazol-2-yl)-2,5-diphenyltetrazolium bromide (MTT) assays. The MTT assay provides information about metabolic activity; however, since this activity is proportional to the number of viable cells, the intensity of the colored product also yields data about cell number and thus cell proliferation [46]. In the absence of SAHA, AGEs at a concentration of 100 µg/mL significantly enhanced proliferation. In the presence of SAHA, AGEs at a concentration of 50 µg/mL but not 100 µg/mL, increased proliferation. Our data also showed no effect of 1 µM SAHA on mouse keratinocyte proliferation (Figure 8).

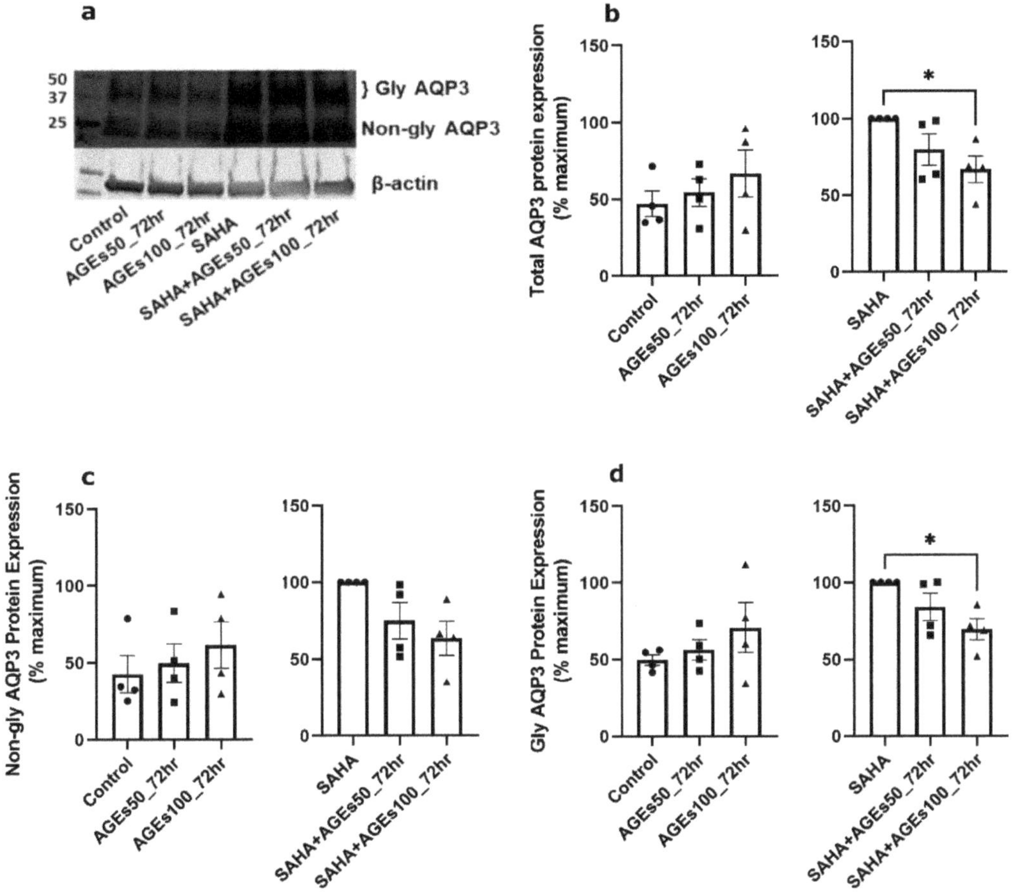

Figure 7. A 72-h Treatment with BioVision AGEs Altered Keratinocyte AQP3 Protein Levels. Western blotting of mouse keratinocytes treated for 72 h with 50 or 100 μg/mL AGEs. (**a**) Representative Western blot showing AQP3 protein expression. (**b**) Left panel: total AQP3 protein expression in the absence of SAHA. Right panel: total AQP3 protein expression in the presence of SAHA. (**c**) Left panel: non-glycosylated AQP3 protein expression in the absence of SAHA. Right panel: non-glycosylated AQP3 protein expression in the presence of SAHA. (**d**) Left panel: glycosylated AQP3 protein expression in the absence of SAHA. Right panel: glycosylated AQP3 protein expression in the presence of SAHA. The data are analyzed as in Figure 3 and expressed as % maximal response; values represent the means ± SEM from four independent experiments, with symbols representing individual experiments. * $p < 0.05$ versus the control value.

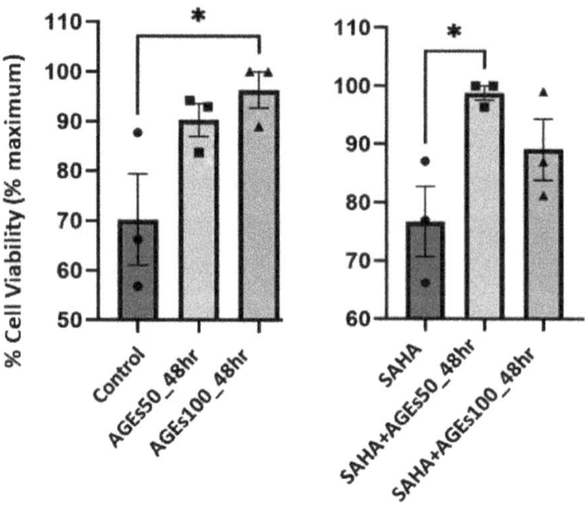

Figure 8. AGEs, not SAHA, IncreasedProliferation in Primary Mouse Keratinocytes. Mouse primary keratinocytes were treated with BioVision AGEs in the presence of absence of 1 μM SAHA for 48 h. Cell proliferation was measured using MTT assays. The data are expressed as % maximal response and represent the means ± SEM from three independent experiments; symbols represent individual experiments. * $p < 0.05$ versus the control value.

3. Discussion

AQP3 is an important protein that regulates keratinocyte proliferation, differentiation, and migration, as well as skin hydration and water permeability repair in vivo, and dysregulation of AQP3 leads to impaired wound healing [10–20] and xerosis [10,22]. The major findings of this study are that TLR2 activation and AGEs (100 μg/mL) upregulated AQP3 protein levels in the absence of SAHA, but TLR2 activation and AGEs (100 μg/mL) downregulated AQP3 protein levels in the presence of SAHA in primary mouse keratinocytes. Interestingly, both TLR2 activation and AGEs (100 μg/mL) had similar effects on AQP3 protein expression, suggesting a common pathway or potential crosstalk between TLR2 and RAGE. The principal mechanism(s) of how AQP3 is regulated by TLR2 activation and AGEs is (are) currently unclear. Nonetheless, both TLR2 activation and AGEs can activate the NF-κB pathway, leading to release of pro-inflammatory cytokines [28,35,47]. A secondary finding of potential significance is our data demonstrating that AGEs that were not contaminated with endotoxin did not activate TLR4, contrary to some reports in the literature [48,49].

In the absence of SAHA, TLR2 stimulation or AGEs (100 μg/mL) treatment of mouse keratinocytes enhanced AQP3 protein expression. The upregulation of AQP3 should enhance the transport function of AQP3, a water, glycerol, and hydrogen peroxide channel, as shown by changes in glycerol transport with endotoxin-contaminated AGEs and SAHA in Supplementary Figure S2. Through glycerol uptake and subsequent cell swelling, AQP3 can promote NLRP3 inflammasome activation and pro-inflammatory cytokine production [50]. The activation of the NLRP3 inflammasome has been shown to enhance skin wound healing by elevating pro-inflammatory cytokine production in the wound site [51,52]. Additionally, increased AQP3 in keratinocytes should enhance their proliferation, differentiation, and migration to contribute to wound healing, as has been shown in various studies in human and mouse keratinocytes (reviewed in [13,53]). Indeed, AGEs (100 μg/mL) enhanced the basal proliferation of primary mouse keratinocytes. This result is consistent with the finding that AGEs (100 μg/mL) enhance wound healing in a human corneal epithelial cell line via RAGE activation [54], and homozygous RAGE null mice exhibit delayed re-epithelialization in mouse corneal wounds [55]. Our results therefore suggest that TLR2 activation and AGEs may have a beneficial effect on acute wound healing in part through upregulation of AQP3 under basal conditions.

On the other hand, in the presence of SAHA, AGEs (100 μg/mL) or the TLR2 agonist PAM reduced AQP3 protein levels in mouse keratinocytes. As SAHA increases the low basal AQP3 protein expression [40], treatment with this agent allows a greater ability to detect decreases in AQP3 levels. SAHA-mediated HDAC inhibition may also mirror the protein hyperacetylation observed in diabetes, although at this point it is unclear whether or not SAHA affects the acetylation of the same or different proteins as in diabetes. In addition, it is likely that HDAC inhibitors have distinctive effects on different cell types. Thus, some studies report that HDAC inhibitors promote β-cell development and proliferation and might potentially therefore be a novel treatment for diabetes [56–58], whereas our data indicated no effect of 1 μM SAHA on mouse epidermal keratinocyte proliferation (Figure 8). The ability of HDAC inhibitors to increase β-cell numbers might suggest their potential therapeutic use in diabetes, despite the increased protein acetylation observed in other tissues in individuals with this disease [30–33]. In addition, HDAC inhibitors exhibit anti-inflammatory effects [59,60] that might be useful in treating diseases like diabetes.

There are limitations to our study. Thus, our study was restricted to in vitro experiments examining cell growth only, and not migration. Another limitation of our study is the fact that we have not defined the mechanisms by which downregulation of AQP3 might affect wound healing in diabetes. However, based on the literature, we can speculate that decreased AQP3 might inhibit reswelling-induced activation of NLRP3-inflammasome to reduce production of pro-inflammatory mediators [50] and hinder the orderly progression of inflammation at the wound site, contributing to delayed wound healing [51,52]. Indeed, multiple lines of evidence support the idea that chronic increases in AGEs in diabetes impair skin wound healing, a function of proliferation and migration, in vivo [61]. Thus, our data support the hypothesis that accumulation of AGEs may downregulate AQP3 expression in diabetes and may explain the findings of Ikarashi et al. [24] that an elevation in blood glucose levels was not sufficient to decrease AQP3 expression but required time to develop (to allow AGEs to form) in diabetic mice. Indeed, AQP3 knockout mice show impaired wound healing accompanied by reduced proliferating keratinocytes, and siRNA-induced AQP3 knockdown slows keratinocyte migration [10]. Our data thus suggest that TLR2 activation and/or AGEs formation in diabetes may be deleterious for wound healing, in part via downregulation of AQP3 protein in this condition.

In this study, we also analyzed the effects of AGEs on glycosylated and non-glycosylated AQP3 separately. Both glycosylated and non-glycosylated AQP3 were upregulated by AGEs in the absence of SAHA but downregulated in the presence of this HDAC inhibitor. The exact role of glycosylation in regulating AQP3 function is unclear but by analogy with AQP2 [62], glycosylation may regulate plasma membrane localization of AQP3. Therefore, this effect of AGEs may impair AQP3's membrane localization and transport function, although further studies are needed to examine this idea.

In conclusion, our study revealed that both TLR2 activation and AGEs treatment (100 μg/mL) of mouse keratinocytes promoted AQP3 protein expression basally, and elevated AQP3 protein should be protective in wound healing. However, with SAHA-mediated HDAC inhibition, TLR2 activation and AGEs decreased AQP3 protein, which is likely detrimental to wound healing, based on the important role of AQP3 in regulating keratinocyte proliferation and migration [10–20]. Our data suggest a possible role for AQP3 in connecting the inflammatory and proliferative phases of wound healing. Further, we speculate that AGEs-induced activation of RAGE acutely following injury may serve to promote wound healing, at least in part through effects on basal AQP3 levels. However, with prolonged elevations in AGEs levels and RAGE activation, as well as other changes accompanying diabetes (e.g., inflammation and enhanced protein acetylation), AGEs instead down-regulate AQP3 to impair skin wound healing. These results suggest that further investigation is warranted to determine the mechanisms by which AQP3 is regulated to exert both beneficial and harmful effects in complex wound healing processes.

4. Materials and Methods

4.1. Cell Culture and Treatment

Mouse keratinocytes were isolated from the skin of newborn ICR CD1 mice and cultured as described previously [63], according to protocols approved by the Institutional Care and Use Committees of Augusta University (Protocol #2017-0915) and the Charlie Norwood VA Medical Center (Protocol #22-04-135). Keratinocytes were treated with 0, 2.0 μg/mL LPS (Sigma-Aldrich, St. Louis, MO, USA), or 2.5 μg/mL Pam_3CSK_4 (EMD Millipore Corporation, Burlington, MA, USA) in the presence or absence of 1 μM SAHA (Sigma-Aldrich, St. Louis, MO, USA) at 37 °C with 5% CO_2 for 24 h; or with 0, 50, or 100 μg/mL AGE-BSA (Sigma-Aldrich or BioVision Incorporated, Milpitas, CA, USA), in the presence or absence of 1 or 2 μM SAHA (based on our previous studies [40]) at 37 °C with 5% CO_2 for 24, 48, and 72 h. Doses of AGEs were selected based on previous literature [54,64–66]. Cells were then harvested for RT-qPCR or Western analysis. Cells treated with SAHA were analyzed separately, to determine whether TLR2/TLR4 activation or AGEs altered AQP3 expression and/or function under basal conditions and settings analogous to diabetes (with enhanced protein acetylation).

4.2. Quantitative RT-PCR

Cells were harvested and RNA isolated using PureLink™ RNA Mini Kits (ThermoFisher Scientific, Waltham, MA, USA). Total RNA was reverse transcribed to cDNA using High-Capacity cDNA Reverse Transcription Kits (ThermoFisher Scientific, St. Louis, MO, USA). Gene expression was analyzed by quantitative RT-PCR using Taqman primer-probe sets purchased from ThermoFisher Scientific according to the supplier's instructions and the delta-delta Ct method with GAPDH as the housekeeping gene, as described previously [40].

4.3. Western Blotting

Western analysis was performed as described previously [40], using a primary anti-AQP3 antibody obtained from Novus Biologicals (Littleton, CO, USA) and secondary IRDye-conjugated antibody purchased from LI-COR Biosciences (Lincoln, NE, USA). Immunoreactive bands were visualized using a LI-COR Odyssey infrared imager and quantified using the LI-COR software. AQP3 levels were normalized to β-actin levels, using an antibody purchased from Sigma-Aldrich (St. Louis, MO, USA), and expressed as the percent maximal response within each experiment and in the presence or absence of SAHA, with results representing the means ± SEM of at least three separate experiments. AQP3 bands were identified in part based on molecular weight and in part on our previous results using AQP3 knockout cells with and without re-expression of AQP3 [16] and up- and down-regulation of the protein [40].

4.4. BSA-AGEs Contamination Test

Contamination of Sigma's BSA-AGEs with TLR4-activated endotoxin was determined using two assays. Initially, a TLR4 reporter cell line was used to monitor TLR4 activation. The HEK-Blue hTLR4 cells (InvivoGen, San Diego, CA, USA) are HEK293 cells engineered to express human TLR4, along with the TLR4 accessory proteins/co-receptors CD14 and MD-2, as well as a secreted embryonic alkaline phosphatase (SEAP) under the control of an interleukin-12 p40 minimal promoter with multiple AP-1 and NF-kB consensus sequences. Thus, activation of TLR4 results in AP-1- and NF-κB-mediated transcription of SEAP, the activity of which can then be measured using the chromogenic substrate in HEK-Blue detection medium. Endotoxin contamination was also measured directly using a Pierce™ Chromogenic Endotoxin Quant Kit (ThermoFisher Scientific), according to the manufacturer's instructions.

4.5. MTT Assay

MTT assays were performed using Roche's Cell Proliferation Kit I according to the manufacturer's instructions. Initially, experiments were performed to determine the initial plating density necessary to ensure that the cells were still proliferating, i.e., they had not achieved confluence, at the time of assay completion. Mouse keratinocytes plated at the determined density in 96-well plates were then treated with 0, 50, or 100 μg/mL AGE-BSA (BioVision Incorporated, CA, USA), in the presence or absence of 1 μM SAHA and incubated at 37 °C in 5% CO_2 for 48 h. Subsequently, MTT (3-[4,5-dimethylthiazol-2-yl]-2,5-diphenyltetrazolium bromide) obtained from Roche Diagnostics GmbH (Mannheim, Germany) was added to each well at a final concentration of 0.5 mg/mL. Cells were incubated for 4 h at 37 °C in 5% CO_2 to allow purple formazan crystals to form in metabolically active cells. An amount of 100 μL of solubilization buffer was then added into each well and incubated overnight at 37 °C with 5% CO_2 to dissolve the formazan crystals. Absorbance at 550 nm was measured using a Gen5 96-well plate reader (BioTek Instruments, Winooski, VT, USA).

4.6. Statistical Analysis

Results represent the means ± SEM of 3-6 separate experiments and are expressed either as a fold over control or as the % maximal response. Group mean values were compared using one-way analysis of variance (ANOVA) with Tukey tests. All statistical analyses were performed using GraphPad software (San Diego, CA, USA).

Supplementary Materials: The following supporting information can be downloaded at: https://www.mdpi.com/article/10.3390/ijms24021376/s1.

Author Contributions: Conceptualization, Y.L., R.U., V.C., M.H., and W.B.B.; formal analysis, Y.L., R.U., S.M., and W.B.B.; funding acquisition, W.B.B.; investigation, Y.L., R.U., V.C., M.H., C.Z., S.M., and X.C.; project administration, W.B.B.; supervision, W.B.B.; writing—original draft, Y.L., R.U., and W.B.B.; writing—review and editing, Y.L., R.U., V.C., M.H., C.Z., S.M., X.C., and W.B.B. All authors have read and agreed to the published version of the manuscript.

Funding: Y.L. was the recipient of a Student Research Program award from the Augusta University Provost's Office, in conjunction with the Translational Research Program of the Department of Medicine of the Medical College of Georgia at Augusta University. M.H. and C.Z. were supported by awards from the Medical College of Georgia Medical Scholars Program. The research was supported in part by an award from the American Legion to W.B.B. W.B.B. was also supported in part by a Veterans Affairs Research Career Scientist Award #BX005691 and a Veterans Affairs Merit Award #CX001357. The contents of this article do not represent the official views of the Department of Veterans Affairs or the United States Government.

Institutional Review Board Statement: The animal study protocol was approved by the Institutional Animal Care and Use Committee of Augusta University (Protocol #2017-0915; approval date 22 October 2020) and Charlie Norwood VA Medical Center (Protocol #22-04-135; approval date 11 April 2022).

Informed Consent Statement: Not applicable.

Data Availability Statement: The data are provided within the figures in the manuscript and Supplementary Materials. The raw data underlying these figures will be provided upon reasonable request.

Acknowledgments: We thank Purnima Merai for excellent technical support in the preparation of primary cultures of mouse keratinocytes.

Conflicts of Interest: The authors declare no conflict of interest.

References

1. Chen, L.; DiPietro, L.A. Toll-Like Receptor Function in Acute Wounds. *Adv. Wound Care* **2017**, *6*, 344–355. [CrossRef]
2. Wilkinson, H.N.; Hardman, M.J. Wound healing: Cellular mechanisms and pathological outcomes. *Open Biol.* **2020**, *10*, 200223. [CrossRef] [PubMed]

3. Seite, S.; Khemis, A.; Rougier, A.; Ortonne, J.P. Importance of treatment of skin xerosis in diabetes. *J. Eur. Acad. Dermatol. Venereol.* **2011**, *25*, 607–609. [CrossRef] [PubMed]
4. Hara-Chikuma, M.; Satooka, H.; Watanabe, S.; Honda, T.; Miyachi, Y.; Watanabe, T.; Verkman, A.S. Aquaporin-3-mediated hydrogen peroxide transport is required for NF-kappaB signalling in keratinocytes and development of psoriasis. *Nat. Commun.* **2015**, *6*, 7454. [CrossRef] [PubMed]
5. Hara-Chikuma, M.; Chikuma, S.; Sugiyama, Y.; Kabashima, K.; Verkman, A.S.; Inoue, S.; Miyachi, Y. Chemokine-dependent T cell migration requires aquaporin-3-mediated hydrogen peroxide uptake. *J. Exp. Med.* **2012**, *209*, 1743–1752. [CrossRef] [PubMed]
6. Miller, E.W.; Dickinson, B.C.; Chang, C.J. Aquaporin-3 mediates hydrogen peroxide uptake to regulate downstream intracellular signaling. *Proc. Natl. Acad. Sci. USA* **2010**, *107*, 15681–15686. [CrossRef]
7. Ecelbarger, C.A.; Terris, J.; Frindt, G.; Echevarria, M.; Marples, D.; Nielsen, S.; Knepper, M.A. Aquaporin-3 water channel localization and regulation in rat kidney. *Am. J. Physiol.* **1995**, *269*, F663–F672. [CrossRef]
8. Echevarria, M.; Windhager, E.E.; Frindt, G. Selectivity of the renal collecting duct water channel aquaporin-3. *J. Biol. Chem.* **1996**, *271*, 25079–25082. [CrossRef]
9. Yang, B.; Verkman, A.S. Water and glycerol permeabilities of aquaporins 1-5 and MIP determined quantitatively by expression of epitope-tagged constructs in Xenopus oocytes. *J. Biol. Chem.* **1997**, *272*, 16140–16146. [CrossRef]
10. Hara-Chikuma, M.; Verkman, A.S. Aquaporin-3 facilitates epidermal cell migration and proliferation during wound healing. *J. Mol. Med.* **2008**, *86*, 221–231. [CrossRef]
11. Hara-Chikuma, M.; Takahashi, K.; Chikuma, S.; Verkman, A.S.; Miyachi, Y. The expression of differentiation markers in aquaporin-3 deficient epidermis. *Arch. Dermatol. Res.* **2009**, *301*, 245–252. [CrossRef] [PubMed]
12. Nakahigashi, K.; Kabashima, K.; Ikoma, A.; Verkman, A.S.; Miyachi, Y.; Hara-Chikuma, M. Upregulation of aquaporin-3 is involved in keratinocyte proliferation and epidermal hyperplasia. *J. Investig. Dermatol.* **2011**, *131*, 865–873. [CrossRef]
13. Bollag, W.B.; Aitkens, L.; White, J.; Hyndman, K.A. Aquaporin-3 in the epidermis: More than skin deep. *Am. J. Physiol. Cell Physiol.* **2020**, *318*, C1144–C1153. [CrossRef] [PubMed]
14. Bollag, W.B.; Xie, D.; Zheng, X.; Zhong, X. A potential role for the phospholipase D2-aquaporin-3 signaling module in early keratinocyte differentiation: Production of a phosphatidylglycerol signaling lipid. *J. Investig. Dermatol.* **2007**, *127*, 2823–2831. [CrossRef] [PubMed]
15. Voss, K.E.; Bollag, R.J.; Fussell, N.; By, C.; Sheehan, D.J.; Bollag, W.B. Abnormal aquaporin-3 protein expression in hyperproliferative skin disorders. *Arch. Dermatol. Res.* **2011**, *303*, 591–600. [CrossRef]
16. Choudhary, V.; Olala, L.O.; Qin, H.; Helwa, I.; Pan, Z.Q.; Tsai, Y.Y.; Frohman, M.A.; Kaddour-Djebbar, I.; Bollag, W.B. Aquaporin-3 re-expression induces differentiation in a phospholipase D2-dependent manner in aquaporin-3-knockout mouse keratinocytes. *J. Investig. Dermatol.* **2015**, *135*, 499–507. [CrossRef]
17. Kim, N.H.; Lee, A.Y. Reduced aquaporin3 expression and survival of keratinocytes in the depigmented epidermis of vitiligo. *J. Investig. Dermatol.* **2010**, *130*, 2231–2239. [CrossRef]
18. Hara-Chikuma, M.; Verkman, A.S. Prevention of skin tumorigenesis and impairment of epidermal cell proliferation by targeted aquaporin-3 gene disruption. *Mol. Cell. Biol.* **2008**, *28*, 326–332. [CrossRef]
19. Guo, L.; Chen, H.; Li, Y.; Zhou, Q.; Sui, Y. An aquaporin 3-notch1 axis in keratinocyte differentiation and inflammation. *PLoS ONE* **2013**, *8*, e80179. [CrossRef]
20. Hara, M.; Ma, T.; Verkman, A.S. Selectively reduced glycerol in skin of aquaporin-3-deficient mice may account for impaired skin hydration, elasticity, and barrier recovery. *J. Biol. Chem.* **2002**, *277*, 46616–46621. [CrossRef]
21. Ma, T.; Song, Y.; Yang, B.; Gillespie, A.; Carlson, E.J.; Epstein, C.J.; Verkman, A.S. Nephrogenic diabetes insipidus in mice lacking aquaporin-3 water channels. *Proc. Natl. Acad. Sci. USA* **2000**, *97*, 4386–4391. [CrossRef] [PubMed]
22. Ikarashi, N.; Mizukami, N.; Pei, C.; Uchino, R.; Fujisawa, I.; Fukuda, N.; Kon, R.; Sakai, H.; Kamei, J. Role of Cutaneous Aquaporins in the Development of Xeroderma in Type 2 Diabetes. *Biomedicines* **2021**, *9*, 104. [CrossRef] [PubMed]
23. Sugimoto, T.; Huang, L.; Minematsu, T.; Yamamoto, Y.; Asada, M.; Nakagami, G.; Akase, T.; Nagase, T.; Oe, M.; Mori, T.; et al. Impaired aquaporin 3 expression in reepithelialization of cutaneous wound healing in the diabetic rat. *Biol. Res. Nurs.* **2013**, *15*, 347–355. [CrossRef] [PubMed]
24. Ikarashi, N.; Mizukami, N.; Kon, R.; Kaneko, M.; Uchino, R.; Fujisawa, I.; Fukuda, N.; Sakai, H.; Kamei, J. Study of the Mechanism Underlying the Onset of Diabetic Xeroderma Focusing on an Aquaporin-3 in a Streptozotocin-Induced Diabetic Mouse Model. *Int. J. Mol. Sci.* **2019**, *20*, 3782. [CrossRef]
25. Goh, S.Y.; Cooper, M.E. Clinical review: The role of advanced glycation end products in progression and complications of diabetes. *J. Clin. Endocrinol. Metab.* **2008**, *93*, 1143–1152. [CrossRef] [PubMed]
26. Gkogkolou, P.; Bohm, M. Advanced glycation end products: Key players in skin aging? *Dermatoendocrinol* **2012**, *4*, 259–270. [CrossRef] [PubMed]
27. Luevano-Contreras, C.; Chapman-Novakofski, K. Dietary advanced glycation end products and aging. *Nutrients* **2010**, *2*, 1247–1265. [CrossRef]
28. Wang, Q.; Zhu, G.; Cao, X.; Dong, J.; Song, F.; Niu, Y. Blocking AGE-RAGE Signaling Improved Functional Disorders of Macrophages in Diabetic Wound. *J. Diabetes Res.* **2017**, *2017*, 1428537. [CrossRef]

29. Goova, M.T.; Li, J.; Kislinger, T.; Qu, W.; Lu, Y.; Bucciarelli, L.G.; Nowygrod, S.; Wolf, B.M.; Caliste, X.; Yan, S.F.; et al. Blockade of receptor for advanced glycation end-products restores effective wound healing in diabetic mice. *Am. J. Pathol.* **2001**, *159*, 513–525. [CrossRef]
30. Kadiyala, C.S.; Zheng, L.; Du, Y.; Yohannes, E.; Kao, H.Y.; Miyagi, M.; Kern, T.S. Acetylation of retinal histones in diabetes increases inflammatory proteins: Effects of minocycline and manipulation of histone acetyltransferase (HAT) and histone deacetylase (HDAC). *J. Biol. Chem.* **2012**, *287*, 25869–25880. [CrossRef]
31. Mosley, A.L.; Ozcan, S. Glucose regulates insulin gene transcription by hyperacetylation of histone h4. *J. Biol. Chem.* **2003**, *278*, 19660–19666. [CrossRef] [PubMed]
32. Mosley, A.L.; Corbett, J.A.; Ozcan, S. Glucose regulation of insulin gene expression requires the recruitment of p300 by the beta-cell-specific transcription factor Pdx-1. *Mol. Endocrinol.* **2004**, *18*, 2279–2290. [CrossRef] [PubMed]
33. Sampley, M.L.; Ozcan, S. Regulation of insulin gene transcription by multiple histone acetyltransferases. *DNA Cell Biol.* **2012**, *31*, 8–14. [CrossRef] [PubMed]
34. Wang, Y.; Zhong, J.; Zhang, X.; Liu, Z.; Yang, Y.; Gong, Q.; Ren, B. The Role of HMGB1 in the Pathogenesis of Type 2 Diabetes. *J. Diabetes Res.* **2016**, *2016*, 2543268. [CrossRef] [PubMed]
35. Dasu, M.R.; Isseroff, R.R. Toll-like receptors in wound healing: Location, accessibility, and timing. *J. Investig. Dermatol.* **2012**, *132*, 1955–1958. [CrossRef]
36. Chen, L.; Guo, S.; Ranzer, M.J.; DiPietro, L.A. Toll-like receptor 4 has an essential role in early skin wound healing. *J. Investig. Dermatol.* **2013**, *133*, 258–267. [CrossRef]
37. Suga, H.; Sugaya, M.; Fujita, H.; Asano, Y.; Tada, Y.; Kadono, T.; Sato, S. TLR4, rather than TLR2, regulates wound healing through TGF-beta and CCL5 expression. *J. Dermatol. Sci.* **2014**, *73*, 117–124. [CrossRef]
38. Dasu, M.R.; Thangappan, R.K.; Bourgette, A.; DiPietro, L.A.; Isseroff, R.; Jialal, I. TLR2 expression and signaling-dependent inflammation impair wound healing in diabetic mice. *Lab. Investig.* **2010**, *90*, 1628–1636. [CrossRef]
39. Dasu, M.R.; Jialal, I. Amelioration in wound healing in diabetic toll-like receptor-4 knockout mice. *J. Diabetes Complicat.* **2013**, *27*, 417–421. [CrossRef]
40. Choudhary, V.; Olala, L.O.; Kagha, K.; Pan, Z.Q.; Chen, X.; Yang, R.; Cline, A.; Helwa, I.; Marshall, L.; Kaddour-Djebbar, I.; et al. Regulation of the Glycerol Transporter, Aquaporin-3, by Histone Deacetylase-3 and p53 in Keratinocytes. *J. Investig. Dermatol.* **2017**, *137*, 1935–1944. [CrossRef]
41. Hodgkinson, C.P.; Laxton, R.C.; Patel, K.; Ye, S. Advanced glycation end-product of low density lipoprotein activates the toll-like 4 receptor pathway implications for diabetic atherosclerosis. *Arterioscler. Thromb. Vasc. Biol.* **2008**, *28*, 2275–2281. [CrossRef] [PubMed]
42. Chen, Y.J.; Sheu, M.L.; Tsai, K.S.; Yang, R.S.; Liu, S.H. Advanced glycation end products induce peroxisome proliferator-activated receptor gamma down-regulation-related inflammatory signals in human chondrocytes via Toll-like receptor-4 and receptor for advanced glycation end products. *PLoS ONE* **2013**, *8*, e66611. [CrossRef]
43. Choudhary, V.; Uaratanawong, R.; Patel, R.R.; Patel, H.; Bao, W.; Hartney, B.; Cohen, E.; Chen, X.; Zhong, Q.; Isales, C.M.; et al. Phosphatidylglycerol Inhibits Toll-Like Receptor-Mediated Inflammation by Danger-Associated Molecular Patterns. *J. Investig. Dermatol.* **2019**, *139*, 868–877. [CrossRef] [PubMed]
44. Choudhary, V.; Griffith, S.; Chen, X.; Bollag, W.B. Pathogen-Associated Molecular Pattern-Induced TLR2 and TLR4 Activation Increases Keratinocyte Production of Inflammatory Mediators and is Inhibited by Phosphatidylglycerol. *Mol. Pharmacol.* **2020**, *97*, 324–335. [CrossRef]
45. Baum, M.A.; Ruddy, M.K.; Hosselet, C.A.; Harris, H.W. The perinatal expression of aquaporin-2 and aquaporin-3 in developing kidney. *Pediatr. Res.* **1998**, *43*, 783–790. [CrossRef]
46. Kamiloglu, S.; Sari, G.; Ozdal, T.; Capanoglu, E. Guidelines for cell viability assays. *Food Front.* **2020**, *1*, 332–349. [CrossRef]
47. Sorci, G.; Riuzzi, F.; Giambanco, I.; Donato, R. RAGE in tissue homeostasis, repair and regeneration. *Biochim. Biophys. Acta* **2013**, *1833*, 101–109. [CrossRef] [PubMed]
48. Erridge, C. Endogenous ligands of TLR2 and TLR4: Agonists or assistants? *J. Leukoc. Biol.* **2010**, *87*, 989–999. [CrossRef]
49. Xing, Y.; Pan, S.; Zhu, L.; Cui, Q.; Tang, Z.; Liu, Z.; Liu, F. Advanced Glycation End Products Induce Atherosclerosis via RAGE/TLR4 Signaling Mediated-M1 Macrophage Polarization-Dependent Vascular Smooth Muscle Cell Phenotypic Conversion. *Oxid. Med. Cell. Longev.* **2022**, *2022*, 9763377. [CrossRef]
50. Da Silva, I.V.; Cardoso, C.; Martinez-Banaclocha, H.; Casini, A.; Pelegrin, P.; Soveral, G. Aquaporin-3 is involved in NLRP3-inflammasome activation contributing to the setting of inflammatory response. *Cell. Mol. Life Sci.* **2021**, *78*, 3073–3085. [CrossRef]
51. Ito, H.; Kanbe, A.; Sakai, H.; Seishima, M. Activation of NLRP3 signalling accelerates skin wound healing. *Exp. Dermatol.* **2018**, *27*, 80–86. [CrossRef] [PubMed]
52. Vinaik, R.; Abdullahi, A.; Barayan, D.; Jeschke, M.G. NLRP3 inflammasome activity is required for wound healing after burns. *Transl. Res.* **2020**, *217*, 47–60. [CrossRef] [PubMed]
53. Qin, H.; Zheng, X.; Zhong, X.; Shetty, A.K.; Elias, P.M.; Bollag, W.B. Aquaporin-3 in keratinocytes and skin: Its role and interaction with phospholipase D2. *Arch. Biochem. Biophys.* **2011**, *508*, 138–143. [CrossRef] [PubMed]
54. Gross, C.; Belville, C.; Lavergne, M.; Choltus, H.; Jabaudon, M.; Blondonnet, R.; Constantin, J.M.; Chiambaretta, F.; Blanchon, L.; Sapin, V. Advanced Glycation End Products and Receptor (RAGE) Promote Wound Healing of Human Corneal Epithelial Cells. *Investig. Ophthalmol. Vis. Sci.* **2020**, *61*, 14. [CrossRef] [PubMed]

55. Nass, N.; Trau, S.; Paulsen, F.; Kaiser, D.; Kalinski, T.; Sel, S. The receptor for advanced glycation end products RAGE is involved in corneal healing. *Ann. Anat.* **2017**, *211*, 13–20. [CrossRef] [PubMed]
56. Christensen, D.P.; Dahllof, M.; Lundh, M.; Rasmussen, D.N.; Nielsen, M.D.; Billestrup, N.; Grunnet, L.G.; Mandrup-Poulsen, T. Histone deacetylase (HDAC) inhibition as a novel treatment for diabetes mellitus. *Mol. Med.* **2011**, *17*, 378–390. [CrossRef]
57. Makkar, R.; Behl, T.; Arora, S. Role of HDAC inhibitors in diabetes mellitus. *Curr. Res. Transl. Med.* **2020**, *68*, 45–50. [CrossRef]
58. Chen, Y.; Du, J.; Zhao, Y.T.; Zhang, L.; Lv, G.; Zhuang, S.; Qin, G.; Zhao, T.C. Histone deacetylase (HDAC) inhibition improves myocardial function and prevents cardiac remodeling in diabetic mice. *Cardiovasc. Diabetol.* **2015**, *14*, 99. [CrossRef]
59. Dokmanovic, M.; Clarke, C.; Marks, P.A. Histone deacetylase inhibitors: Overview and perspectives. *Mol. Cancer Res.* **2007**, *5*, 981–989. [CrossRef]
60. Hancock, W.W.; Akimova, T.; Beier, U.H.; Liu, Y.; Wang, L. HDAC inhibitor therapy in autoimmunity and transplantation. *Ann. Rheum. Dis.* **2012**, *71* (Suppl. 2), i46–i54. [CrossRef]
61. Van Putte, L.; De Schrijver, S.; Moortgat, P. The effects of advanced glycation end products (AGEs) on dermal wound healing and scar formation: A systematic review. *Scars Burn Heal.* **2016**, *2*, 2059513116676828. [CrossRef] [PubMed]
62. Hendriks, G.; Koudijs, M.; van Balkom, B.W.; Oorschot, V.; Klumperman, J.; Deen, P.M.; van der Sluijs, P. Glycosylation is important for cell surface expression of the water channel aquaporin-2 but is not essential for tetramerization in the endoplasmic reticulum. *J. Biol. Chem.* **2004**, *279*, 2975–2983. [CrossRef] [PubMed]
63. Bailey, L.J.; Choudhary, V.; Merai, P.; Bollag, W.B. Preparation of primary cultures of mouse epidermal keratinocytes and the measurement of phospholipase D activity. *Methods Mol. Biol.* **2014**, *1195*, 111–131. [CrossRef]
64. Tian, M.; Lu, S.; Niu, Y.; Xie, T.; Dong, J.; Cao, X.; Song, F.; Jin, S.; Qing, C. Effects of advanced glycation end-products (AGEs) on skin keratinocytes by nuclear factor-kappa B (NF-κB) activation. *Afr. J. Biotechnol.* **2012**, *11*, 11132–11142.
65. Zhu, P.; Yang, C.; Chen, L.-H.; Ren, M.; Lao, G.-j.; Yan, L. Impairment of human keratinocyte mobility and proliferation by advanced glycation end products-modified BSA. *Arch. Dermatol. Res.* **2011**, *303*, 339–350. [CrossRef]
66. Zhu, P.; Chen, C.; Wu, D.; Chen, G.; Tan, R.; Ran, J. AGEs-induced MMP-9 activation mediated by Notch1 signaling is involved in impaired wound healing in diabetic rats. *Diabetes Res. Clin. Pract.* **2022**, *186*, 109831. [CrossRef]

International Journal of
Molecular Sciences

MDPI

Article

Mechanical Compression by Simulating Orthodontic Tooth Movement in an In Vitro Model Modulates Phosphorylation of AKT and MAPKs via TLR4 in Human Periodontal Ligament Cells

Charlotte E. Roth [1,†], Rogerio B. Craveiro [1,*,†], Christian Niederau [1], Hanna Malyaran [2,3], Sabine Neuss [2,3], Joachim Jankowski [4] and Michael Wolf [1]

1 Department of Orthodontics, Dental Clinic, University of Aachen, 52074 Aachen, Germany; croth@ukaachen.de (C.E.R.); cniederau@ukaachen.de (C.N.); michwolf@ukaachen.de (M.W.)
2 Helmholtz Institute for Biomedical Engineering, BioInterface Group, RWTH Aachen University, 52056 Aachen, Germany; hmalyaran@ukaachen.de (H.M.); sneuss-stein@ukaachen.de (S.N.)
3 Institute of Pathology, RWTH Aachen University, 52056 Aachen, Germany
4 Institute for Molecular Cardiovascular Research (IMCAR), University Hospital RWTH Aachen, 52074 Aachen, Germany; jjankowski@ukaachen.de
* Correspondence: rcraveiro@ukaachen.de
† These authors contributed equally to this work.

Abstract: Mechanical compression simulating orthodontic tooth movement in in vitro models induces pro-inflammatory cytokine expression in periodontal ligament (PDL) cells. Our previous work shows that TLR4 is involved in this process. Here, primary PDL cells are isolated and characterized to better understand the cell signaling downstream of key molecules involved in the process of sterile inflammation via TLR4. The TLR4 monoclonal blocking antibody significantly reverses the upregulation of phospho-AKT, caused by compressive force, to levels comparable to controls by inhibition of TLR4. Phospho-ERK and phospho-p38 are also modulated in the short term via TLR4. Additionally, moderate compressive forces of 2 g/cm^2, a gold standard for static compressive mechanical stimulation, are not able to induce translocation of Nf-kB and phospho-ERK into the nucleus. Accordingly, we demonstrated for the first time that TLR4 is also one of the triggers for signal transduction under compressive force. The TLR4, one of the pattern recognition receptors, is involved through its specific molecular structures on damaged cells during mechanical stress. Our findings provide the basis for further research on TLR4 in the modulation of sterile inflammation during orthodontic therapy and periodontal remodeling.

Keywords: monoclonal antibody TLR4; compression force; MAPKs; AKT; human PDL; sterile inflammation

Citation: Roth, C.E.; Craveiro, R.B.; Niederau, C.; Malyaran, H.; Neuss, S.; Jankowski, J.; Wolf, M. Mechanical Compression by Simulating Orthodontic Tooth Movement in an In Vitro Model Modulates Phosphorylation of AKT and MAPKs via TLR4 in Human Periodontal Ligament Cells. *Int. J. Mol. Sci.* **2022**, *23*, 8062. https://doi.org/10.3390/ijms23158062

Academic Editors: Oliver Planz and Ralf Kircheis

Received: 6 July 2022
Accepted: 20 July 2022
Published: 22 July 2022

1. Introduction

Mechanical forces are used as a therapeutic tool enabling orthodontic tooth movement (OTM) to improve functional and aesthetical malocclusions [1,2]. The mechanical stimulation is transmitted from the teeth to the periodontal ligament (PDL) where it is translated into a biochemical reaction, that leads to the remodeling of the alveolar bone [2–5]. The tooth is anchored to the surrounding alveolar bone via the periodontium, which absorbs the various shocks associated with mastication and provides tooth stability by continuously remodeling its extracellular matrix, the periodontal ligament (PDL) [6]. It represents a specialized connective soft tissue with viscoelastic properties, mainly comprised of fibroblasts and extracellular matrix (ECM) with the following main functions: anchoring teeth by Sharpey's fibers into the alveolar cavity [4], supply of tooth nutrients, homeostasis and tissue remodeling. Mechanical stimuli, initially directed to the ECM of the

PDL, are transduced via mechanosensitive receptors, ion channels [7] and pattern recognition receptors (PRR) to biochemical responses. This leads to a local sterile inflammation with pro-inflammatory cytokines, e.g., interleukins (IL-6, IL-8) [8–10], vascular endothelial growth factor (VEGF) [11] and matrix metalloproteinases (MMP8 and 9) [10,12]. MMPs are a group of enzymes that in concert are responsible for the degradation of most extracellular matrix proteins during organogenesis, growth and normal tissue turnover and increase significantly in various pathological conditions [13] and serve to accurately predict the level of inflammation during OTM [14]. These factors trigger an immune response, resulting in migration, adhesion and differentiation of monocytes and osteoclasts [15–17]. The monocytes dislodge cellular debris and allow for the remodeling of periodontal architecture in the initial phase and in the later phase during the tooth movement [18,19]. This is followed by alveolar bone resorption in the compression area and bone formation in the tension area of the periodontal ligament.

As shown in in vitro and in vivo studies, Toll-like receptor 4 (TLR4), a type of pattern recognition receptors (PRR) family, is involved in osteoclastogenesis [20–22]. Furthermore, TLR4 has been considered an important player in the initiation and progression of diverse inflammatory diseases. Our previous study has shown that the TLR4 expression is significantly increased due to mechanical stimulation in periodontal cells [8]. TLR4 seems to play an important role in two common inflammation types of the periodontium. Firstly, the sterile inflammation resulting from trauma or injuries is characterized by the release of endogenous molecules termed damage-associated molecular patterns (DAMPs). Secondly, the infectious inflammation, triggered by conserved structural motifs found in microorganisms, called pathogen-associated molecular patterns (PAMPs), leads to periodontal diseases, e.g., periodontitis. Furthermore, TLR4 has been well characterized to be involved in the transduction of innate and adaptive host immune responses to microbial pathogens, such as lipopolysaccharides (LPS) [23]. Nowadays, many DAMPs are reported as TLR4 ligands, such as High Mobility Group Box 1 (HMGB1) which is described in PDL fibroblasts during OTM [8,24,25]. Thus, HMGB1 can be defined as an accessory regulator for bone remodeling during OTM. HMGB1 is a damage protein that is released in the extracellular medium through traumatic events [26,27]. However, the effect of mechanical cell stress in human PDL cells depends on variables such as force magnitude, duration and constancy [25,27]. These findings underline the need for further investigation of the signal transduction cascades involved in the mediation of mechanical cell stress into host immune responses.

We hypothesized that the expression of TLR4 in human PDL cells may be involved in the initial inflammation process in orthodontic tooth movement. In this study, we want to evaluate the role of TLR4 in human periodontal ligament cells in an in vitro compression model for OTM and TLR4 signaling during sterile inflammation. To this end, we investigated the modulation of TLR4 under mechanical stress and its involvement in key downstream molecules such as AKT, also known as protein kinase B, and the highly conserved mitogen-activated protein kinase (MAPK) family, with its three important members (extracellular signal-regulated kinase (ERK), p38 and c-Jun N-terminal kinase (JNK)). The TLR4 modulation regulates the cells by transducing extracellular into cellular responses and enables a better understanding of the molecular mechanisms underlining sterile periodontal remodeling.

2. Results

2.1. Characterization of Primary Human PDL Cells

At first, three isolated PDL cells from the upper jaw were evaluated for stem cell characteristics via surface characterization according to the Guidelines of the International Society for Stem Cell Research. The percentages of specific stem cell markers were analyzed by flow cytometry. All donors express stem cell markers (CD34−, CD45−, CD73+, CD90+ and CD105+), sharing in addition morphological and phenotypical features specific for multipotent adult mesenchymal stem cells (MSCs). To demonstrate the multipotency of hPDLs, when cultured under the appropriate conditions, the mesodermal differentiation towards adipocytes, osteoblasts and chondrocytes was induced in three separate donors. The adipogenic induction results in the formation of lipid droplets, which were visualized by Oil red O staining, on the culture plate. The first lipid droplets were observed after 14 days. hPDLs, which were cultured in standard media showed no formation of lipid vacuole formation during the differentiation process. Chondrocytes produce large amounts of extracellular matrix composed of collagen, proteoglycan, and elastin. The differentiation towards chondrocytes was induced with CIM. Staining with Toluidine blue revealed excessive production of extracellular matrix components and proteoglycans during the culture process with CIM. In comparison with the control, the staining was darker and hyaline structures appeared more organized. Osteogenic differentiation was induced with OIM. hPDLs were seeded at a density of 31.000 cells/cm^2 and stained with Alizarin red after 14 days in culture. Alizarin red stains calcium depositions that are characteristic of the osteogenic cell fate. Dark red formations represent the calcium depositions. Over the whole culture period, no red staining was observed in the control group (Figure 1).

Furthermore, as shown in previous studies, PDL cells are characterized by their ability to express inflammatory markers under mechanical compression [9,11]. Here, we also observed the upregulation of inflammatory cytokines under compressive force. The pro-inflammatory cytokines interleukin-6 (IL-6), interleukin-8 (IL-8) and cyclooxygenase-2 (Cox2) and the vascular endothelial growth factor A (VEGFA), a mediator of inflammation and angiogenesis, are significantly upregulated (Figure 2A,B).

2.2. Changes in the TLR4 Production under Compressive Forces

To investigate the effect of compressive forces on the TLR4 protein production, the ligand HMGB1 (100 ng/mL) was used as a stimulator, and TLR4 blocking antibody (5 μg/mL) as an inhibitor for the TLR4 downstream signaling. The TLR4 protein level was clearly upregulated after a short-term culture (3 h) of mechanical stimulation and downregulated after a long-term culture (24 h) (Figure 3). The same pattern was observed for all three donors (Figure 3). TLR4 block antibody alone did not affect the basal TLR4 production, but in combination with 3 h and 24 h compressive force, it was downregulated. The activation of TLR4 with HMGB1 resulted in a slightly higher production after 3 h without having an effect after 24 h. MyD88, a downstream adapter molecule, that plays a pivotal role in immune activation through TLRs, showed just a slight decrease for all conditions, except for a significant downregulation by compressive forces for 24 h (Figure 3).

2.3. Phospho-AKT Was Upregulated by Compressive Force

To investigate downstream signaling of TLR4, AKT and its phosphorylation status was analyzed by Western blots under the same conditions. Under all conditions, AKT showed no differences, while the phosphorylated AKT was significantly upregulated with compressive forces (CF) for 3 h and 24 h. In particular, compressive forces with additional blocking TLR4 monoclonal antibody led to a significant downregulation comparable to the control (Figure 4).

2.4. Phospho-ERK and Phospho-p38 Were Significantly Upregulated under Compressive Force

Further, the phosphorylation status and production of MAP-Kinase (ERK, p38 and JNK) were investigated by Western blots. The MAPK did not change in all conditions. However, the phosphorylation of ERK (3 h) and p38 (3 and 24 h) was significantly upregulated under CF (Figure 5). The phosphorylation of ERK and p38 under 3 h compressive forces with additional blocking TLR4 monoclonal antibody was reduced. The phosphorylated form of JNK was not detected for PDL cells in any condition. The validation of antibody phosphorylation of JNK, however, was confirmed by a positive control (Figure 5C).

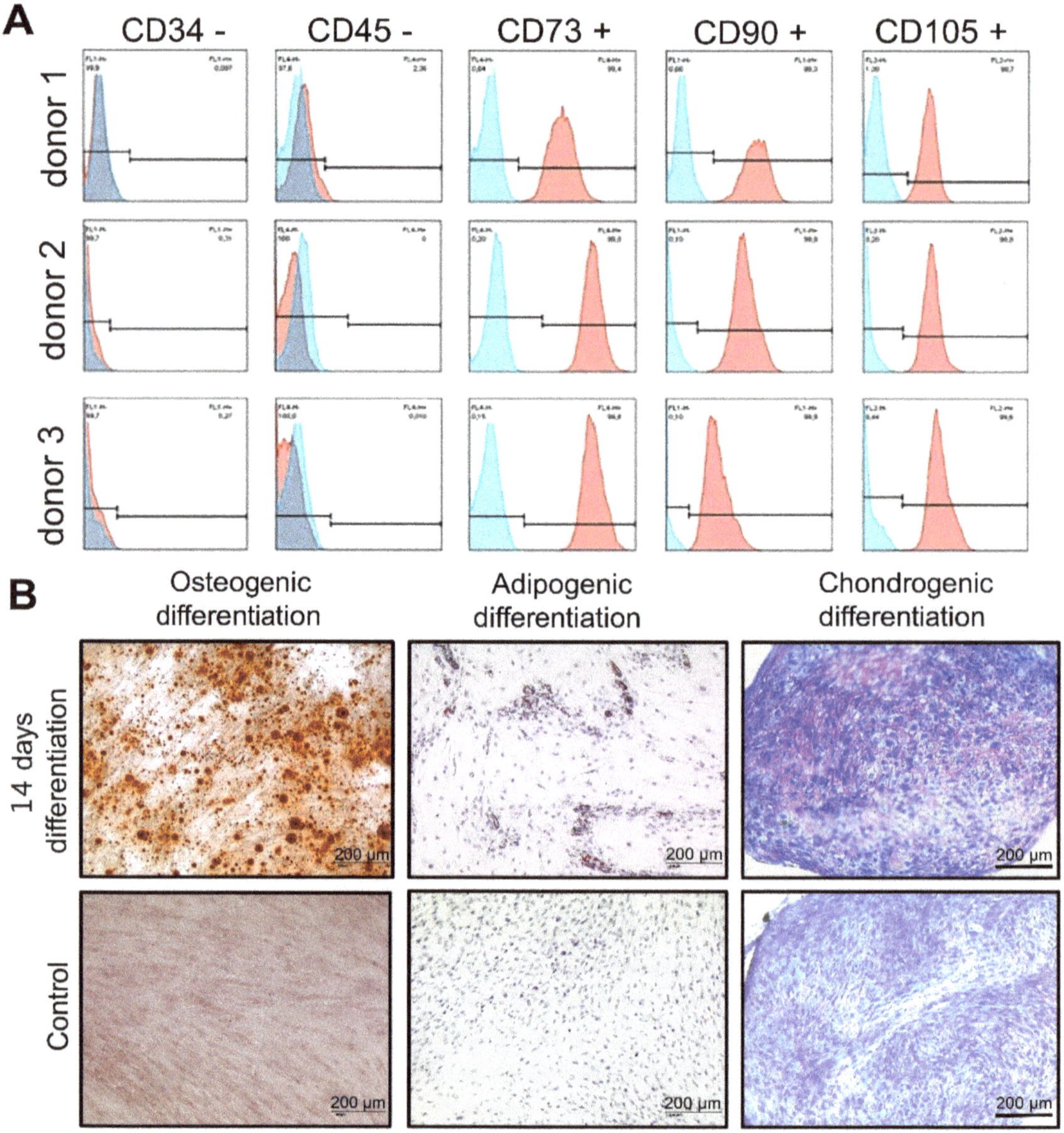

Figure 1. Characterization of primary human PDL cells. (**A**) hPDL cells were first characterized by flow cytometry analysis of cell surface markers. The expression pattern for all donors is CD34−, CD45−, CD73+, CD90+ and CD105+ (blue color: isotype control, red color: surface marker) and (**B**) hPDLs differentiation towards adipocytes, chondrocytes and osteoblasts was performed. hPDL cells were grown in osteogenic, adipogenic and chondrogenic differentiation media for 14 days and then stained with alizarin red, Oil red O and Toluidin blue, respectively. The red color represented mineralized calcium depositions after PDL cells differentiated into osteoblast-like cells. A large number of orange lipid vacuoles were seen in PDL cells after culture in AIM and staining with Oil red O. Staining with Toluidine blue revealed excessive production of extracellular matrix components and proteoglycans during the culture process with CIM (scale: 200 µm).

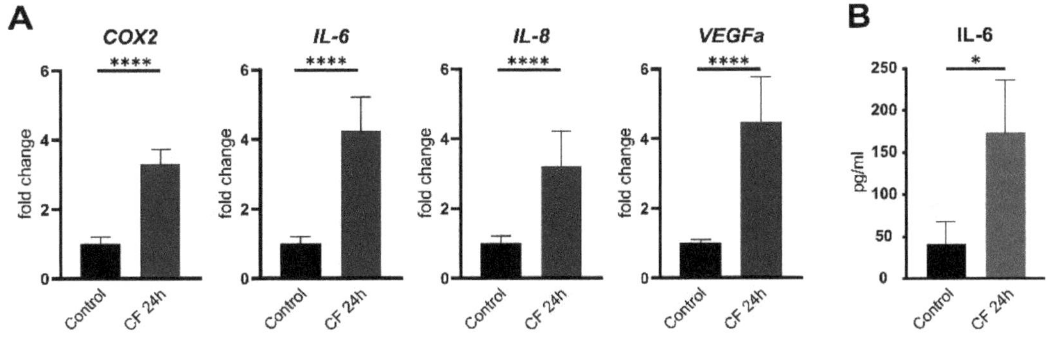

Figure 2. Inflammatory markers are upregulated by compressive forces in hPDL after 24 h. To analyze the regulation of inflammatory cytokines under compressive forces, the expression of different markers was analyzed on mRNA level (**A**) and protein levels (**B**) after mechanical stimulation. The pro-inflammatory cytokines interleukin-6 (IL-6), interleukin-8 (IL-8) and cyclooxygenase-2 (Cox2) are significantly upregulated. The vascular endothelial growth factor A (VEGFA), a mediator of inflammation and angiogenesis, is also significantly upregulated. The data represent two independent experiments in triplicates; normalization by ddC$_t$ method to RPL22 and control 100% statistical data were tested for normal distribution by Shapiro–Wilk test and a *t*-test was performed. Statistically significant differences are marked by an asterisk (* $p < 0.05$, **** $p < 0.0001$).

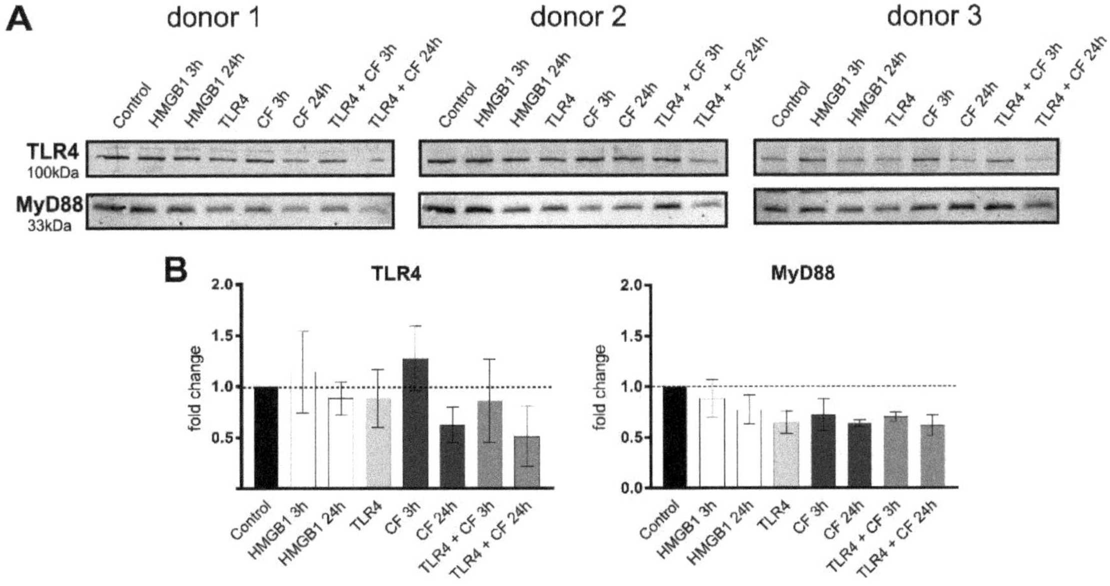

Figure 3. Compressive forces and TLR4 blocking antibody modulate TLR4 production and MyD88 in primary PDL cells. (**A**) Protein levels of TLR4 and MyD88 were determined by Western blot in different conditions: HMGB1 (100 ng/mL), TLR4 blocking antibody (5 µg/mL) (TLR4: TLR4 blocking antibody) and compressive force (CF) 2 g/cm^2 for 3 and 24 h. Three different donors showed similar patterns with reduction of TLR4 antibody. TLR4 production was upregulated under 3 h compression forces. MyD88 production was reduced in all conditions. (**B**) Quantification of three donors, normalized to the control with stain-free technology. Data were tested for normal distribution by Shapiro–Wilk test. Afterward, a one-way analysis of variance (ANOVA) followed by Tukeys' post hoc test was performed.

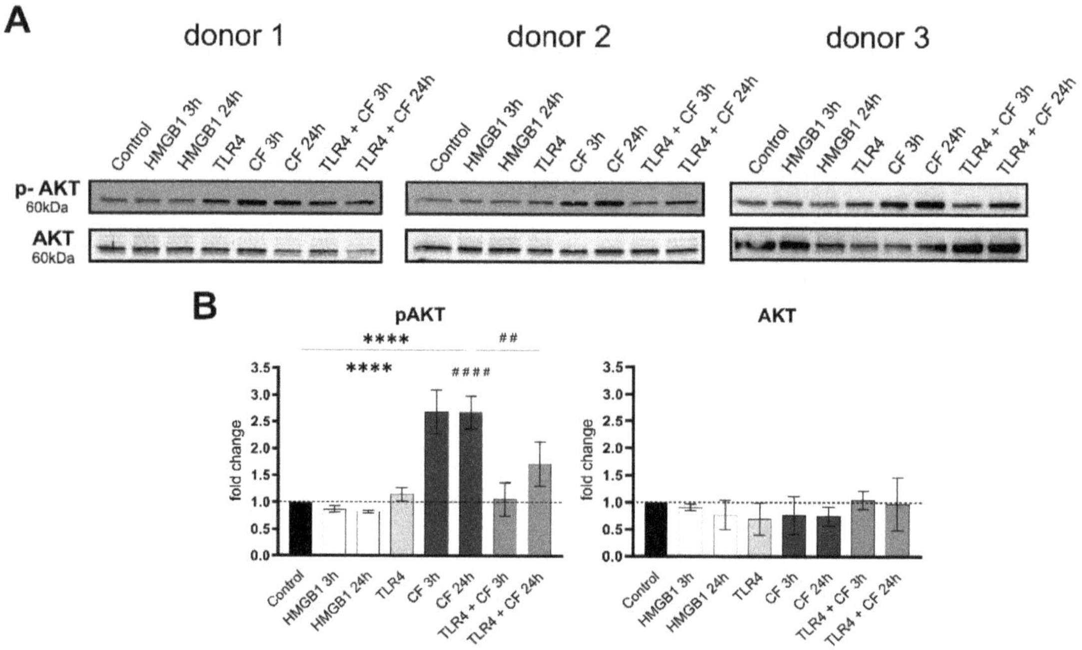

Figure 4. Compressive forces upregulate phosphor AKT in primary hPDL cells. (**A**) Protein production levels and phospho-AKT were determined by Western blot in different conditions: HMGB1 (100 ng/mL), TLR4 blocking antibody (5 μg/mL) (TLR4: TLR4 blocking antibody) and compressive force (CF) 2 g/cm^2 at 3 and 24 h. Three different donors showed similar patterns. Under all conditions, AKT showed no differences, but the phosphorylated AKT was significantly upregulated with compressive forces (CF) for 3 h and 24 h. Compressive forces with additional blocking TLR4 monoclonal antibody led to significant downregulation comparable to the control. (**B**) Quantification of three donors, normalized to the control with stain-free technology. CF conditions without TLR4 blocking antibody showed a significant upregulation. Data were tested for normal distribution by Shapiro–Wilk test. Afterward, a one-way analysis of variance (ANOVA) followed by Tukeys' post hoc test was performed. Statistically significant differences to control are marked by asterisks (**** $p < 0.0001$) and hashtag show significant differences between CF and TLR4 +CF (## $p < 0.01$; #### $p < 0.0001$).

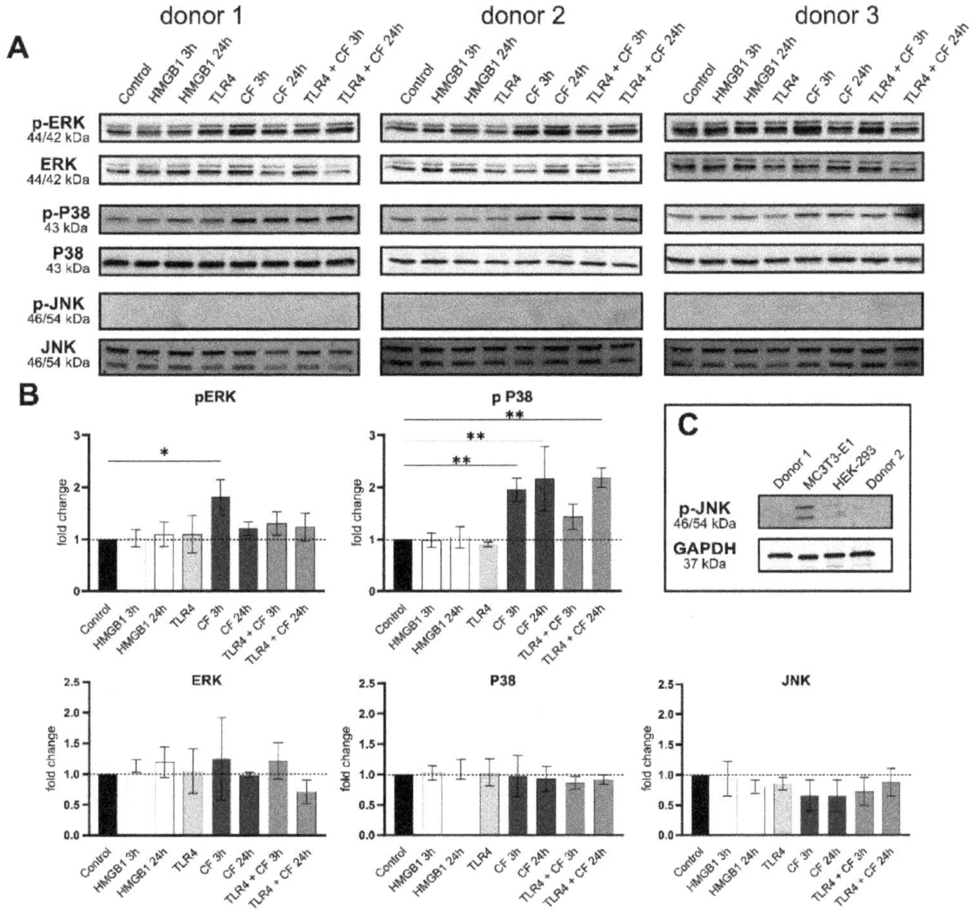

Figure 5. Compressive forces upregulate pERK and p P38 in primary hPDL cells. (**A**) Protein levels and phospho-MAPK (ERK, p38 and JNK) were determined by Western blot in different conditions: HMGB1 (100 ng/mL), TLR4 blocking antibody (5 µg/mL) (TLR4: TLR4 blocking antibody) and compressive force 2 g/cm^2 at 3 and 24 h. Three different donors showed similar patterns. Under all conditions, ERK, p38 and JNK showed no differences, but the phosphorylated ERK and p38 were significantly upregulated with compressive forces (CF) as follows: ERK for 3 h and p38 for 3 h and 24 h. (**B**) Quantification of three donors, normalized to the control with stain-free technology. CF conditions without TLR4 blocking antibody showed a significant upregulation for phospho-ERK and phospho-p38. (**C**) HEK-293 and MC3T3 cells were used as a positive control for phospho-JNK antibody. Data were tested for normal distribution by Shapiro–Wilk test. Afterward, a one-way analysis of variance (ANOVA) followed by Tukeys' post hoc test was performed. Statistically significant differences to control are marked by asterisks (* $p < 0.05$; ** $p < 0.01$).

2.5. Moderate Compressive Forces on PDL Cells Is Not Able to Translocate NF-kB and ERK to the Nucleus

Inflammation is a process coordinated by the local secretion of adhesion molecules, chemotactic factors and cytokines [28]. The nuclear factor-kB (NF-kB), a transcription factor, is an important mediator for the activation of the IL-6 gene [29]. As shown in Figure 2, PDL cells are able to express inflammatory markers, in particular IL6, under mechanical compression. Based on this, we investigated, whether PDL cells under compressive force are able to translocate NF-kB or phospho-ERK to the nucleus and whether a TLR4 blocking antibody plays a role in this process. To study the translocation of NF-kB, fluorescence images and Western blot from cytoplasmic and nuclear fractions were performed. Fluorescence images for all conditions (control and 3 h compressive force with and without TLR4 blocking antibody) were not able to detect NF-kB translocation (Figure 6A). These results correspond to the data of the fractionated Western blot. NF-κB, as well as phosho-ERK did not lead to any changes in the cytosol and nuclear fraction (Figure 6B).

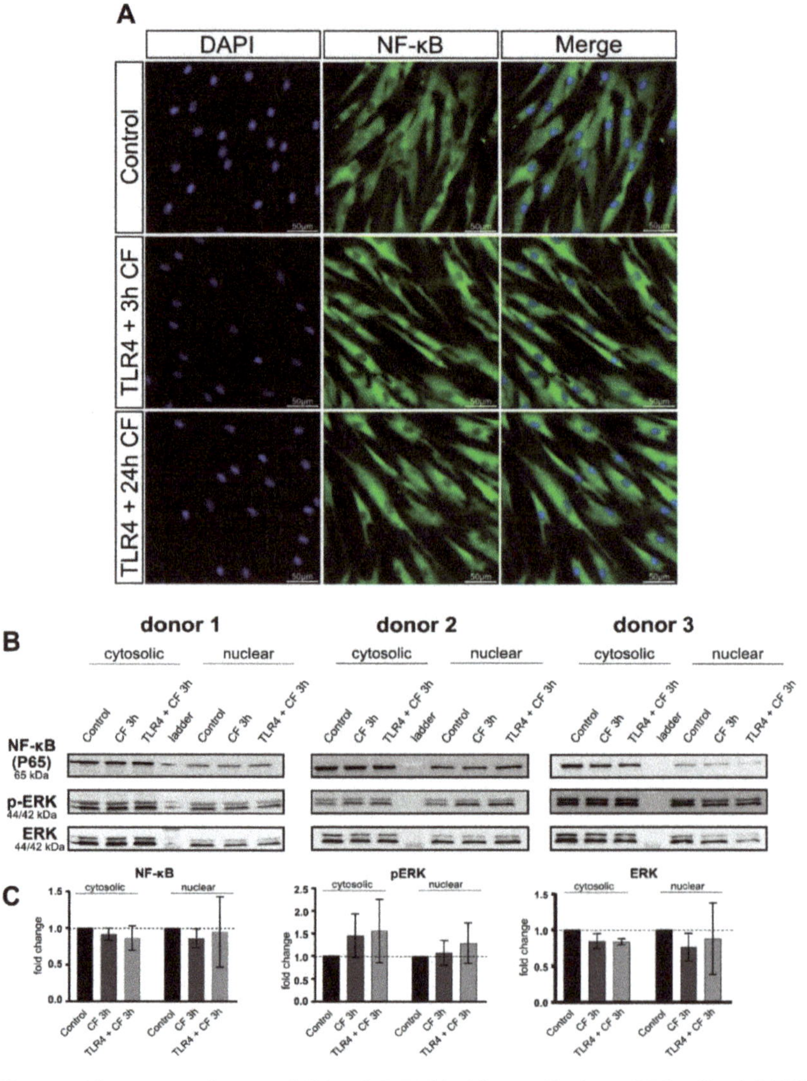

Figure 6. Fluorescence images of CF and TLR4 blocking antibody on PDL cells. (**A**) PDL cells treaded with and without TLR4 blocking antibody (5 µg/mL) and compressive force 2 g/cm^2 for 24 h. NF-kB was stained green (Alexa 488), blue areas represent nuclei (DAPI); scalebar 50 µm, CF: compressive force NF-kB was not translocated to the nucleus in any condition. (**B**) Protein production of NF-kB, ERK and phospho-ERK were determined by Western blot. (**C**) Quantification of three donors, normalized to the control with stain-free technology. CF conditions without TLR4 antibody showed a significant upregulation for phospho-ERK and phospho-p38.

3. Discussion

The level of inflammatory markers in affected periodontal tissues during the orthodontic tooth movement (OTM) plays an important role as a viable diagnosis tool in monitoring the progression of the periodontium [14]. In our previous work, we reported that human PDL cells modify gene expression and protein production of pro-inflammatory cytokines via TLR4 signaling in an in vitro model, which represents the compressed periodontal ligament in the initial phase of OTM [8]. Hence in this study, we further explored the possible cell signaling downstream key molecules modulation via TLR4 caused by a sterile inflammation in three new isolated and characterized human PDL cells. To this end, we used the gold standard model for static compressive mechanical stimulation induced by static compression forces of 2 g/cm^2 [9,11,30]. Some studies use this model with PDL cells to investigate mechanotransduction, i.e., the ability of a cell to actively sense, integrate, and convert mechanical stimuli into biochemical signals that result in intracellular changes, such as ion concentrations, activation of signaling pathways and transcriptional regulation [31–33]. Concerning mechanotransduction, this model can be used to explore, e.g., the mechanosensory protein complex and focal adhesion kinase (FAK) involved in mechanotransduction and subsequent YAP/TAZ translocation into the nucleus [34] or Wnt signaling responsive to mechanical loading [35], or mechanical strain modulating the amount of the matrix metalloproteinase MMP-13 [36]. Additionally, to this, we want to deepen the understanding of signal transduction cascades involved in mediating mechanical cell stress into host immune responses via TLR4 in a sterile inflammation by the release of damage-associated molecular patterns (DAMPs), resulting from injuries during OTM. TLR4 is very well investigated as a receptor involved in pathogenic models (LPS) PAMPs described in PDL cells [37,38]. Although TLR4 has been considered an important player in the initiation and progression of several inflammatory pathological conditions, we hypothesized that there might be a strong interplay between TLR4 and cell signaling in bacterial and sterile inflammation.

Differently from our previous work using a commercial primary cell—HPdLF [8], the presented data are based on self-isolated human PDL cells from the upper jaw of three different donors. Using only the samples from the upper jaw can help to avoid possible variability from primary PDL cells and support the level of reliability of our results. With newly isolated cells, we could confirm the gene expression and protein production under 24 h compressive force with results identical to our previous work. In our present study, it was observed that the TLR4 production increased after 3 h, and not after 24 h.

The PDL cells showed similar characteristics in the evaluation of the stem cell character as described in the literature for periodontal ligament stem cells [39]. In terms of morphology, surface epitopes and differentiation capacity towards adipocytes, osteoblasts and chondrocytes, PDLs behave comparably to MSCs. Moreover, the protein levels and phosphorylation status of different proteins analyzed in this study show very similar profiles in all Western blotting analyses for all three donors.

On the basis of our previous work about gene expression, similar results could be observed in this study, with TLR4 protein production increasing under compressive force and reducing with TLR4 monoclonal blocking antibody in a short-term culture. Moreover, the downstream signaling from TLR4 has clearly changed.

TLR4 can be activated by HMGB1 [40] and regulated under mechanical stress in PDL cells [41,42]. Furthermore, the HMGB1–TLR pathway is linked to the MyD88-mediated NF-κB pathway and activates downstream signaling pathways which in turn leads to the induction of innate immune responses by producing inflammatory cytokines and other mediators [43,44]. However, our results with HMGB1 show no significant changes in TLR4 and downstream molecules. The activation of TLR4 via HMGB1 might better work in a higher concentration or may be dependent on the complexes it forms with other molecules, immunostimulatory complexes, or with the binding of cytokines and other molecules [40,45–47]. HMGB1 binds to TLR4 with remarkably less affinity than LPS, and it activates gene expression of distinct signaling patterns after stimulation. Both HMGB1 and LPS significantly increase the nuclear translocation of NF-κB [40].

MyD88 used in this work as a downstream activation of TLR4 showed in all conditions a similar regulation pattern to the regulation of TLR4. After ligand binding, TLRs interact with adaptor proteins as myeloid differentiation primary response gene 88 (MyD88) or TIR domain-containing adaptor proteins and initiate signal transduction pathways that activate NF-κB or MAP kinases.

AKT is also known to be regulated by mechanical stimulation leading to activated FGF-2 production and to the involvement of FGF-2 in the PI3K/Akt or Rho pathway. In addition, AKT can be related to pro-inflammatory responses [48,49]. Our data confirmed that AKT is phosphorylated in response to a compressive force, being significantly upregulated with the compressive force for 3 h and 24 h. What is more, we could very clearly demonstrate that inhibition of TLR4 signaling via TLR4 monoclonal blocking antibody led to significant downregulation comparable to control.

In addition to AKT, mitogen-activated protein kinases (MAPKs), and other key molecules involved in signal transduction and localized downstream of TLR4, were evaluated. Recently, it has been published, that MAPKs are involved in mechanotransduction and resulting in inflammatory responses [50–54]. Our data also confirmed the modulation of phosphorylation of ERK and p38 by mechanical stimulation in human PDL cells. Additional blocking of TLR4 with a specific antibody reduced in a short-term culture, 3 h under compressive force, the phosphorylation of these MAPKs following mechanical stress. Interestingly, JNK does not seem to be regulated at all, which contrasts to several studies that showed regulation of phosphor-JNK in PDL cells due to mechanical compression. Of note, a phospho-JNK antibody recognized specifically the twice phosphorylated epitope at Thr183 and Tyr185. This indicates that if phospho-JNK binds only in one phospho residue, e.g., Tyr185, this antibody is not able to detect the phosphorylation of JNK.

LPS from Porphyromonas gingivalis leads to the activation of the TLR4/MyD88 complex, triggering the secretion of pro-inflammatory cytokine cascades as: IL-1α, IL-8, TNF-α and β and Eotaxin. Moreover, the upregulation of pERK/ERK signaling pathways and Nf-kB nuclear translocation was evident [55]. In our study, we could see that the activation of TLR4/MyD88 complex and modulation of phosphorylation of ERK and p38 triggering the secretion of pro-inflammatory cytokines work in a sterile inflammation during mechanical stress too. The strong correlation between these two common inflammation types of the periodontium triggered by PAMP or DAMP molecules might confirm that—as hypothesized [56]—the activation of pattern recognition receptor TLR4 shares certain biochemical actors that can regulate the inflammatory process in both types of inflammation. Based on this, the translocation of Nf-kB and phospho-ERK to the nucleus was analyzed by immunofluorescence as well as cytoplasmic and nuclear fractions by Western blotting. In a short and long compressive force time with and without TLR4 blocking antibody, no translocation of Nf-kB by immunofluorescence was observed. The same was observed in the nuclear and cytoplasmatic fractions of Nf-kB and phospho-ERK. Taking both results together, we could conclude that a moderate compressive force of 2 g/cm^2 is not able to induce translocation of Nf-kB and phospho-ERK into the nucleus. However, despite this, inflammatory cytokines in gene expression, as well as protein levels of IL-6, were observed in our experiments. The activation of TLR4 through damage proteins during mechanical compression in this model may require less affinity compared with LPS. Another reason could be that the effect of mechanical cell stress on TLR production in human PDL cells is dependent on variables such as force magnitude, duration and constancy [25].

We could clearly demonstrate that in PDL cells under compressive forces, the inflammatory markers, as well as IL-6 protein production, are upregulated—as expected in line with other publications. Further, we could see that the use of TLR4 monoclonal blocking antibody significantly reverses the upregulation of phospho-AKT, caused by compressive force, to levels comparable to controls. The inhibition of TLR4 may indicate that this modulation is induced not only by mechanotransduction from mechanoreceptors as described by now. However, this study must be seen in the light of certain limitations since we could not describe or suggest a cell signaling mechanism independent from NF-kB or phosphor-ERK translocation for explaining the upregulation of inflammatory markers.

4. Materials and Methods

4.1. Reagents and Methods

Primary antibodies p38 MAPK (8690S), Phospho-p38 MAPK (9216S), p44/42 MAPK (Erk1/2) (3A7) (9107S), Phospho-p44/42MAPK (Erk1/2) (Thr202/Tyr204) (D13.14.4E) (4370S), SAPK/JNK (9252), Phospho-SAPK/JNK (Thr183/Tyr185) (G9) (9255S), Akt (pan) (40D4) (2920S), Phospho-Akt (Ser473) (D9E) (4060S) and MyD88 (D80F5) (4283) were purchased from Cell Signaling and TLR4 (MA5-16216) from Thermo Fisher Scientific, Waltham, MA, USA. Secondary antibodies StarBright Blue 700 (12004158) and StarBright Blue 520 (12005869) were purchased from Bio-Rad, Hercules, CA, USA. For flow cytometry APC, FITC and PE-labeled antibodies were purchased from eBiosciences, San Diego, CA, USA (FITC-labeled anti-human CD34 (11-0349-42), APC-labeled anti-human CD45 (17-0459-42), APC-labeled anti-human CD73 (17-0739-42), FITC-labeled anti-human CD90 (11-0909-42), PE-labeled anti-human CD105 (12-1057-42)). For Western blotting, RIPA buffer (Thermo Fisher Scientific, Waltham, MA, USA) complemented with cOmplete Tablets Mini and PhosStop (Roche, Basel, Switzerland), Neutralization monoclonal blocking antibody TLR-4 (HTA125) (14-9917-82) 5 µg/mL from Thermo Fisher Scientific, Recombinant Human HMGB1 Protein (1690-HMB) R&D Systems was used.

4.2. Primary Human Periodontal Ligament (hPDL) Cell Isolation

Primary human periodontal ligament (hPDL) cells were cultured from periodontal connective tissue, isolated from the middle root section of healthy human teeth. Only decay-free teeth from healthy donors, which needed to be extracted for medical reasons were used for human PDL cell isolation. Immediately after extraction, the teeth were transferred into an isolation medium of DMEM high-glucose (Gibco, Billings, MT, USA), 10% Fetal Bovine Serum (FBS), qualified heat-inactivated (Gibco), 50 mg/L ascorbic acid (Sigma, Saint Louis, MO, USA), antibiotic-antimycotic (Gibco) and stored at RT until isolation was started within 24 h after extraction. For isolation, residual tissue was mechanically scraped off and incubated in a solution of 3 mg/mL collagenase type I (Worthington Biochem, Freehold, NJ, USA) for 1 h at 37 °C. After incubation, cells were plated into 6-well plates. After the first splitting, the isolation medium was replaced by a culture medium of DMEM high-glucose (Gibco), 10 % FBS (Gibco), 50 mg/L ascorbic acid (Sigma), Penicillin/Streptomycin (Gibco). Human PDL cells (upper jaw, third molar, 28) from three different patients were used (1 male, 2 females, age: 19–22 years). Collection and usage of hPDL cells from discarded patient samples were approved by the ethics committee of the RWTH Aachen, Germany (approval number EK374/19), and all experiments were carried out in accordance with the relevant guidelines and regulations. Informed consent was obtained from all participants and/or their legal guardian/s. In this work, human PDL cells were used from passages two to five.

4.3. Characterization of hPDL Cells Flow Cytometry Analysis of Cell Surface Markers

To characterize the cells from the three different donors, the expression of specific stem cell markers was analyzed by flow cytometry. Briefly, 250,000 isolated hPDL cells were trypsinized and resuspended in FACS buffer (PBS + 0.1% FCS) and centrifuged for 5 min at 300× *g* at 4 °C. Afterwards, the cells were incubated for 30 min in the dark with primary PE/FITC/APC-coupled antibody CD34, CD45, CD73, CD90, and CD105 in a concentration of 0.5 µg, 0.06 µg, 0.125 µg and 1 µg per 250,000 cells, respectively. Cells were centrifuged for 5 min at 300× *g* and resuspended in 300 µL FACS buffer before measurement with BD FACSCalibur™ Flow Cytometer (BD Science, Franklin Lakes, NJ, USA). Analysis was performed using BD Cell Quest Pro Software from BD Science, USA.

4.4. PDL Cell Differentiation

Differentiation towards adipocytes: To induce the differentiation into an adipogenic phenotype, PDL cells were seeded in a density of 80.000 cells/cm^2 on TCPS incubated at 37 °C with 5% CO_2 in a humidified atmosphere for 14 days. The medium was changed the next day to an adipogenic induction medium (AIM) containing DMEM high-glucose (Gibco, Darmstadt, Germany), 10% FCS (PAN-Biotech, Aidenbach, Germany) and the following supplements: 1 μM Dexamethasone, 0.2 μM Indomethacin, 0.5 mM IBMX, 0.01 mg/mL human Insulin (all Sigma Aldrich, Steinheim, DE, Germany) and 1% LGPS (80 U/mL Penicillin; 80 μg/mL Streptomycin; 1.6 mM L-Glutamine; Gibco, Germany). Medium change was performed two times a week, alternating with adipogenic induction medium (AIM) and adipogenic maintenance medium (AMM) containing DMEM high-glucose (Gibco, Germany), 10% FCS (PAN-Biotech, Germany), 0.01 mg/mL human Insulin (Sigma Aldrich, Germany) and 1% LGPS (80 U/mL Penicillin; 80 μg/mL Streptomycin; 1.6 mM L-Glutamine).

Oil red O Staining: To prove the success of adipogenic differentiation, Oil red O staining was performed. The staining solution was prepared by mixing 35 mL 0.2% (*w/v*) Oil red O powder (Sigma Aldrich, Darmstadt, Germany) in 100% ethanol (Merck, Darmstadt, Germany) with 10 mL 1 M NaOH (Merck, Germany) and filtrated. Cells were fixed in 50% ice-cold ethanol (4 °C) for 30 min and stained with Oil red O for 10 min. The supernatant was removed and cells were rinsed with 50% ethanol and aqua ad iniectabilia (B. Braun, Melsungen, Germany). Cell nuclei were stained blue by hemalum (Abcam, Cambridge, UK) and rinsed with tap water afterward.

Differentiation towards chondrocytes: Differentiation toward chondrocytes was induced with a chondrogenic induction medium (CIM) containing DMEM high-glucose (Gibco, Germany) and the following supplements: 100 nM Dexamethasone, 0.17 mM L-Ascorbic-Acid-2-phosphate, 100 μg/mL Sodium pyruvate, 4 μg/mL L-Proline (all Sigma Aldrich, Germany), 10 ng/mL TGF-β3 (R&D Systems, Wiesbaden, Germany), 1% LGPS (80 U/mL Penicillin; 80 μg/mL Streptomycin; 1.6 mM L-Glutamine; Gibco, Germany) and 5% ITS-Plus Premix (6.25 μg/mL Bovine Insulin, 6.25 μg/mL Transferrin, 6.25 μg/mL Selenium acid, 6.25 μg/mL Linoleic acid, 6.25 μg/mL BSA; Life Technologies, Darmstadt, Germany). Cells were seeded as pellet cultures in a density of 250,000 cells/0.5 mL and transferred to a 15 mL centrifuge tube, centrifuged for 7 min at 500 g and incubated at 37 °C with 5% CO2 in a humidified atmosphere for 14 days. The medium was changed from standard medium to CIM the next day after seeding and changed three times per week. The growth factor TGF-β3 was added freshly to the medium (0.5 μL/mL).

Toluidine blue staining: Cell pellets were fixed in 4% formalin (Morphisto, Karlsruhe, Germany) overnight at 4 °C and embedded in 3% agarose (Sigma Aldrich, Germany), dehydrated in an ascending ethanol series, treated with xylol and histoplast in a tissue processor. Pellets were embedded in paraffin and sectioned with a rotating microtome into slices of 2 μm thickness. Afterward, paraffin sections were incubated at 60 °C for 10 min. Paraffin residues were removed by placing the slides in xylene for 5–10 min and a descending alcohol series was performed (10 min each in 100%, 96%, 70% ethanol and aqua dest. (B. Braun, Germany)). Pellets were stained with toluidine blue (2 g toluidine blue powder (Sigma Aldrich, Germany) in acetate buffer (Merck, Germany), pH 4.66) for 2 min. Next, an ascending alcohol series was applied (96%, 100% ethanol and xylene for 1 min each). Finally, the samples were encapsulated in Vitro-Clud® (R. Langenbrinck, Emmendingen, Germany) and investigated under a light microscope.

Differentiation towards osteoblasts: Induction towards osteoblasts was performed by osteogenic induction medium (OIM) containing DMEM low glucose (Gibco, Germany), 10% FCS (PAN-Biotech, Germany) and the following supplements: 100 nM Dexamethason, 10 mM Sodium-β-glycerophosphate, 0.05 mM L-Ascorbic-Acid-2-phosphate (all Sigma Aldrich, Germany) and 1% LGPS (80 U/mL Penicillin; 80 μg/mL Streptomycin; 1.6 mM L-Glutamine; Gibco, Germany). Cells were seeded in a density of 31.000 cells/cm^2 on TCPS and cultured at 37 °C with 5% CO_2 in a humidified atmosphere for 14 days. The medium was changed three times a week. Cells were quantified with Alizarin red to visualize possible calcium deposits as a result of the osteogenic differentiation.

Alizarin red staining: Staining solution was prepared by dissolving 1.37 g Alizarin red powder (Sigma Aldrich, Germany; 342.3 g/mol) in 100 mL aqua ad iniectabilia (B. Braun, Germany) and filtrated. Cells were fixed for 1 h at room temperature with 70% ethanol and washed three times with aqua ad iniectabilia for five minutes. Samples were stained with alizarin red for 10 min and washed three times with PBS (Gibco, Germany) for 5 min before visualization by light microscopy.

4.5. Isolation and Purification of RNA

For RNA isolation, cells in each well were first washed with 2 mL phosphate-buffered saline (PBS; Gibco) and the cells were harvested with 0.5 mL TRIzolTM Reagent (Thermo Fisher Scientific, USA), two wells were pooled. This leads to biological triplicates for each condition. After isolation according to the manufacturers' instructions, the RNA yield of each sample was verified photometrically at 280 nm and 260 nm (Nanodrop OneTM, Thermo Fisher Scientific, USA). Afterward, RNA purification was performed with Quick-RNA MicroPrep kit (Zymo Research Europe GmbH, Freiburg, Germany) following the producers' protocol including an on-column DNA digestion. In order to control the success of the purification and to ensure a uniform cDNA synthesis, each sample was measured again (Nanodrop OneTM).

4.6. Quantitative Realtime-RT-PCR Analysis (RT-qPCR)

The RNA was transcribed into cDNA (SuperScript III RT, Thermo Fisher Scientific, USA) with a final concentration of 25 ng/μL. All steps from RNA isolation to cDNA synthesis were performed in parallel for all samples of each experiment in order to avoid experimental variations. RT-qPCR was performed in technical duplicates using 2.5 ng/μL cDNA in each reaction and a primer concentration of 0.5 μM. The qTower3 (Analytik Jena, Jena, Germany), High Green Mastermix (Thermo Fisher Scientific, USA), qPCRSoft 3 (Analytik Jena, Germany) and self-designed intron spanning primers were used (Eurofins Genomics, Luxembourg). Primers were designed by using Primer-BLAST (NCBI, Bethesda, MD, USA) followed by a PCR-Check (Eurofins Oligo Analyse Tool, Luxembourg) to ensure in silico PCR specificity. RT-qPCR protocol was performed as follows: 2 min 50 °C for 2 min, 95 °C for 10 min followed by 40 cycles of 95 °C/15 s, 60 °C/30 s and 72 °C/30 s. After 95 °C for 15 s as the last step, a melting curve (60–95 °C). Gene, primer and target/amplicon information for the reference and target genes are displayed in Table 1.

Table 1. Primer information: RT-qPCR gene, primer and target/amplicon information for the reference gene RPL22 and investigated target genes. Tm: melting temperature of primer/specific qPCR product (amplicon), %GC: guanine/cytosine content, bp: base pairs.

Gene Symbol	Gene Name (Mus Musculus)	Gene Function	Accession Number (NCBI Gene Bank)	Chromosoma Location (Length)	5′-Forward Primer-3′ (Length/Tm/GC)	5′reverse Primer-3′ (Length/Tm/GC)	Primer Location	Amplicon Length	Amplicon Location (bp of Start/Stop)	Intron-Flanking (Length)	Variants Targeted (Transcript/Splice)
RPL22	ribosomal protein L22	translation of mRNA in protein	NM_000983	1; 1p36.31 (2061 bp)	tgattgcacccaccctgtag (20 bp/59.67 °C/55%GC)	ggttcccagcttttccgttc (20 bp/59.4 °C/55%GC)	Exon 2/ Exon 3	98	91/188	yes	yes
IL-6	Interleukin 6	important role in bone metabolism; os-teoclastogenesis	NM_000600	7; 7p15.3 (1127 bp)	catcctcgacggcatctcag (20 bp/60.32 °C/60%GC)	tcaccaggcaagtctcctca (20 bp/60.47 °C/55%GC)	Exon 2/ Exon 4	164	240/403	yes	yes
IL-8	Interleukin 8	important role in bone metabolism; os-teoclastogenesis	NM_000584	4; 4q13.3 (1642 bp)	catactccaaacctttccacc (21 bp/57.9 °C/47,6%GC)	cttcaaaaacttctccacaacc (22 bp/56.9 °C/40.9%GC)	Exon 2/ Exon 3	167	206/372	yes	yes
VEGF A	vascular endothelial growth factor A	induces proliferation and migration of vascular endothelial cells	NM 001171623	6p21.1 (3660 bp)	GGAGGGCAGAATCATCACGAA (21 bp/60.1 °C/52.3%GC)	GGTACTCCTGGAAGATGTCCAC (22 bp/59.8 °C/54.5%GC)	Exon 2/ Exon 3	100	1153/1211	yes	yes
PTGS2 COX2	prostaglandin-endoperoxide synthase 2	involved in prostaglandin synthesis	NM_000963	1q31.1 (4510 pb)	GATGATTGCCCGACTCCCTT (20 bp/59.8 °C/55%GC)	GGCCCTCGCTTATGATCTGT (20 pb/59.6 °C/55%GC)	Exon 4/ Exon 5	185	560/725	yes	yes

4.7. ELISA

To analyze the level of IL-6, a commercially available enzyme-linked immunosorbent assay (ELISA) kit for IL-6 (CSB-E04638h, Cusabio Wuhan Huamei Biotech Co., Wuhan, China) was used following manufacturers' instructions with fresh cell culture supernatant.

4.8. In Vitro Compressive Stimulation Model

The hPDL cells were cultured in high-glucose Dulbecco's Modified Eagle's Medium (DMEM; Gibco, USA), containing 100 units/mL of penicillin, 100 g/mL of streptomycin (Gibco, Gaithersburg, MD, USA), 10% FCS (Gibco, USA) and 50 mg/l L-ascorbic acid (Sigma-Aldrich, USA) at humidified 37 °C and 5% CO_2. Cells were trypsinized and centrifuged at 300× *g* and 90,000 cells were seeded into 6-well plates and cultured for 4 days till 90% cell density was reached. After the incubation of 24 h, a static force of 2 g/cm^2 (=0.02 N/cm^2) was applied to the monolayer with sterile round-glass cylinders (34 mm Ø; 18 g), as described and established by Kanzaki et al. [27].

4.9. Isolation of Total Protein Respective Cytoplasmic and Nuclear Fractions

After the treatment, cells were analyzed by immunoblotting. Cells were washed and lysed with (100 µL/well) RIPA buffer. Alternatively, to acquire protein fractions from the cytoplasm and nucleus, the NE-PER™ Nuclear and Cytoplasmic Extraction Reagents kit (neper TM Thermo Fischer, nuclear and cytoplasmic extraction reagents Thermo Fisher, USA) was used, according to manufacturers´ instructions. The protein amount was quantified by Bradford assay (Bio-Rad, USA) and 25 µg total protein was used for gel electrophoresis, respective 10 µg for fractionated gel electrophoresis.

4.10. Immunoblotting Analysis

The protein lysates from hPDL cells were separated by gel electrophoresis (TGX Stain-FreeTM FastCast™, 12%, Bio-Rad, Hercules, CA, USA) and transferred to nitrocellulose membranes (Trans-Blot Turbo ®Turbo TM RTA Transfer kit, Nitrocellulose, Bio-Rad, USA). Membranes were blocked for 1 h at RT in 1× Tris-buffered saline containing 0.05% Tween-20 (TBST) supplemented with 5% BSA. Incubation with the primary antibodies occurred overnight at 4 °C and subsequently with the respective secondary antibody for 1 h at RT. Immunoblot was detected by fluorescence, quantified and normalized by means of a ChemiDoc MP Imaging System (Bio-Rad, USA) with Stain-Free technology and Image Lab™ Software (Version6.01 Bio-Rad, Hercules, CA, USA).

4.11. NF-kB/DAPI Staining

To investigate the translocation of Nf-kB, cells with and without compressive mechanical stimulation, as well as with HMGB1 and TLR4 AB were fixed with 3.7% formaldehyde suspension (Carl Roth, Karlsruhe, Germany). Afterward, the cells were permeabilized with PBS, supplemented with 0.1% Triton X-100. Subsequently, samples were blocked in PBS containing 1% BSA (Carl Roth, Karlsruhe, Germany), and incubated overnight with NF-kB antibody (#8242, Cell Signaling Technology, USA) followed by ProLong Gold Antifade Mountant with DAPI (Thermo Fisher Scientific, USA), according to the manufacturer's protocol. Afterward, the monolayer was covered for preservation with coverslips and examined by immunofluorescence imaging microscopy (Observer 7, Zeiss, Germany).

4.12. Statistical Analysis

Graphs show mean ± standard deviations (SD). Data were tested for normal distribution by the Shapiro–Wilk test. Afterward, a *t*-test or one-way analysis of variance (ANOVA) followed by Tukeys' post hoc test was performed in GraphPad Prism (version 9.0; San Diego, CA, USA). A p-value < 0.05 was considered statistically significant.

5. Conclusions

In this work, it has been shown for the first time that mechanical compression modulates phosphorylation of AKT and MAPKs (phopho-ERK and phospho-p38) via TLR4 in an in vitro model simulating orthodontic tooth movement in human periodontal ligament cells. Furthermore, the inhibition of TLR4 through a TLR4 monoclonal blocking antibody hinders the upregulation of phospho-AKT caused by compressive force to levels comparable to the control. Therefore, it can be assumed that these signaling pathways are modulated not only by mechanotransduction from mechanoreceptors as described in other works but also by TLR4 via transduction of pattern recognition receptors through specific molecular structures on damaged senescent cells during mechanical stress. Additionally, a moderate compressive force is not able to induce the translocation of Nf-kB and phospho-ERK into the nucleus and the signaling of inflammatory cytokine IL-6 may be induced differently. The present findings provide evidence that TLR4 modulating strategies seem to be effective to regulate clinical orthodontic therapy and might have the potential to treat its inflammatory-related side effects such as mechanical or trauma-induced tooth root resorption and periodontal degeneration.

Author Contributions: Conceptualization, M.W. and R.B.C.; performing experiments, C.E.R., H.M., R.B.C. and C.N.; data analysis, C.E.R., C.N. and H.M.; statistical analysis, C.E.R., R.B.C. and C.N.; data interpretation, C.E.R., R.B.C., S.N. and J.J.; visualization, C.E.R. and C.N.; writing—original draft preparation, C.E.R.; writing—review and editing, R.B.C., S.N., J.J. and M.W.; funding acquisition, M.W. and S.N.; project administration, R.B.C. and M.W. All authors have read and agreed to the published version of the manuscript.

Funding: This work was supported by a grant from the Interdisciplinary Centre for Clinical Research within the faculty of Medicine at the RWTH Aachen University (OC1-2, 3 and 11/IA) and Deutsche Forschungsgemeinschaft (DFG, 490932300).

Institutional Review Board Statement: Collection and usage of hPDL cells from discarded patient samples were approved by the ethics committee of the RWTH Aachen, Germany (approval number EK374/19), and all experiments were carried out in accordance with the relevant guidelines and regulations. Informed consent was obtained from all participants and/or their legal guardian/s.

Informed Consent Statement: Informed consent was obtained from all subjects involved in the study.

Data Availability Statement: The data that support the findings of this study are available from the corresponding author, R.B.C., upon reasonable request.

Conflicts of Interest: The authors declare no conflict of interest.

References

1. Richmond, S.; Shaw, W.C.; O'brien, K.D.; Buchanan, I.B.; Jones, R.; Stephens, C.D.; Roberts, C.T.; Andrews, M. *The Development of the PAR Index (Peer Assessment Rating): Reliability and Validity*; European Orthodontic Society: London, UK, 1992; Volume 14.
2. Meikle, M.C. The Tissue, Cellular, and Molecular Regulation of Orthodontic Tooth Movement: 100 Years after Carl Sandstedt. *Eur. J. Orthod.* **2006**, *28*, 221–240. [CrossRef] [PubMed]
3. Beertsen, W.; Mcculloch, C.A.G.; Sodek, J. The Periodontal Ligament: A Unique, Multifunctional Connective Tissue. *Periodontol. 2000* **1997**, *13*, 20–40. [CrossRef] [PubMed]
4. Bresin, A.; Kiliaridis, S.; Strid, K.G. Effect of Masticatory Function on the Internal Bone Structure in the Mandible of the Growing Rat. *Eur. J. Oral Sci.* **1999**, *107*, 35–44. [CrossRef] [PubMed]
5. Silva, M.A.J.; Merzel, J. Alveolar Bone Sharpey Fibers of the Rat Incisor in Normal and Altered Functional Conditions Examined by Scanning Electron Microscopy. *Anat. Rec. Part A Discov. Mol. Cell. Evol. Biol.* **2004**, *279*, 792–797. [CrossRef]
6. Pagella, P.; de Vargas Roditi, L.; Stadlinger, B.; Moor, A.E.; Mitsiadis, T.A. A Single-Cell Atlas of Human Teeth. *iScience* **2021**, *24*, 102405. [CrossRef]
7. Hlaing, E.E.H.; Ishihara, Y.; Wang, Z.; Odagaki, N.; Kamioka, H. Role of Intracellular Ca2+–Based Mechanotransduction of Human Periodontal Ligament Fibroblasts. *FASEB J.* **2019**, *33*, 10409–10424. [CrossRef]
8. Marciniak, J.; Lossdörfer, S.; Knaup, I.; Bastian, A.; Craveiro, R.B.; Jäger, A.; Wolf, M. Orthodontic Cell Stress Modifies Proinflammatory Cytokine Expression in Human PDL Cells and Induces Immunomodulatory Effects via TLR-4 Signaling in Vitro. *Clin. Oral Investig.* **2020**, *24*, 1411–1419. [CrossRef]

9. Brockhaus, J.; Craveiro, R.B.; Azraq, I.; Niederau, C.; Schröder, S.K.; Weiskirchen, R.; Jankowski, J.; Wolf, M. In Vitro Compression Model for Orthodontic Tooth Movement Modulates Human Periodontal Ligament Fibroblast Proliferation, Apoptosis and Cell Cycle. *Biomolecules* **2021**, *11*, 932. [CrossRef]
10. Schröder, A.; Käppler, P.; Nazet, U.; Jantsch, J.; Proff, P.; Cieplik, F.; Deschner, J.; Kirschneck, C. Effects of Compressive and Tensile Strain on Macrophages during Simulated Orthodontic Tooth Movement. *Mediat. Inflamm.* **2020**, *2020*, 2814015. [CrossRef]
11. Weider, M.; Schröder, A.; Docheva, D.; Rodrian, G.; Enderle, I.; Seidel, C.L.; Andreev, D.; Wegner, M.; Bozec, A.; Deschner, J.; et al. A Human Periodontal Ligament Fibroblast Cell Line as a New Model to Study Periodontal Stress. *Int. J. Mol. Sci.* **2020**, *21*, 7961. [CrossRef]
12. Könönen, E.; Gursoy, M.; Gursoy, U.K. Periodontitis: A Multifaceted Disease of Tooth-Supporting Tissues. *J. Clin. Med.* **2019**, *8*, 1135. [CrossRef] [PubMed]
13. Sorsa, T.; Tja¨derhane, L.T.; Salo, T. Matrix Metalloproteinases (MMPs) in Oral Diseases. *Oral Dis.* **2004**, *10*, 311–318. [CrossRef] [PubMed]
14. Luchian, I.; Moscalu, M.; Goriuc, A.; Nucci, L.; Tatarciuc, M.; Martu, I.; Covasa, M. Using Salivary MMP-9 to Successfully Quantify Periodontal Inflammation during Orthodontic Treatment. *J. Clin. Med.* **2021**, *10*, 379. [CrossRef] [PubMed]
15. Garlet, T.P.; Coelho, U.; Silva, J.S.; Garlet, G.P. Cytokine Expression Pattern in Compression and Tension Sides of the Periodontal Ligament during Orthodontic Tooth Movement in Humans. *Eur. J. Oral Sci.* **2007**, *115*, 355–362. [CrossRef]
16. Konermann, A.; Beyer, M.; Deschner, J.; Allam, J.P.; Novak, N.; Winter, J.; Jepsen, S.; Jäger, A. Human Periodontal Ligament Cells Facilitate Leukocyte Recruitment and Are Influenced in Their Immunomodulatory Function by Th17 Cytokine Release. *Cell. Immunol.* **2012**, *272*, 137–143. [CrossRef]
17. Konermann, A.; Stabenow, D.; Knolle, P.A.; Held, S.A.E.; Deschner, J.; Jäger, A. Regulatory Role of Periodontal Ligament Fibroblasts for Innate Immune Cell Function and Differentiation. *Innate Immun.* **2012**, *18*, 745–752. [CrossRef]
18. Bartold, P.M.; McCulloch, C.A.G.; Narayanan, A.S.; Pitaru, S. Tissue Engineering: A New Paradigm for Periodontal Regeneration Based on Molecular and Cell Biology. *Periodontol. 2000* **2000**, *24*, 253–269. [CrossRef]
19. Shimono, M.; Ishikawa, T.; Ishikawa, H.; Matsuzaki, H.; Hashimoto, S.; Muramatsu, T.; Shima, K.; Matsuzaka, K.I.; Inoue, T. Regulatory Mechanisms of Periodontal Regeneration. *Microsc. Res. Tech.* **2003**, *60*, 491–502. [CrossRef]
20. Hayashi, S.I.; Tsuneto, M.; Yamada, T.; Nose, M.; Yoshino, M.; Shultz, L.D.; Yamazaki, H. Lipopolysaccharide-Induced Osteoclastogenesis in Src Homology 2-Domain Phosphatase-1-Deficient Viable Motheaten Mice. *Endocrinology* **2004**, *145*, 2721–2729. [CrossRef]
21. AlQranei, M.S.; Senbanjo, L.T.; Aljohani, H.; Hamza, T.; Chellaiah, M.A. Lipopolysaccharide- TLR-4 Axis Regulates Osteoclastogenesis Independent of RANKL/RANK Signaling. *BMC Immunol.* **2021**, *22*, 23. [CrossRef]
22. Yim, M. The Role of Toll-like Receptors in Osteoclastogenesis. *J. Bone Metab.* **2020**, *27*, 227–235. [CrossRef] [PubMed]
23. Kim, H.M.; Park, B.S.; Kim, J.I.; Kim, S.E.; Lee, J.; Oh, S.C.; Enkhbayar, P.; Matsushima, N.; Lee, H.; Yoo, O.J.; et al. Crystal Structure of the TLR4-MD-2 Complex with Bound Endotoxin Antagonist Eritoran. *Cell* **2007**, *130*, 906–917. [CrossRef] [PubMed]
24. Wolf, M.; Lossdörfer, S.; Römer, P.; Kirschneck, C.; Küpper, K.; Deschner, J.; Jäger, A. Short-Term Heat Pre-Treatment Modulates the Release of HMGB1 and pro-Inflammatory Cytokines in HPDL Cells Following Mechanical Loading and Affects Monocyte Behavior. *Clin. Oral Investig.* **2016**, *20*, 923–931. [CrossRef]
25. Kanzaki, H.; Nakamura, Y. Orthodontic Tooth Movement and HMGB1. *J. Oral Biosci.* **2018**, *60*, 49–53. [CrossRef]
26. Scaffidi, P.; Misteli, T.; Bianchi, M.E. Release of Chromatin Protein HMGB1 by Necrotic Cells Triggers Inflammation. *Nature* **2002**, *418*, 191–195. [CrossRef] [PubMed]
27. Kanzaki, H.; Chiba, M.; Shimizu, Y.; Mitani, H. Periodontal Ligament Cells under Mechanical Stress Induce Osteoclastogenesis by Receptor Activator of Nuclear Factor KB Ligand Up-Regulation via Prostaglandin E2 Synthesis. *J. Bone Miner. Res.* **2002**, *17*, 210–220. [CrossRef]
28. Brasier, A.R. The Nuclear Factor-B-Interleukin-6 Signalling Pathway Mediating Vascular Inflammation. *Cardiovasc. Res.* **2010**, *86*, 211–218. [CrossRef]
29. Kircheis, R.; Haasbach, E.; Lueftenegger, D.; Heyken, W.T.; Ocker, M.; Planz, O. NF-KB Pathway as a Potential Target for Treatment of Critical Stage COVID-19 Patients. *Front. Immunol.* **2020**, *11*, 598444. [CrossRef]
30. Janjic, M.; Docheva, D.; Trickovic Janjic, O.; Wichelhaus, A.; Baumert, U. In Vitro Weight-Loaded Cell Models for Understanding Mechanodependent Molecular Pathways Involved in Orthodontic Tooth Movement: A Systematic Review. *Stem Cells Int.* **2018**, *2018*, 3208285. [CrossRef]
31. Dupont, S.; Morsut, L.; Aragona, M.; Enzo, E.; Giulitti, S.; Cordenonsi, M.; Zanconato, F.; Le Digabel, J.; Forcato, M.; Bicciato, S.; et al. Role of YAP/TAZ in Mechanotransduction. *Nature* **2011**, *474*, 179–184. [CrossRef]
32. Chukkapalli, S.S.; Lele, T.P. Periodontal Cell Mechanotransduction. *Open Biol.* **2018**, *8*, 180053. [CrossRef] [PubMed]
33. Du, J.; Li, M. Functions of Periostin in Dental Tissues and Its Role in Periodontal Tissues' Regeneration. *Cell. Mol. Life Sci.* **2017**, *74*, 4279–4286. [CrossRef] [PubMed]
34. Belgardt, E.; Steinberg, T.; Husari, A.; Dieterle, M.P.; Hülter-Hassler, D.; Jung, B.; Tomakidi, P. Force-Responsive Zyxin Modulation in Periodontal Ligament Cells Is Regulated by YAP Rather than TAZ. *Cell Signal* **2020**, *72*, 109662. [CrossRef] [PubMed]
35. Premaraj, S.; Souza, I.; Premaraj, T. Mechanical Loading Activates β-Catenin Signaling in Periodontal Ligament Cells. *Angle Orthod.* **2011**, *81*, 592–599. [CrossRef] [PubMed]

36. Ziegler, N.; Alonso, A.; Steinberg, T.; Woodnutt, D.; Kohl, A.; Müssig, E.; Schulz, S.; Tomakidi, P. Mechano-Transduction in Periodontal Ligament Cells Identifies Activated States of MAP-Kinases P42/44 and P38-Stress Kinase as a Mechanism for MMP-13 Expression. *BMC Cell Biol.* **2010**, *11*, 10. [CrossRef]
37. Hatakeyama, J.; Tamai, R.; Sugiyama, A.; Akashi, S.; Sugawara, S.; Takada, H. Contrasting Responses of Human Gingival and Periodontal Ligament Fibroblasts to Bacterial Cell-Surface Components through the CD14/Toll-like Receptor System. *Oral Microbiol. Immunol.* **2003**, *18*, 14–23. [CrossRef] [PubMed]
38. Tang, L.; Zhou, X.; Wang, Q.; Zhang, L.; Wang, Y.; Li, X.; Huang, D. Expression of TRAF6 and Pro-Inflammatory Cytokines through Activation of TLR2, TLR4, NOD1, and NOD2 in Human Periodontal Ligament Fibroblasts. *Arch Oral Biol.* **2011**, *56*, 1064–1072. [CrossRef]
39. Zhu, W.; Liang, M. Periodontal Ligament Stem Cells: Current Status, Concerns, and Future Prospects. *Stem Cells Int.* **2015**, *2015*, 972313. [CrossRef]
40. Andersson, U.; Tracey, K.J. HMGB1 Is a Therapeutic Target for Sterile Inflammation and Infection. *Annu. Rev. Immunol.* **2011**, *29*, 139–162. [CrossRef]
41. Wolf, M.; Lossdörfer, S.; Küpper, K.; Jäger, A. Regulation of High Mobility Group Box Protein 1 Expression Following Mechanical Loading by Orthodontic Forces in Vitro and in Vivo. *Eur. J. Orthod.* **2014**, *36*, 624–631. [CrossRef]
42. Wolf, M.; Lossdörfer, S.; Römer, P.; Bastos Craveiro, R.; Deschner, J.; Jäger, A. Anabolic Properties of High Mobility Group Box Protein-1 in Human Periodontal Ligament Cells in Vitro. *Mediat. Inflamm.* **2014**, *2014*, 347585. [CrossRef] [PubMed]
43. Brubaker, S.W.; Bonham, K.S.; Zanoni, I.; Kagan, J.C. Innate Immune Pattern Recognition: A Cell Biological Perspective. *Annu. Rev. Immunol.* **2015**, *33*, 257. [CrossRef] [PubMed]
44. Janeway, C.A.; Medzhitov, R. Innate Immune Recognition. *Annu. Rev. Immunol.* **2002**, *20*, 197–216. [CrossRef] [PubMed]
45. Sha, Y.; Zmijewski, J.; Xu, Z.; Abraham, E. HMGB1 Develops Enhanced Proinflammatory Activity by Binding to Cytokines. *J. Immunol.* **2008**, *180*, 2531–2537. [CrossRef] [PubMed]
46. Bianchi, M.E. HMGB1 Loves Company. *J. Leukoc. Biol.* **2009**, *86*, 573–576. [CrossRef]
47. Rouhiainen, A.; Tumova, S.; Valmu, L.; Kalkkinen, N.; Rauvala, H. Analysis of Proinflammatory Activity of Highly Purified Eukaryotic Recombinant HMGB1 (Amphoterin). *J. Leukoc. Biol.* **2007**, *81*, 49–58. [CrossRef]
48. Zegeye, M.M.; Lindkvist, M.; Fälker, K.; Kumawat, A.K.; Paramel, G.; Grenegård, M.; Sirsjö, A.; Ljungberg, L.U. Activation of the JAK/STAT3 and PI3K/AKT Pathways Are Crucial for IL-6 Trans-Signaling-Mediated pro-Inflammatory Response in Human Vascular Endothelial Cells. *Cell Commun. Signal.* **2018**, *16*, 1–10. [CrossRef]
49. Huang, C.Y.; Deng, J.S.; Huang, W.C.; Jiang, W.P.; Huang, G.J. Attenuation of Lipopolysaccharide-Induced Acute Lung Injury by Hispolon in Mice, through regulating the TLR_4/PI_3K/Akt/mTOR and $Keap_1$/Nrf_2/HO_{-1} pathways, and suppressing oxidative stress-mediated ER stress-induced apoptosis and autophagy. *Nutrients* **2020**, *12*, 1742. [CrossRef]
50. Hülter-Hassler, D.; Wein, M.; Schulz, S.D.; Proksch, S.; Steinberg, T.; Jung, B.A.; Tomakidi, P. Biomechanical Strain-Induced Modulation of Proliferation Coincides with an ERK1/2-Independent Nuclear YAP Localization. *Exp. Cell Res.* **2017**, *361*, 93–100. [CrossRef]
51. Yuan, F.; Chen, J.; Sun, P.P.; Guan, S.; Xu, J. Wedelolactone Inhibits LPS-Induced pro-Inflammation via NF-KappaB Pathway in RAW 264.7 Cells. *J. Biomed. Sci.* **2013**, *20*, 84. [CrossRef]
52. Huang, H.; Yang, R.; Zhou, Y.-H. Mechanobiology of Periodontal Ligament Stem Cells in Orthodontic Tooth Movement. *Stem Cells Int.* **2018**, *2018*, 6531216. [CrossRef] [PubMed]
53. Scheller, J.; Chalaris, A.; Schmidt-Arras, D.; Rose-John, S. The Pro- and Anti-Inflammatory Properties of the Cytokine Interleukin-6. *Biochim. Biophys. Acta-Mol. Cell Res.* **2011**, *1813*, 878–888. [CrossRef] [PubMed]
54. Lee, S.I.; Park, K.H.; Kim, S.J.; Kang, Y.G.; Lee, Y.M.; Kim, E.C. Mechanical Stress-Activated Immune Response Genes via Sirtuin 1 Expression in Human Periodontal Ligament Cells. *Clin. Exp. Immunol.* **2012**, *168*, 113–124. [CrossRef] [PubMed]
55. Diomede, F.; Zingariello, M.; Cavalcanti, M.F.X.B.; Merciaro, I.; Pizzicannella, J.; de Isla, N.; Caputi, S.; Ballerini, P.; Trubiani, O. MyD88/ERK/NFkB Pathways and pro-Inflammatory Cytokines Release in Periodontal Ligament Stem Cells Stimulated by Porphyromonas Gingivalis. *Eur. J. Histochem.* **2017**, *61*, 122–127. [CrossRef]
56. Aveic, S.; Craveiro, R.B.; Wolf, M.; Fischer, H. Current Trends in In Vitro Modeling to Mimic Cellular Crosstalk in Periodontal Tissue. *Adv. Healthc. Mater.* **2021**, *10*, 2001269. [CrossRef]

International Journal of *Molecular Sciences*

MDPI

Article

Identification of an Optimal TLR8 Ligand by Alternating the Position of 2′-O-Ribose Methylation

Marina Nicolai [1], Julia Steinberg [2], Hannah-Lena Obermann [1], Francisco Venegas Solis [1], Eva Bartok [3], Stefan Bauer [1] and Stephanie Jung [2,*]

1 Institute for Immunology, Philipps-University Marburg, 35043 Marburg, Germany
2 Institute of Cardiovascular Immunology, University Hospital Bonn, University of Bonn, 53127 Bonn, Germany
3 Institute of Experimental Haematology and Transfusion Medicine, University Hospital Bonn, University of Bonn, 53127 Bonn, Germany
* Correspondence: sjung@uni-bonn.de

Abstract: Recognition of RNA by receptors of the innate immune system is regulated by various posttranslational modifications. Different single 2′-O-ribose (2′-O-) methylations have been shown to convert TLR7/TLR8 ligands into specific TLR8 ligands, so we investigated whether the position of 2′-O-methylation is crucial for its function. To this end, we designed different 2′-O-methylated RNA oligoribonucleotides (ORN), investigating their immune activity in various cell systems and analyzing degradation under RNase T2 treatment. We found that the 18S rRNA-derived TLR7/8 ligand, RNA63, was differentially digested as a result of 2′-O-methylation, leading to variations in TLR8 and TLR7 inhibition. The suitability of certain 2′-O-methylated RNA63 derivatives as TLR8 agonists was further demonstrated by the fact that other RNA sequences were only weak TLR8 agonists. We were thus able to identify specific 2′-O-methylated RNA derivatives as optimal TLR8 ligands.

Keywords: TLR7; TLR8; 2′-O-ribose-methylation; RNase T2; immune activation

Citation: Nicolai, M.; Steinberg, J.; Obermann, H.-L.; Solis, F.V.; Bartok, E.; Bauer, S.; Jung, S. Identification of an Optimal TLR8 Ligand by Alternating the Position of 2′-O-Ribose Methylation. *Int. J. Mol. Sci.* **2022**, *23*, 11139. https://doi.org/10.3390/ijms231911139

Academic Editors: Oliver Planz and Ralf Kircheis

Received: 3 August 2022
Accepted: 17 September 2022
Published: 22 September 2022

1. Introduction

As the first line of host defense, the innate immune system recognizes molecular danger signals and subsequently triggers signal transduction and secretion of interferons (IFN) and proinflammatory cytokines [1,2]. This function is enabled by pattern-recognition receptors (PRRs), which recognize pathogen-associated molecular patterns (PAMPs) characteristic of a particular group of pathogens. Among these PAMPs are bacterial and viral nucleic acids, which are recognized not only by their structure and composition, but also by their intracellular localization [3]. For example, single-stranded RNA (ssRNA) in human endosomes is recognized by the two related Toll-like receptors (TLRs) 7 and 8, inducing diverse cytokine patterns in different cell types and species. In humans, TLR7 is strongly expressed in plasmacytoid dendritic cells (pDCs) and B cells, where its activation leads to B cell activation and IFN-α release from pDCs [3–7]. In contrast, TLR8 is primarily expressed in the myeloid compartment and its activation induces proinflammatory cytokine and IFN-β release [3–5,8]. In mice, TLR7 also functions similarly in pDCs and B cells [9–11]. However, it is further expressed in the myeloid compartment, where it induces a similar cytokine profile to human TLR8 [12,13]. In contrast, murine TLR8 does not induce cytokine release, and its function remains largely unknown [9,10].

Early reports stating that TLR7 and TLR8 preferentially recognize uridine-rich sequences were followed by structural analyses showing that both receptors bind single nucleosides and short ssRNA sequences at two distinct binding sites [14–19]. Recent studies have shown that the activities of endosomal ribonuclease (RNase) T2 and RNase 2 are essential for TLR8 activation; RNase activity generates both single uridines and short ssRNA fragments as degradation products that bind TLR8 and initiate signal transduction [20–22]. It was also reported that 2′-O-ribose methylation of phosphodiester ssRNA impairs RNase digestion, and thus, influences TLR8 activation [20]. This observation necessitates further research on the influence of 2′-O-methylations on TLR8 activity, as the importance of therapeutic TLR8 ligands has greatly increased in recent years [23–26]. For this purpose, the use of 2′-O-ribose methylation may be useful, as it can both stabilize RNA and lead to adequate control of the immune response.

Several groups have reported that RNA methylations can prevent activation of TLR7 [27–30]. Remarkably, a single 2′-O-ribose methylation is sufficient to prevent TLR7 signaling and convert a TLR7/TLR8 ligand into a sole TLR8 ligand; however, the influence of the positioning of 2′-O-methylation has not yet been explored in much detail [31–33]. Therefore, it is important to determine whether the position of 2′-O-methylation matters, as it may be of both technical and immunological relevance for the most efficient generation of TLR8 ligands.

The aim of this study was to investigate whether a known TLR8 ligand could be further optimized by alternating the position of the 2′-O-ribose methylation. Taking the importance of RNase activity into account, 2′-O-ribose methylation may alter the preferred RNA cleavage site to different positions, thus generating RNA degradation products with strong differential effects on TLR7 and TLR8 activity. As TLR7 activation should be avoided, we focused only on guanosine methylations, because binding of both single guanosine nucleosides and guanosine-rich sequences is essential for TLR7 activation [14,16,18,19,34]. To this end, we designed different oligoribonucleotides (ORNs) that were methylated at different guanosine positions and derived from different naturally occurring RNA sequences. Following RNase digestion, we monitored the effect of 2′-O-methylations on RNA degradation patterns. TLR8 activation by methylated and unmethylated ORNs was investigated by immunostimulations with primary human immune cells, murine pDCs, and TLR8-transfected HEK293 reporter cells. We observed a flexibility in TLR8 activation with respect to the position of 2′-O-methylation on its ligands. The results presented here contribute to the generation of optimal TLR8 ligands under different requirements and for diverse clinical applications.

2. Results

2.1. 2′-O-Ribose Methylation Prevents TLR7 Activation Independent of the Position

We aimed to investigate whether the conversion of a TLR7 and TLR8 ligand into a sole TLR8 ligand was a universal effect of 2′-O-ribose methylation or whether this effect was mediated only by a small group of naturally occurring methylation patterns, located at specific positions. Therefore, we first employed the previously characterized TLR7 and TLR8 ligand, RNA63, derived from 18S rRNA (position 1488–1499), as its immunostimulatory properties have already been confirmed in previous publications (Figure 1a) [30,32]. Under natural conditions, this ligand is 2′-O-ribose methylated at the first guanosine (RNA63M1) [35]. To investigate the influence of the position of the methyl group on preferential RNase cleavage sites and immune activation, we designed four different RNA63 derivatives, each with a 2′-O-ribose methylation at one of the four guanosines (RNA63M1 to RNA63M4). RNaseT2 activity has been shown to be necessary for TLR8 activation and affected by 2′O-methylation [20,21]. Thus, we compared methylated and unmethylated RNA63 with and without RNase T2 treatment and observed that RNA methylation indeed resulted in altered RNA fragmentation patterns (Figures 1b and A1). In particular, RNase T2-digested RNA63M1 and RNA63M3 showed a slightly differential fragment pattern compared with RNA63M2 and RNA63M4. Specifically, RNase digests of RNA63M1 and RNA63M3 demonstrated at least five fragments with a prominent double band (marked by an arrow 🡸). In contrast, the analysis of RNA63M2 and RNA63M4 digests showed an altered pattern in this region resembling a single band or two co-migrating bands. Furthermore, digested RNA63M1 and RNA63M3 showed two additional small bands (compared with a single smaller band in digested RNA63M2 and RNA63M4), which is marked with a line (——). Overall, these patterns suggest a difference in ORN fragments created by RNase T2 digestion.

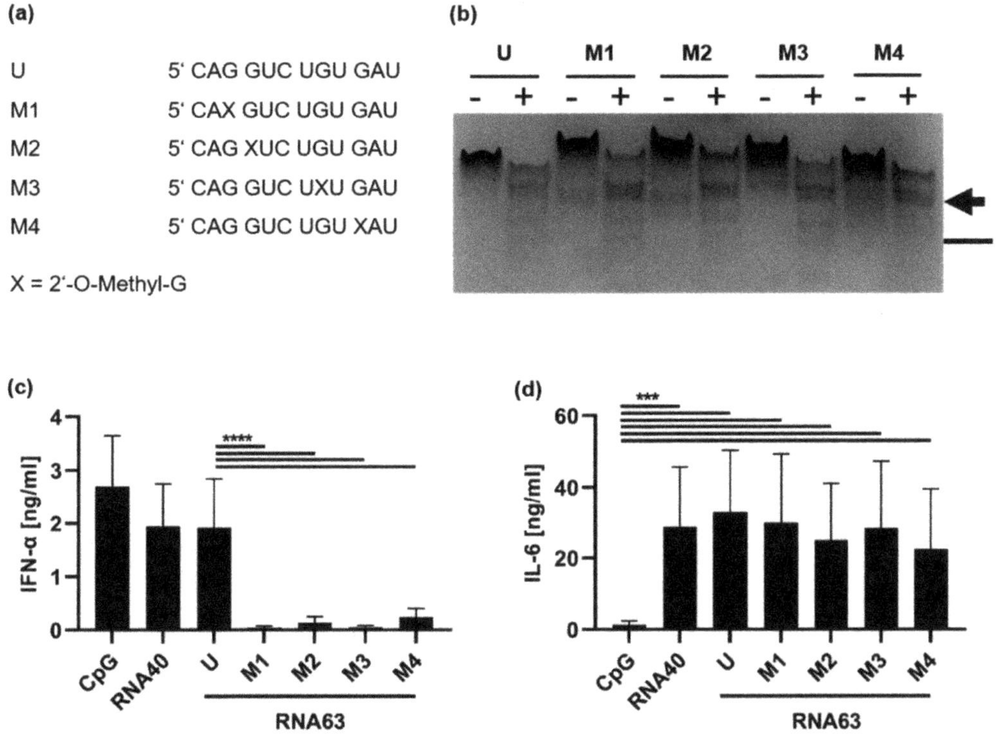

Figure 1. Methylation by 2′-O-ribose prevents IFN-α but not IL-6 secretion, independent of the position. (**a**) Sequences of unmethylated RNA63 (U) or methylated derivatives (M1, M2, M3, and M4): 2′-O-ribose-methylated guanosines are indicated with an X. (**b**) RNA63 derivatives were either treated with 0.005 units of RNase T2/μg RNA (+) or mock treated (−), and RNA fragmentation was visualized in polyacrylamide gel electrophoresis. Position of the RNA fragments of interest are indicated with an arrow (🡄) and a line (——). (**c**,**d**) PBMCs were stimulated with RNA63 derivatives at a final concentration of 10 μg/mL: 1 μM CpG ODN 2216 or 5 μg/mL RNA40. Supernatants were harvested 20 h post-stimulation (h p. s.) and both (**c**) IFN-α and (**d**) IL-6 concentrations were measured in ELISA. Graphs depict six independent experiments with PBMCs obtained from six individual donors, each in biological duplicates (twelve measurements per data point, mean + S.D). Data were analyzed using paired *t*-tests. **** $p < 0.0001$, *** $p < 0.001$.

It has been reported that human TLR7 induces both IFN-α and interleukin-6 (IL-6) release, whereas human TLR8 leads to IL-6 release but not IFN-α [5,15,36,37]. We then tested whether 2′-O-methylations at different positions on the RNA would lead to changes in the observed cytokine response. To this end, we stimulated human peripheral blood mononuclear cells (PBMCs) with the different ORNs, and both IFN-α and IL-6 release was measured by ELISA (Figure 1c,d). The previously described unmethylated oligodeoxynucleotide (CpG ODN 2216), which is a well-known TLR9 ligand that contains CG dinucleotide, was used as a TLR7-independent positive control for pDC activation [6,7,38]. RNA40, a well characterized TLR7 and TLR8 ligand, was used as a specific TLR7/8 control agonist [14]. In human PBMC, CpG2216 induced IFN-α release, but not IL-6. This is consistent with previously descriptions [38] of the functional activity of the pDCs present within the isolated PBMC [6,7]. In contrast, RNA40 induced both IFN-α and IL-6 release, in line with previous studies on TLR7 and TLR8 in human pDCs and monocytes [14]. In contrast, we observed that while RNA63 induced IFN-α secretion, each 2′-O-ribose-methylation that was applied strongly decreased IFN-α release (Figure 1c). In contrast, RNA63M1-M4 derivatives induced a robust IL-6 signal (Figure 1d). Comparing the differentially methylated RNA63 derivatives, we observed that RNA63M2 and RNA63M4 still induced slight IFN-α release (Figure 1c) and caused a slightly weaker IL-6 induction than both RNA63M1 and RNA63M3 (Figure 1d). Nonetheless, our data suggest that human TLR7 activation can be inhibited by guanosine methylation in RNA63, irrespective of the position, whereas TLR8 activation remains unaffected

In mice, TLR7 can also induce both IFN-α and IL-6 release, whereas TLR8 activation does not induce these cytokines [9,10,12,13]. To examine murine TLR7 activity, we generated FMS-like tyrosine kinase 3 ligand-differentiated dendritic cells (FLT3L DCs) from the bone marrow of wildtype (wt) and Tlr7-deficient ($Tlr7^{-/-}$) mice and stimulated these cells with both RNA63 and methylated RNA63M1-M4 derivatives (Figure 2). As in human PBMC, induction of both IFN-α and IL-6 secretion by CpG ODN 2216 confirmed the activity and responsiveness of FLT3L DCs, in both wildtype and $Tlr7^{-/-}$, to nucleic acid stimuli. In contrast, IFN-α and IL-6 secretion was completely abrogated in $Tlr7^{-/-}$ after RNA40, demonstrating the expected functional activity of this ligand in mice.

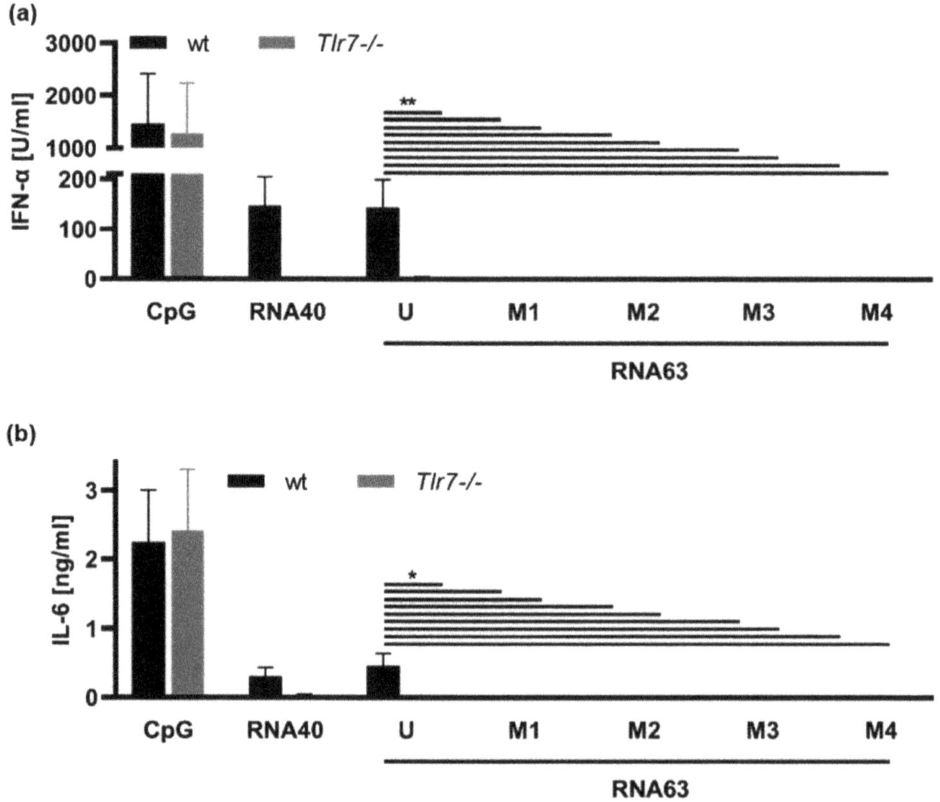

Figure 2. Methylation by 2′-O-ribose of a different guanosine prevents TLR7 activation. FLT3L DCs from wt or $Tlr7^{-/-}$ mice were stimulated with 1 μM CpG ODN 2216 and 5 μg/mL RNA40 or RNA63 derivatives (U, M1, M2, M3, or M4) at a final concentration of 10 μg/mL. Supernatants were harvested 20 h p. s. and concentrations of (**a**) IFN-α and (**b**) IL-6 were determined by ELISA. Graphs depict six independent experiments each in biological duplicates (twelve measurements per data point, mean + S.D). Data were analyzed using Wilcoxon tests. ** $p < 0.01$, * $p < 0.05$.

Whereas RNA63 induced a significant IFN-α and IL-6 response in FLT3L DCs of wt mice, the Tlr7 knockout completely abolished RNA63-dependent cytokine secretion (Figure 2). Of note, each 2′-O-ribose-methylation of RNA63 abrogated both IFN-α and IL-6-release. Consequently, murine TLR7 activation was inhibited by guanosine methylations of RNA63, independent of the position of the modification.

2.2. *Impact of 2′-O-Ribose Methylation on TLR7 Activation by Naturally Methylated and Unmethylated Sequences*

We then investigated whether the generation of a strong TLR8 ligand could only be achieved by 2′-O-methylation of specific RNA sequences such as the optimized RNA63, or whether this effect could be universally achieved using GU-containing ORNs. We used two additional RNA sequences in their unmethylated and methylated derivatives: RNA28, which is derived from 28S rRNA (position 2409–2420 on 28S rRNA, XR_007090848.1) and carries a 2′-O-methylation at the first guanosine under natural conditions, and RNA66, which is derived from 18S rRNA (position 773–784 on 18S rRNA, NR 003286.2) and is not methylated under natural conditions (Figure 3a) [35]. The TLR7/TLR8 stimulatory activities of these RNA sequences have not been previously described, and these particular sequences were selected due to their high uridine content, natural occurrence in ribosomal RNA, and presence of a guanosine at position 3. PBMCs were stimulated with the ORNs, and RNA-induced cytokine release was determined by ELISA (Figure 3b,c).

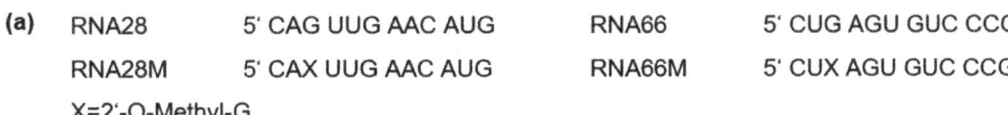

(a)

RNA28	5' CAG UUG AAC AUG	RNA66	5' CUG AGU GUC CCG
RNA28M	5' CAX UUG AAC AUG	RNA66M	5' CUX AGU GUC CCG

X=2'-O-Methyl-G

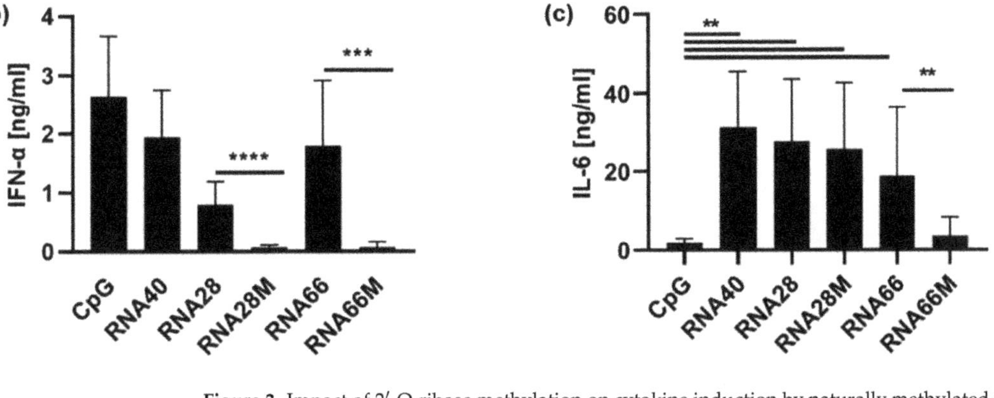

Figure 3. Impact of 2′-O-ribose methylation on cytokine induction by naturally methylated and unmethylated sequences. (**a**) Sequences of applied RNA derivatives. 2′-O-ribose methylated guanosine are indicated with an X. (**b**,**c**) PBMCs were stimulated with RNA66 or RNA28 derivatives at a final concentration of 10 μg/mL, with 1 μM CpG ODN 2216 or 5 μg/mL RNA40. Supernatants were harvested 20 h p. s. and both (**b**) IFN-α and (**c**) IL-6 concentrations were measured by ELISA. Graphs depict six independent experiments with PBMCs obtained from six individual healthy donors each in biological duplicates (twelve measurements per data point, mean + S.D). Data were analyzed using paired t-tests; IL-6-data of RNA66 and RNA66M were analyzed using Wilcoxon tests. **** $p < 0.0001$, *** $p < 0.001$, ** $p < 0.01$.

As shown in Figure 1, CpG2216 induced IFN-α, providing a TLR7-independent control of pDC activity. Moreover, RNA40, as well as both RNA28 and RNA66, induced IFN-α and IL-6 release, as was observed for RNA63 (Figure 3b,c, compare Figure 1). In contrast, RNA28M induced the release of IL-6 only, as observed for the methylated derivatives of RNA63, whereas RNA66M did not induce IFN-α or IL-6 release, indicating that this particular 2′-O-methylation blocked both TLR7 and TLR8 activation.

To examine RNA28- and RN66-induced immune activation in murine cells, we studied TLR7 using wt and $Tlr7^{-/-}$ murine dendritic cells.

In murine FLT3L DCs, as shown in Figure 2, CpG2216 and RNA40 acted as TLR7-independent and TLR7-dependent controls, respectively. However, both RNA28 and RNA66 were weak agonists of TLR7 in comparison with RNA40 and RNA63 (Figure 4a, compare Figure 2). Nonetheless, IFN-α release was induced by RNA28 and RNA66 in a TLR7-dependent manner. Moreover, in line with what was observed for RNA63M1-4, neither RNA28M nor RNA66M induced IFN-α (Figure 4a). RNA28 also only induced low levels of IL-6, which were inhibited by 2′-O-methylation (RNA28M), as seen for RNA63M1-4. However, RNA66 stimulation did not induce IL-6 release from murine FLT3L DCs (Figure 4b).

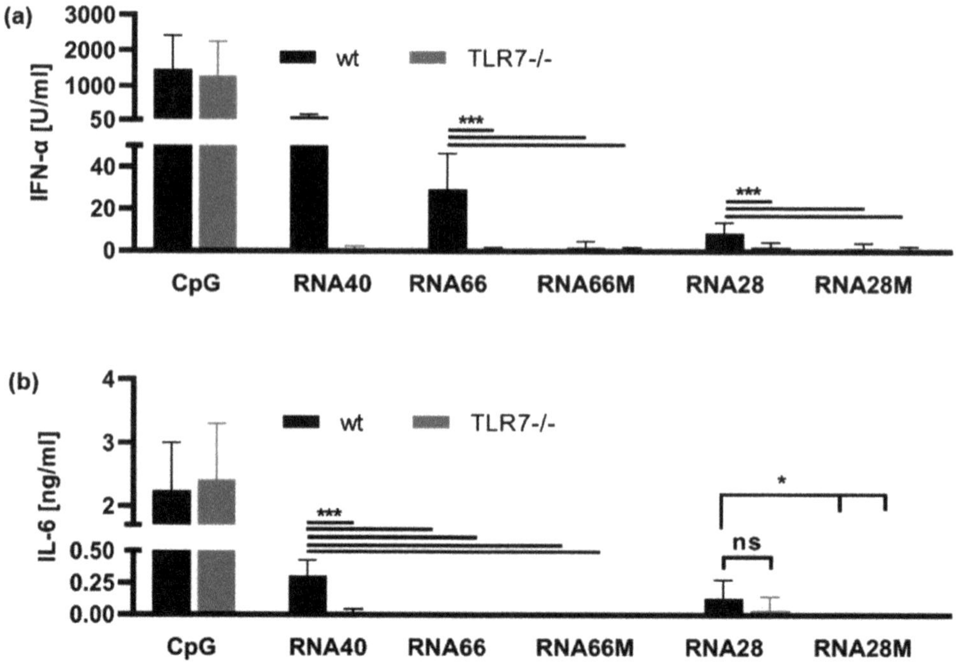

Figure 4. Methylation by 2′-O-ribose of different sequences prevents TLR7 activation. FLT3L DCs from wt or $Tlr7^{-/-}$ mice were stimulated with 1 μM CpG ODN 2216, 5 μg/mL RNA40, and RNA66 derivatives or RNA28 derivatives at a final concentration of 10 μg/mL. Supernatants were harvested 20 h p. s. and concentrations of (**a**) IFN-α and (**b**) IL-6 were determined in ELISA. Graphs depict six independent experiments each in biological duplicates (twelve measurements per data point, mean + S.D). Data were analyzed using Wilcoxon tests. *** $p < 0.001$, * $p < 0.05$, ns = not significant.

2.3. *Activation of TLR8 by 2′-O-Ribose Methylated ORNs*

Our previous data demonstrating IL-6 but not IFN-α induction after stimulation with 2′-O-ribose methylated ORNs only provide an indirect indication that TLR8 is activated by 2′-O-ribose methylated ORNs. To provide direct evidence of TLR8 activation, we used a HEK reporter cell line stably overexpressing TLR8 (HEK-TLR8) and stimulated it with the different methylated and unmethylated ORNs.

Immune stimulation of TLR8 reporter cells resulted in activation of the transcription factor nuclear factor ′kappa-light-chain-enhancer′ of activated B-cells (NF-κB), which resulted in the release of secreted embryonic alkaline phosphatase (SEAP) and was normalized to the SEAP activity induced by the synthetic TLR8-specific agonist TL8-506 [39] (Figure 5). Although mock treatment of cells (mock) without ORN did not lead to NF-κB-activation, transfection of all methylated RNA63 derivatives upregulated SEAP activity (Figure 5a). However, the strength of immune activation differed between the individual RNA63 derivatives, and 2′-O-methylation at position 3 (RNA63M1) resulted in the weakest signal transduction in TLR8 reporter cells. In contrast, TLR8 activation by the other ORNs was different; although RNA66 and RNA28 led to distinct TLR8 activation, RNA66M and RNA28M only led to weak NF-κB activation (Figure 5b,c).

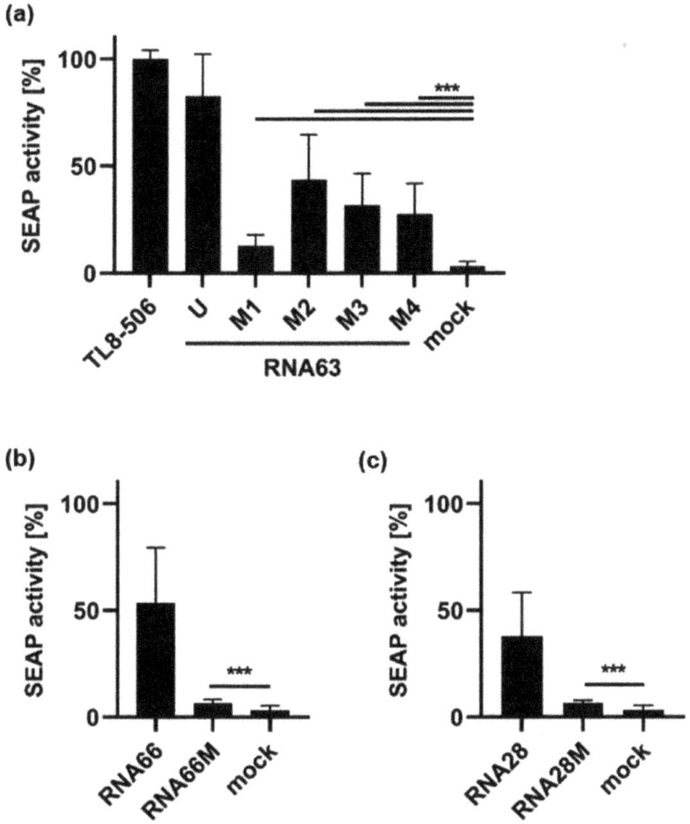

Figure 5. Oligoribonucleotides with 2′-O-ribose methylation activate TLR8. HEK-TLR8 cells were stimulated with TLR8-specific TL8-506 at 0.5 μg/mL and different derivatives of (**a**) RNA63 (U = unmethylated), (**b**) RNA66, or (**c**) RNA28 at a final concentration of 20 μg/mL or mock-treated. Cell supernatant was mixed with QUANTI-Blue™ medium at 20 h p. s. for 15 min at 37 °C. Secreted embryonic alkaline phosphatase (SEAP) activity, reflecting NF-κB activation, was determined by colorimetric measurement at 650 nm and results were normalized to TL8-506-induced NFκB activation. Graphs depict six independent experiments each in biological duplicates (twelve measurements per data point, mean + S.D). Data were analyzed using Wilcoxon tests. *** $p < 0.001$.

In summary, a single 2′-O-methylation, tested at different positions within the RNA, converted RNA63 from a TLR7/TLR8 ligand into an exclusive TLR8 ligand. However, 2′-O-methylations at different guanosine positions resulted in different strengths of TLR7 inhibition and TLR8 activation, although it was observed for all positions tested. In addition, the switch from TLR7 and TLR8 activation to sole TLR8 activation worked particularly well for the optimized RNA63 derivatives and seemed to be a coordinated process, as some methylated sequences showed only weak TLR8-activating properties.

3. Discussion

The application of RNA as a therapeutic agent has become increasingly important in recent years [23–26]. Just as important as the production of stable ORNs that lead to sufficient immune activation is the targeting of the innate immune response to avoid an excessive or misdirected danger signal. For this reason, it is essential to design optimal PRR-ligands and gain knowledge about the optimal localization of ORN methylation, especially for cases with technical limitations. Consequently, we aimed to design an optimized TLR8 ligand by investigating the influence of 2′-O-methylations at different positions on TLR8 activity. We first investigated different derivatives of the previously described ORN RNA63 [32]. This sequence was converted from a TLR7/TLR8 ligand to a TLR8 ligand by a single ribose methylation at position 3. We designed several RNA63 derivatives that were methylated at different guanosine positions and applied them in PBMC stimulations (Figure 1). All RNA63 derivatives with a single ribose methylation strongly decreased or abrogated IFN-α release from human PBMCs but still induced robust IL-6 release. Of note, RNA63M2 and RNA63M4, which showed residual IFN-α induction and weaker IL-6 release from human PBMCs, also exhibited a different RNA fragmentation pattern than RNA63M1 and RNA63M3 derivatives, which induced a strong IL-6 signal only. Therefore, we speculate that a single 2′-O-methylation may indeed direct RNase activity, thereby aiding the development of optimal TLR8 ligands essential for TLR8 activation [20–22].

To demonstrate the prevention of TLR7-dependent immune activation by 2′-O-ribose methylation, we stimulated DCs differentiated from the bone marrow of both wt and *Tlr7*$^{-/-}$ mice (Figure 2). These experiments proved that unmodified RNA63 is a TLR7 ligand, and TLR7 activity was inhibited by 2′-O-methylation at all positions tested. Thus, we demonstrated the flexibility of TLR7 inhibition with respect to the position of 2′-O-methylation.

As our data did not allow us to draw conclusions about whether the conversion from TLR7 to TLR8 by 2′-O-methylation was transferable to other sequences, we designed two additional ORNs, RNA28 and RNA66, as well as their methylated derivatives. The RNA28 sequence, similar to RNA63, is methylated under natural conditions [35]. Unmethylated RNA28 appears to be only a weak TLR7 activator, as stimulation of both human PBMCs and murine FLT3L-differentiated DCs with RNA28 resulted in only weak IFN-α release. Nevertheless, RNA28M proved to be a good TLR8 inducer in primary human immune cells, as it led to a comparably strong IL-6 induction as its unmethylated derivate and RNA63. In contrast, RNA66 but not RNA66M led to marked cytokine release. This provided initial evidence that RNA66M is not a strong TLR8 ligand.

We used murine FLT3-differentiated DCs to confirm the TLR7 dependence of RNA28 and RNA66. Indeed, both sequences were shown to be TLR7 ligands (Figure 4). The observation of weak RNA28-mediated cytokine release in the murine system supports the hypothesis that RNA28 is a low-potency TLR7 activator (compare Figure 3). RNA66, on the other hand, was a strong IFN-α inducer in the murine system, as it is in humans, but did not induce IL-6. At this point, the question arises why the RNA66 sequence, which was already a weak IL-6 inducer in human PBMCs and is not methylated under natural conditions [35], was not recognized as a pattern that requires a proinflammatory danger signal. For this reason, it seemed necessary to directly verify which RNA sequences were TLR8 ligands.

Therefore, we tested the TLR8 specificity of the different RNA sequences in TLR8-expressing HEK reporter cells (Figure 5). Indeed, all unmethylated and methylated ORNs induced significant TLR8-dependent NF-κB activation, which was weak in the cases of RNA63M1, RNA66M, and RNA28M. Nevertheless, we assume that RNA63M1 and RNA28M are TLR8 ligands because they induced a robust IL-6 response in PBMCs (compare Figure 1). In contrast, RNA66M-induced IL-6 release was significantly reduced in

PBMCs compared with RNA66, so we hesitate to conclude that RNA66M is also a TLR8 ligand. With respect to our objective of identifying an optimal and specific TLR8 ligand, we would choose the newly designed RNA63M3. the previously described RNA63M1 would be a second choice, as it showed no TLR7 activity and robust TLR8 activity in all systems tested. The fact that randomly selected ORNs such as RNA28M and RNA66M only had weak TLR8 activity or lacked it completely highlights the importance of RNA63M3.

The differences in TLR7 and TLR8 activity of RNAs methylated at different positions are probably because 2′-O-methylation can protect against RNase digestion of the phosphodiester bond between the methylated and 3′ nucleotide [20,40]. In our case, 2′-O-methylation occurred at only a single variable position, leaving wide stretches of ssRNA accessible to RNase digestion, which was reported to be essential for TLR8 activation [20–22]. However, 2′-O-methylations at different positions within the RNA most likely also directed preferential RNA cleavage to different positions, resulting in an RNA fragmentation pattern that was also variable between individual RNA63 derivatives. Consequently, both short single-stranded RNA stretches and single nucleosides reported to be essential for TLR7 and TLR8 activation can be generated from all methylated derivatives, but methylation at the optimal site presumably results in the generation of RNA degradation products that activate TLR8 and not TLR7 [18–22,34]. Here, the base composition of the resulting RNA fragments and their methylation status both play a role, as 2′-O-methylations have been reported to prevent TLR7 and TLR8 activation by ssRNA [20,30,41]. However, a balance between stability and accessibility to RNases seems to be required, as it has also been reported that some ORNs lead to better TLR7 and TLR8 activation when made more resistant to nuclease-mediated digestion by modification [42].

Consistent with our data, specific TLR8 activation by ORNs with single methylations has been reported previously [27,31,32]. However, these publications only investigated naturally occurring 2′-O-methylations, and not the influence of position, which we have investigated here.

To verify the specificity of our potential optimal TLR8 ligands in our systems, we also examined other naturally methylated and unmethylated RNA segments for their TLR7 and TLR8 activity. An influence of the sequence context of the 2′-O-methylation on the silencing of TLR7 and TLR8 has already been discussed in previous publications [28,29]. In this context, we observed that RNA66, compared with RNA63 and RNA28, was no longer a strong TLR8 ligand after ribose methylation. However, unlike RNA63 and RNA28, RNA66 does not contain AU segments, which were reported to be important for a TLR8 ligand [16]. However, it should be noted that all ORNs tested (including RNA66) contained GU segments, which have been described to induce TLR8 activation [14]. Of note, we observed strong TLR8 activity after ribose methylation in ORNs RNA63M1–M4 and RNA28M. Therefore, we conclude that only RNA sequences which are strong TLR8 ligands per se still induce NF-κB activation after a single 2′-O-methylation.

In summary, we have shown that a single 2′-O-methylation at any position can convert a TLR7/8 ligand to an exclusive TLR8 ligand, but with varying efficiency. Here, the RNA63M3 derivative turned out to be a particularly suitable TLR8 ligand, as 2′-O-methylation at other positions or other methylated sequences resulted in less potent TLR8 ligands or less potent inhibition of TLR7 signaling. These findings could facilitate the generation of therapeutic TLR8 ligands.

4. Materials and Methods

4.1. Kits and Reagents

TL8-506 was purchased from Invivogen (Toulouse, France) and RNA40 (GCCCGUCUGUUGUGUGACUC) was purchased from IBA (Göttingen, Germany). Phosphodiester-linked ORNs with the following sequences were synthesized by Metabion (Planegg, Germany): RNA63 (caggucugugau), RNA63M1 (caxgucugugau; "x" depicts 2′-O-methylguanosine), RNA63M2 (cagxucugugau), RNA63M3 (caggucuxugau), RNA63M4 (caggucuguxau), RNA28 (caguugaacaug), RNA28M (caxuugaacaug), RNA66 (cugagugucccg), and RNA66M (cuxagugucccg). CpG ODN 2216 was obtained from Biomol (Berlin, Germany). Poly-dT-phosphorothioate (PTO) was provided by Metabion (Martinsried, Germany). Recombinant human FMS-like tyrosine kinase 3 ligand (FLT3L) was obtained from H. Hochrein (Bavarian Nordic GmbH, Martinsried, Germany).

4.2. Cells

Human peripheral blood mononuclear cells (PBMCs) were purified from buffy coats by Ficoll gradient centrifugation (PAA Laboratories GmbH, Cölbe, Germany) and cultivated in RPMI1640 (PAN Biotech GmbH, Aidenbach, Germany), supplemented with 2 mM glutamine (Merck, Darmstadt, Germany), 100 units/mL penicillin and 100 µg/mL streptomycin (Merck), non-essential amino acids (Merck), 1 mM sodium pyruvate (Merck), and 2% serum of AB positive male donors (Merck). Differentiation of dendritic cells (DCs) from the bone marrow of both wt and $Tlr7^{-/-}$ mice was induced by stimulation with FLT-3 ligand, as described in [43], and cells were maintained in Opti-MEM (Thermo-Fisher, Waltham, Massachusetts) with 10% fetal calf serum (FCS) (Biochrom AG, Berlin, Germany), 2 mM glutamine, 100 units/mL penicillin, 100 µg/mL streptomycin, and 0.05 mM β-Mercaptoethanol (Sigma-Aldrich, St. Louis, MO, USA).

4.3. Cell Stimulation

For cell stimulation, PBMCs were seeded at 3×10^5 cells/well and murine DCs were seeded at 2×10^5 cells/well in 100 µL growth medium in a 96-well flat-bottom plate. Cells were stimulated with CpG ODN 2216 at 1 µM without a transfection reagent as stimulation control. For stimulation with ORNs, RNA40 was applied at 5 µg/mL, and RNA63, RNA66 and RNA28 ORNs were applied at 10 µg/mL. The stimulation mixture was prepared as follows: RNA stimuli were combined with 50 µL Opti-MEM and 1.5 µL DOTAP (Roche, Mannheim, Germany) per well. Samples were incubated for 5 min at room temperature, then 50 µL of the medium was added without serum and used to stimulate cells at 100 µL. Supernatants were harvested 20 h post stimulation (h p. s.) and cytokine secretion was determined by ELISA analysis.

4.4. Enzyme-Linked Immunosorbent Assay (ELISA)

Murine IL-6, murine IFN-α, and human IL-6 concentrations were measured by ELISA according to the manufacturer's instructions (R&D Systems, Minneapolis, MN, USA for murine IL-6; PBL Interferon Source, Piscataway, NJ, USA for murine IFN-α; BD Biosciences, Heidelberg, Germany for human IL-6). Human IFN-α concentrations were measured using capture antibody mouse monoclonal anti-human IFN-α (PBL Interferon Source) and detection antibody anti-human IFN-α HRP-conjugate (eBioscience, San Diego, CA, USA). Recombinant human IFN-α (PeproTech, Cranbury, NY, USA) was used as a standard.

4.5. Genetic Complementation Assay

HEK-Blue™ hTLR8 (Invivogen) cells expressing the human TLR8 gene and a NF-κB/AP-1 inducible SEAP (secreted embryonic alkaline phosphatase) reporter gene were used as a TLR8-specific reporter system. Upon TLR8-dependent NFκB activation, SEAP activity was monitored using QUANTI-Blue™ (Invivogen). Cells were seeded in a 96-well flat-bottom plate at 6×10^4 cells/well. Cells were stimulated 12 h after seeding with final concentrations of 0.5 µg/mL TL8-506 or 5 µg/mL RNA40, and 20 µg/mL RNA63, RNA66, and RNA28 ORN complexed to 5 µL DOTAP/well. To enhance TLR8-activation, Poly-dT-PTO was added at 1.5 µM final concentration. The readout was performed with QUANTI-Blue™ medium, as described in the manufacturer's protocol. Optical density was measured with a Berthold luminometer (Pforzheim, Germany) at a wavelength of 650 nm. Results were normalized to percent NF-κb induction by TL8-506.

4.6. Mice

For experiments with Tlr7-deficient mice, we used the Tlr7-deficient mouse line established by Hemmi et al. [12]. Mice were bred under specific pathogen-free conditions at the animal facility of Philipps University Marburg.

4.7. RNase T2 Digestion Assay

An amount of 500 ng of each ORN was incubated with 0.0025 U RNase T2 (Aspergillus oryzae, MoBiTec GmbH, Göttingen, Germany) for 1.5 h at 37 °C. RNA fragments were separated in a 25% polyacrylamide gel with the following running conditions: 3–4 h at 250 V, 8 °C. Staining of RNA bands was performed in a 0.05% toluidine blue aqueous solution with 20% methanol and 2% glycerol for 1 h. Destaining and visualization of band patterns was performed in a destain solution containing 20% methanol and 2% glycerol for at least 2 h. Documentation was performed on a Chemi Doc Imaging System (Bio-Rad, Hercules, CA, USA).

4.8. Statistical Analysis

Experiments tested for statistical significance were performed six times. Data were tested for normality using D'Agostino-Pearson and Kolmogorov-Smirnov tests and are presented as arithmetic means + S.D. Statistical analyses of normally distributed data were based on paired two-tailed *t*-tests. Non-normally distributed data were analyzed using Wilcoxon tests.

Author Contributions: Conceptualization, S.J.; methodology, M.N., F.V.S., S.B. and S.J.; validation, M.N., J.S., H.-L.O. and S.J.; formal analysis, S.B. and S.J.; investigation, M.N., J.S., H.-L.O. and S.J.; resources, E.B., S.B. and S.J.; writing—original draft preparation, S.J. and E.B.; writing—review and editing, J.S., E.B., S.B. and S.J.; visualization, S.J.; supervision, S.B. and S.J.; project administration, S.J.; funding acquisition, E.B., S.B. and S.J. All authors have read and agreed to the published version of the manuscript.

Funding: J.S., E.B. and S.J. were funded by the Deutsche Forschungsgemeinschaft (DFG, German Research Foundation) under Germany's Excellence Strategy-EXC2151-390873048. M.N. was funded by the Deutsche Forschungsgemeinschaft (DFG)-KFO325-Project number 329116008-Project SP7. F.V.S. was funded by the Deutsche Forschungsgemeinschaft (DFG)-TRR84-Project number 114933180-Project C10. H.-L.O. and S.B. were funded by the Deutsche Forschungsgemeinschaft (DFG)-TRR237-Project number 369799452-Project A02 and E.B by Project A06.

Institutional Review Board Statement: Ethical review and approval were waived for the use of buffy coats in this study, due to non-identifiability of blood samples. For experiments with murine pDCs, mice were sacrificed, and bone marrow was harvested. The experiments were performed according to National German welfare law §4 (3) TierSchG and §2 and Annex 2 (TierSchVerV) of the National Order for the use of animals in research and did not require approval by a local ethics committee. The number of mice used was reported to the Animal Welfare Officer of Philipps-University Marburg according to the regulations.

Informed Consent Statement: Not applicable.

Data Availability Statement: Not applicable.

Acknowledgments: We thank H. Hackstein and G. Bein, Institute for Clinical Immunology and Transfusion Medicine, Justus-Liebig-University Giessen, for providing human buffy coats. The authors thank Timo Wadenpohl for excellent technical support. Furthermore, the authors thank Thomas Zillinger, Damien Bertheloot, and Eicke Latz for cells, reagents, protocols, and fruitful discussions.

Conflicts of Interest: The authors declare no conflict of interest.

Appendix A

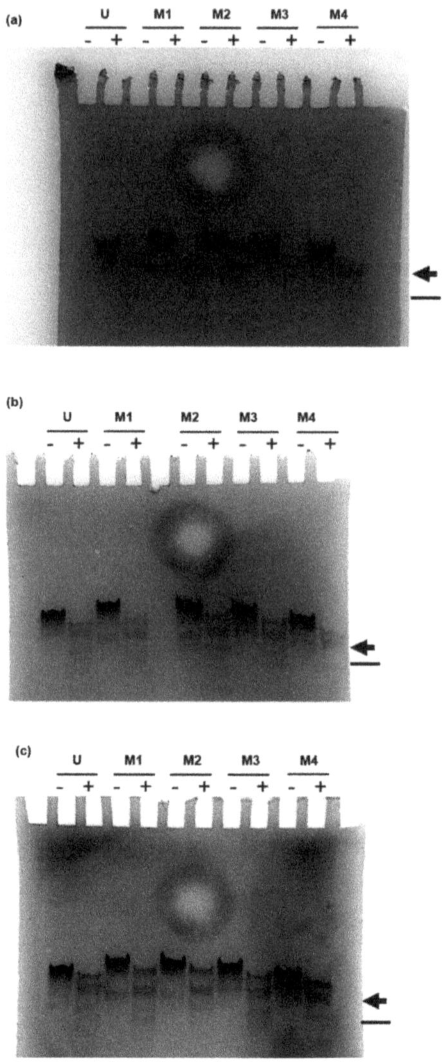

Figure A1. RNase T2 digestion patterns of RNA63 derivatives are influenced by the position of 2′-O-ribose methylation. (**a**–**c**) Unmethylated RNA63 (U) or methylated derivatives (M1, M2, M3, and M4) were treated with 0.005 units RNase T2/µg RNA (+) or mock-treated (−). RNA fragmentation was visualized in polyacrylamide gel electrophoresis. Position of RNA fragments of interest are indicated with an arrow (🡸) and a line (▬). Three independent replicates of RNase T2 digestion are shown.

References

1. Chan, Y.K.; Gack, M.U. Viral evasion of intracellular DNA and RNA sensing. *Nat. Rev. Microbiol.* **2016**, *14*, 360–373. [CrossRef]
2. Murphy, K.; Weaver, C. *Janeway's Immunobiology*; Garland Science: New York, NY, USA, 2017.
3. Lind, N.A.; Rael, V.E.; Pestal, K.; Liu, B.; Barton, G.M. Regulation of the nucleic acid-sensing Toll-like receptors. *Nat. Rev. Immunol.* **2022**, *22*, 224–235. [CrossRef]

4. Cervantes, J.L.; Weinerman, B.; Basole, C.; Salazar, J.C. TLR8: The forgotten relative revindicated. *Cell Mol. Immunol.* **2012**, *9*, 434–438. [CrossRef]
5. Bender, A.T.; Tzvetkov, E.; Pereira, A.; Wu, Y.; Kasar, S.; Przetak, M.M.; Vlach, J.; Niewold, T.B.; Jensen, M.A.; Okitsu, S.L. TLR7 and TLR8 Differentially Activate the IRF and NF-κB Pathways in Specific Cell Types to Promote Inflammation. *Immunohorizons* **2020**, *4*, 93–107. [CrossRef]
6. Hornung, V.; Guenthner-Biller, M.; Bourquin, C.; Ablasser, A.; Schlee, M.; Uematsu, S.; Noronha, A.; Manoharan, M.; Akira, S.; de Fougerolles, A.; et al. Sequence-specific potent induction of IFN-alpha by short interfering RNA in plasmacytoid dendritic cells through TLR7. *Nat. Med.* **2005**, *11*, 263–270. [CrossRef] [PubMed]
7. Ablasser, A.; Poeck, H.; Anz, D.; Berger, M.; Schlee, M.; Kim, S.; Bourquin, C.; Goutagny, N.; Jiang, Z.; Fitzgerald, K.A.; et al. Selection of molecular structure and delivery of RNA oligonucleotides to activate TLR7 versus TLR8 and to induce high amounts of IL-12p70 in primary human monocytes. *J. Immunol.* **2009**, *182*, 6824–6833. [CrossRef] [PubMed]
8. de Marcken, M.; Dhaliwal, K.; Danielsen, A.C.; Gautron, A.S.; Dominguez-Villar, M. TLR7 and TLR8 activate distinct pathways in monocytes during RNA virus infection. *Sci. Signal.* **2019**, *12*, eaaw1347. [CrossRef]
9. Bartok, E.; Hartmann, G. Immune Sensing Mechanisms that Discriminate Self from Altered Self and Foreign Nucleic Acids. *Immunity* **2020**, *53*, 54–77. [CrossRef]
10. Hartmann, G. Nucleic Acid Immunity. *Adv. Immunol.* **2017**, *133*, 121–169. [CrossRef]
11. Barchet, W.; Wimmenauer, V.; Schlee, M.; Hartmann, G. Accessing the therapeutic potential of immunostimulatory nucleic acids. *Curr. Opin. Immunol.* **2008**, *20*, 389–395. [CrossRef]
12. Hemmi, H.; Kaisho, T.; Takeuchi, O.; Sato, S.; Sanjo, H.; Hoshino, K.; Horiuchi, T.; Tomizawa, H.; Takeda, K.; Akira, S. Small anti-viral compounds activate immune cells via the TLR7 MyD88-dependent signaling pathway. *Nat. Immunol.* **2002**, *3*, 196–200. [CrossRef]
13. Jurk, M.; Heil, F.; Vollmer, J.; Schetter, C.; Krieg, A.M.; Wagner, H.; Lipford, G.; Bauer, S. Human TLR7 or TLR8 independently confer responsiveness to the antiviral compound R-848. *Nat. Immunol.* **2002**, *3*, 499. [CrossRef]
14. Heil, F.; Hemmi, H.; Hochrein, H.; Ampenberger, F.; Kirschning, C.; Akira, S.; Lipford, G.; Wagner, H.; Bauer, S. Species-specific recognition of single-stranded RNA via toll-like receptor 7 and 8. *Science* **2004**, *303*, 1526–1529. [CrossRef]
15. Krüger, A.; Oldenburg, M.; Chebrolu, C.; Beisser, D.; Kolter, J.; Sigmund, A.M.; Steinmann, J.; Schäfer, S.; Hochrein, H.; Rahmann, S.; et al. Human TLR8 senses UR/URR motifs in bacterial and mitochondrial RNA. *EMBO Rep.* **2015**, *16*, 1656–1663. [CrossRef] [PubMed]
16. Forsbach, A.; Nemorin, J.G.; Montino, C.; Müller, C.; Samulowitz, U.; Vicari, A.P.; Jurk, M.; Mutwiri, G.K.; Krieg, A.M.; Lipford, G.B.; et al. Identification of RNA sequence motifs stimulating sequence-specific TLR8-dependent immune responses. *J. Immunol.* **2008**, *180*, 3729–3738. [CrossRef] [PubMed]
17. Zhang, Z.; Ohto, U.; Shibata, T.; Taoka, M.; Yamauchi, Y.; Sato, R.; Shukla, N.M.; David, S.A.; Isobe, T.; Miyake, K.; et al. Structural Analyses of Toll-like Receptor 7 Reveal Detailed RNA Sequence Specificity and Recognition Mechanism of Agonistic Ligands. *Cell Rep.* **2018**, *25*, 3371–3381.e5. [CrossRef] [PubMed]
18. Zhang, Z.; Ohto, U.; Shibata, T.; Krayukhina, E.; Taoka, M.; Yamauchi, Y.; Tanji, H.; Isobe, T.; Uchiyama, S.; Miyake, K.; et al. Structural Analysis Reveals that Toll-like Receptor 7 Is a Dual Receptor for Guanosine and Single-Stranded RNA. *Immunity* **2016**, *45*, 737–748. [CrossRef]
19. Shibata, T.; Ohto, U.; Nomura, S.; Kibata, K.; Motoi, Y.; Zhang, Y.; Murakami, Y.; Fukui, R.; Ishimoto, T.; Sano, S.; et al. Guanosine and its modified derivatives are endogenous ligands for TLR7. *Int. Immunol.* **2016**, *28*, 211–222. [CrossRef]
20. Ostendorf, T.; Zillinger, T.; Andryka, K.; Schlee-Guimaraes, T.M.; Schmitz, S.; Marx, S.; Bayrak, K.; Linke, R.; Salgert, S.; Wegner, J.; et al. Immune Sensing of Synthetic, Bacterial, and Protozoan RNA by Toll-like Receptor 8 Requires Coordinated Processing by RNase T2 and RNase 2. *Immunity* **2020**, *52*, 591–605.e6. [CrossRef] [PubMed]
21. Greulich, W.; Wagner, M.; Gaidt, M.M.; Stafford, C.; Cheng, Y.; Linder, A.; Carell, T.; Hornung, V. TLR8 Is a Sensor of RNase T2 Degradation Products. *Cell* **2019**, *179*, 1264–1275.e13. [CrossRef]
22. Tanji, H.; Ohto, U.; Shibata, T.; Taoka, M.; Yamauchi, Y.; Isobe, T.; Miyake, K.; Shimizu, T. Toll-like receptor 8 senses degradation products of single-stranded RNA. *Nat. Struct. Mol. Biol.* **2015**, *22*, 109–115. [CrossRef] [PubMed]
23. Damase, T.R.; Sukhovershin, R.; Boada, C.; Taraballi, F.; Pettigrew, R.I.; Cooke, J.P. The Limitless Future of RNA Therapeutics. *Front. Bioeng. Biotechnol.* **2021**, *9*, 628137. [CrossRef]
24. Auderset, F.; Belnoue, E.; Mastelic-Gavillet, B.; Lambert, P.H.; Siegrist, C.A. A TLR7/8 Agonist-Including DOEPC-Based Cationic Liposome Formulation Mediates Its Adjuvanticity Through the Sustained Recruitment of Highly Activated Monocytes in a Type I IFN-Independent but NF-κB-Dependent Manner. *Front. Immunol.* **2020**, *11*, 580974. [CrossRef] [PubMed]
25. Mackman, R.L.; Mish, M.; Chin, G.; Perry, J.K.; Appleby, T.; Aktoudianakis, V.; Metobo, S.; Pyun, P.; Niu, C.; Daffis, S.; et al. Discovery of GS-9688 (Selgantolimod) as a Potent and Selective Oral Toll-Like Receptor 8 Agonist for the Treatment of Chronic Hepatitis B. *J. Med. Chem.* **2020**, *63*, 10188–10203. [CrossRef] [PubMed]
26. Amin, O.E.; Colbeck, E.J.; Daffis, S.; Khan, S.; Ramakrishnan, D.; Pattabiraman, D.; Chu, R.; Micolochick Steuer, H.; Lehar, S.; Peiser, L.; et al. Therapeutic Potential of TLR8 Agonist GS-9688 (Selgantolimod) in Chronic Hepatitis B: Remodeling of Antiviral and Regulatory Mediators. *Hepatology* **2021**, *74*, 55–71. [CrossRef]

27. Freund, I.; Buhl, D.K.; Boutin, S.; Kotter, A.; Pichot, F.; Marchand, V.; Vierbuchen, T.; Heine, H.; Motorin, Y.; Helm, M.; et al. 2′-O-methylation within prokaryotic and eukaryotic tRNA inhibits innate immune activation by endosomal Toll-like receptors but does not affect recognition of whole organisms. *RNA* **2019**, *25*, 869–880. [CrossRef] [PubMed]
28. Freund, I.; Eigenbrod, T.; Helm, M.; Dalpke, A.H. RNA Modifications Modulate Activation of Innate Toll-Like Receptors. *Genes* **2019**, *10*, 92. [CrossRef]
29. Schmitt, F.C.F.; Freund, I.; Weigand, M.A.; Helm, M.; Dalpke, A.H.; Eigenbrod, T. Identification of an optimized 2′-O-methylated trinucleotide RNA motif inhibiting Toll-like receptors 7 and 8. *RNA* **2017**, *23*, 1344–1351. [CrossRef]
30. Hamm, S.; Latz, E.; Hangel, D.; Muller, T.; Yu, P.; Golenbock, D.; Sparwasser, T.; Wagner, H.; Bauer, S. Alternating 2′-O-ribose methylation is a universal approach for generating non-stimulatory siRNA by acting as TLR7 antagonist. *Immunobiology* **2009**, *215*, 559–569. [CrossRef]
31. Jockel, S.; Nees, G.; Sommer, R.; Zhao, Y.; Cherkasov, D.; Hori, H.; Ehm, G.; Schnare, M.; Nain, M.; Kaufmann, A.; et al. The 2′-O-methylation status of a single guanosine controls transfer RNA-mediated Toll-like receptor 7 activation or inhibition. *J. Exp. Med.* **2012**, *209*, 235–241. [CrossRef]
32. Jung, S.; von Thülen, T.; Laukemper, V.; Pigisch, S.; Hangel, D.; Wagner, H.; Kaufmann, A.; Bauer, S. A single naturally occurring 2′-O-methylation converts a TLR7- and TLR8-activating RNA into a TLR8-specific ligand. *PLoS ONE* **2015**, *10*, e0120498. [CrossRef]
33. Gehrig, S.; Eberle, M.E.; Botschen, F.; Rimbach, K.; Eberle, F.; Eigenbrod, T.; Kaiser, S.; Holmes, W.M.; Erdmann, V.A.; Sprinzl, M.; et al. Identification of modifications in microbial, native tRNA that suppress immunostimulatory activity. *J. Exp. Med.* **2012**, *209*, 225–233. [CrossRef]
34. Lee, J.; Chuang, T.H.; Redecke, V.; She, L.; Pitha, P.M.; Carson, D.A.; Raz, E.; Cottam, H.B. Molecular basis for the immunostimulatory activity of guanine nucleoside analogs: Activation of Toll-like receptor 7. *Proc. Natl. Acad. Sci. USA* **2003**, *100*, 6646–6651. [CrossRef]
35. Piekna-Przybylska, D.; Decatur, W.A.; Fournier, M.J. The 3D rRNA modification maps database: With interactive tools for ribosome analysis. *Nucleic Acids Res.* **2008**, *36*, D178–D183. [CrossRef]
36. Fabbri, M.; Paone, A.; Calore, F.; Galli, R.; Gaudio, E.; Santhanam, R.; Lovat, F.; Fadda, P.; Mao, C.; Nuovo, G.J.; et al. MicroRNAs bind to Toll-like receptors to induce prometastatic inflammatory response. *Proc. Natl. Acad. Sci. USA* **2012**, *109*, E2110–E2116. [CrossRef]
37. Sarvestani, S.T.; Williams, B.R.; Gantier, M.P. Human Toll-like receptor 8 can be cool too: Implications for foreign RNA sensing. *J. Interferon Cytokine Res.* **2012**, *32*, 350–361. [CrossRef]
38. Krug, A.; Rothenfusser, S.; Hornung, V.; Jahrsdörfer, B.; Blackwell, S.; Ballas, Z.K.; Endres, S.; Krieg, A.M.; Hartmann, G. Identification of CpG oligonucleotide sequences with high induction of IFN-alpha/beta in plasmacytoid dendritic cells. *Eur. J. Immunol.* **2001**, *31*, 2154–2163. [CrossRef]
39. Lu, H.; Dietsch, G.N.; Matthews, M.A.; Yang, Y.; Ghanekar, S.; Inokuma, M.; Suni, M.; Maino, V.C.; Henderson, K.E.; Howbert, J.J.; et al. VTX-2337 is a novel TLR8 agonist that activates NK cells and augments ADCC. *Clin. Cancer Res.* **2012**, *18*, 499–509. [CrossRef]
40. Czauderna, F.; Fechtner, M.; Dames, S.; Aygün, H.; Klippel, A.; Pronk, G.J.; Giese, K.; Kaufmann, J. Structural variations and stabilising modifications of synthetic siRNAs in mammalian cells. *Nucleic Acids Res.* **2003**, *31*, 2705–2716. [CrossRef]
41. Obermann, H.L.; Lederbogen, I.I.; Steele, J.; Dorna, J.; Sander, L.E.; Engelhardt, K.; Bakowsky, U.; Kaufmann, A.; Bauer, S. RNA-Cholesterol Nanoparticles Function as Potent Immune Activators via TLR7 and TLR8. *Front. Immunol.* **2021**, *12*, 658895. [CrossRef]
42. Lan, T.; Putta, M.R.; Wang, D.; Dai, M.; Yu, D.; Kandimalla, E.R.; Agrawal, S. Synthetic oligoribonucleotides-containing secondary structures act as agonists of Toll-like receptors 7 and 8. *Biochem. Biophys. Res. Commun.* **2009**, *386*, 443–448. [CrossRef]
43. Spies, B.; Hochrein, H.; Vabulas, M.; Huster, K.; Busch, D.H.; Schmitz, F.; Heit, A.; Wagner, H. Vaccination with plasmid DNA activates dendritic cells via Toll-like receptor 9 (TLR9) but functions in TLR9-deficient mice. *J. Immunol.* **2003**, *171*, 5908–5912. [CrossRef]

International Journal of
Molecular Sciences

MDPI

Article

Breast Cancer Vaccine Containing a Novel Toll-like Receptor 7 Agonist and an Aluminum Adjuvant Exerts Antitumor Effects

Shuquan Zhang, Yu Liu †, Ji Zhou, Jiaxin Wang, Guangyi Jin and Xiaodong Wang *

School of Pharmaceutical Sciences, Nation-Regional Engineering Lab for Synthetic Biology of Medicine, International Cancer Center, Shenzhen University Health Science Center, Shenzhen 518060, China
* Correspondence: wangxiaodong@szu.edu.cn
† Current Address: Department of General Practice, The 3rd Affiliated Hospital of Shenzhen University, Shenzhen 518006, China.

Abstract: Mucin 1 (MUC1) has received increasing attention due to its high expression in breast cancer, in which MUC1 acts as a cancer antigen. Our group has been committed to the development of small-molecule TLR7 (Toll-like receptor 7) agonists, which have been widely investigated in the field of tumor immunotherapy. In the present study, we constructed a novel tumor vaccine (SZU251 + MUC1 + Al) containing MUC1 and two types of adjuvants: a TLR7 agonist (SZU251) and an aluminum adjuvant (Al). Immunostimulatory responses were first verified in vitro, where the vaccine promoted the release of cytokines and the expression of costimulatory molecules in mouse BMDCs (bone marrow dendritic cells) and spleen lymphocytes. Then, we demonstrated that SZU251 + MUC1 + Al was effective and safe against a tumor expressing the MUC1 antigen in both prophylactic and therapeutic schedules in vivo. The immune responses in vivo were attributed to the increase in specific humoral and cellular immunity, including antibody titers, $CD4^+$, $CD8^+$ and activated $CD8^+$ T cells. Therefore, our vaccine candidate may have beneficial effects in the prevention and treatment of breast cancer patients.

Keywords: TLR7 (Toll-like receptor 7); MUC1 (Mucin 1); aluminum adjuvant; tumor vaccine; immunotherapy

Citation: Zhang, S.; Liu, Y.; Zhou, J.; Wang, J.; Jin, G.; Wang, X. Breast Cancer Vaccine Containing a Novel Toll-like Receptor 7 Agonist and an Aluminum Adjuvant Exerts Antitumor Effects. *Int. J. Mol. Sci.* **2022**, *23*, 15130. https://doi.org/10.3390/ijms232315130

Academic Editors: Oliver Planz and Ralf Kircheis

Received: 7 September 2022
Accepted: 28 November 2022
Published: 1 December 2022

1. Introduction

Toll-like receptors (TLRs) are a family of integral membrane proteins, primarily localized in immune cells such as dendritic cells (DCs) and macrophages. TLRs, recognizing pathogen-associated molecular patterns (PAMPs) in a variety of viruses and microorganisms, are the bridge between innate and adaptive immunity [1,2]. Among all of the TLRs, only TLR7/8 have small-molecule ligands, which are easier to obtain and modify than other biomacromolecules such as TLR4 and TLR9 ligands [3]. Recently, small-molecule TLR7 agonists have been widely investigated in the field of tumor immunotherapy through inducing tumor-specific immune responses and reducing the tumor growth [4–6]. Our group has been committed to the development of TLR7 agonists, and SZU101 is one of the representative compounds. We have previously proved that SZU101 enhances the therapeutic efficacy of doxorubicin in a mouse lymphoma model via the generation of systemic immune responses. SZU101 has been used in the form of chemical conjugation with targeted anticancer drugs, including JQ1 (a BET inhibitor) and ibrutinib (a BTK inhibitor) [7]. Furthermore, SZU101 is a successful adjuvant in the development of tumor vaccines, having been conjugated with multiple tumor-associated antigens [8].

Mucin 1 (MUC1) is a membrane-associated glycoprotein overexpressed in many different kinds of epithelial tumor tissues, making it an attractive target for tumor immunotherapy [9]. MUC1 contains a variable number of tandem repeat (VNTR) sequence of twenty amino acids, and five potential O-glycosylation sites are located on the threonine and serine residues of each repeat [10,11]. The peptides from the VNTR sequence are often used as the antigens for tumor vaccines targeting MUC1 [12]. A tumor vaccine has been constructed by covalent attachment of an MUC1 glycopeptide and a T-helper epitope, inducing both humoral and cellular immune responses [13]. Our group developed a novel TLR7 agonist-conjugated MUC1 peptide vaccine that elicited effective immune responses and robust antitumor effects in a mouse breast cancer model [14]. However, appropriate immune stimulators still need to be studied due to the weak immunogenicity of MUC1 peptides.

In this study, we construct a novel tumor vaccine (SZU251 + MUC1 + Al) with three components: a small-molecule TLR7 agonist synthesized for the first time (SZU251), an MUC1-related peptide (MUC1), and an aluminum adjuvant (Al). It is widely accepted that the glycosylated MUC1 peptide is more immunogenic than the non-glycosylated one [13]. However, we already demonstrated the immunogenicity of the non-glycosylated MUC1 peptide in previous work, and the combination of the naked MUC1 sequence and the TLR7 agonist (T7 + MUC1) could be an effective tumor vaccine [14]. Therefore, we have chosen the same naked sequence here in order to verify the improvement of this three-component vaccine (T7 + MUC1 + Al) compared to the previous two-component vaccine (T7 + MUC1). Immunostimulatory effects of the vaccine are first determined in vitro by the detection of cytokines and costimulatory molecules. Then, the humoral and cellular immune responses are verified in a mouse breast cancer model in both prophylactic and therapeutic vaccination schedules. Finally, the vaccine is proved to be well tolerated in the mice by HE staining (hematoxylin–eosin staining) of the major organs and blood tests.

2. Results

2.1. Chemical Synthesis of the Vaccine Candidates

MUC1 peptide was synthesized by the solid phase method, containing a well-documented MUC1-derived epitope (SAPDTRPAP) and a murine MHC class II restricted T-helper epitope (KLFAVWKITYKDT) (Figure 1A). SZU251 was synthesized for the first time by our group according to the methods shown in Figure 1B. The above-mentioned compounds were confirmed by NMR spectrometry, mass spectrometry or high-performance liquid chromatography (Figures S1 and S2). SZU251 + MUC1 was a mixed formulation at a 3:1 ratio of SZU251 and MUC1, and the 3:1 ratio was selected according to our previous study [14]. All of the vaccine candidates, especially for the in vivo experiments, were mixed with the aluminum adjuvant (Figure 1C). Moreover, the ability of SZU251 for TLR activation was determined on HEK-Blue hTLR7 and hTLR8 cells. SZU251 and R848 (TLR7/8 agonists) induced much more potent TLR7 activation than imiquimod (a TLR7 agonist), while SZU251 and imiquimod could not induce TLR8 activation (Figure S3). Therefore, SZU251 is a novel TLR7 agonist with specificity and selectivity.

2.2. In Vitro Immunostimulatory Responses of the Vaccine Candidates

To access the immunostimulatory activity of the vaccine candidates, the release of cytokines was first decided by the ELISA method in mouse spleen lymphocytes and BMDCs (bone marrow dendritic cells). As shown in Figure 2A,B, MUC1 rarely stimulated cytokine production due to its poor immunogenicity. SZU251 and SZU251 + MUC1 displayed similar and remarkable trends of increase in the relevant cytokines in a dose-dependent manner (TNFα, IFNγ, IL6 and IL12 in spleen lymphocytes; TNFα, IFNγ and IL6 in BMDCs). IL12 in BMDCs was also detected but is not displayed here owing to the low secretion levels. Next, maturation of BMDCs was accessed by flow cytometry by detecting the upregulated expression of CD40, CD80 and CD86, the costimulatory molecules necessary for T cell activation. The flow cytometry results and statistical analyses are shown in Figure 2C. SZU251 and SZU251 + MUC1 increased the levels of CD40, CD80 and CD86, while SZU251 + MUC1 exhibited an even better stimulatory capacity than SZU251. However, significant differences between SZU251 and SZU251 + MUC1 could be observed for CD40 and CD86 but not CD80, which may be due to the different sensitivity of the costimulatory molecules.

Moreover, SZU251 + MUC1 promoted the expression of the costimulatory molecules basically in a dose-dependent manner, similar to the results of cytokine levels (Figure S4).

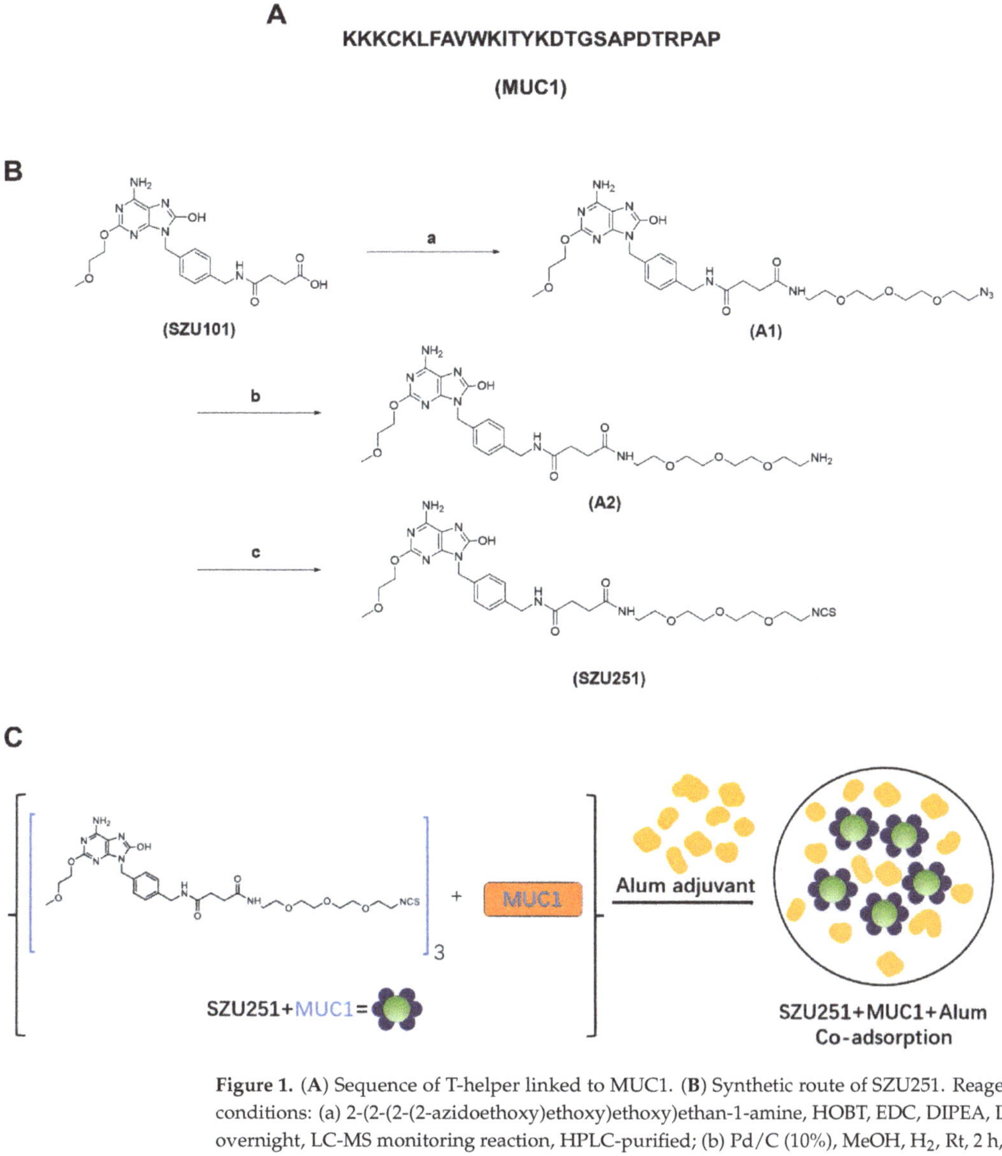

Figure 1. (**A**) Sequence of T-helper linked to MUC1. (**B**) Synthetic route of SZU251. Reagents and conditions: (a) 2-(2-(2-(2-azidoethoxy)ethoxy)ethoxy)ethan-1-amine, HOBT, EDC, DIPEA, DMF, Rt, overnight, LC-MS monitoring reaction, HPLC-purified; (b) Pd/C (10%), MeOH, H_2, Rt, 2 h, LC-MS monitoring reaction, MeOH spin-dried; (c) (1) CS_2, TEA, DMF, Rt, overnight; (2) TsCl, Rt, 5 h, LC-MS monitoring reaction, HPLC-purified. (**C**) Scheme of SZU251 + MUC1 + Al.

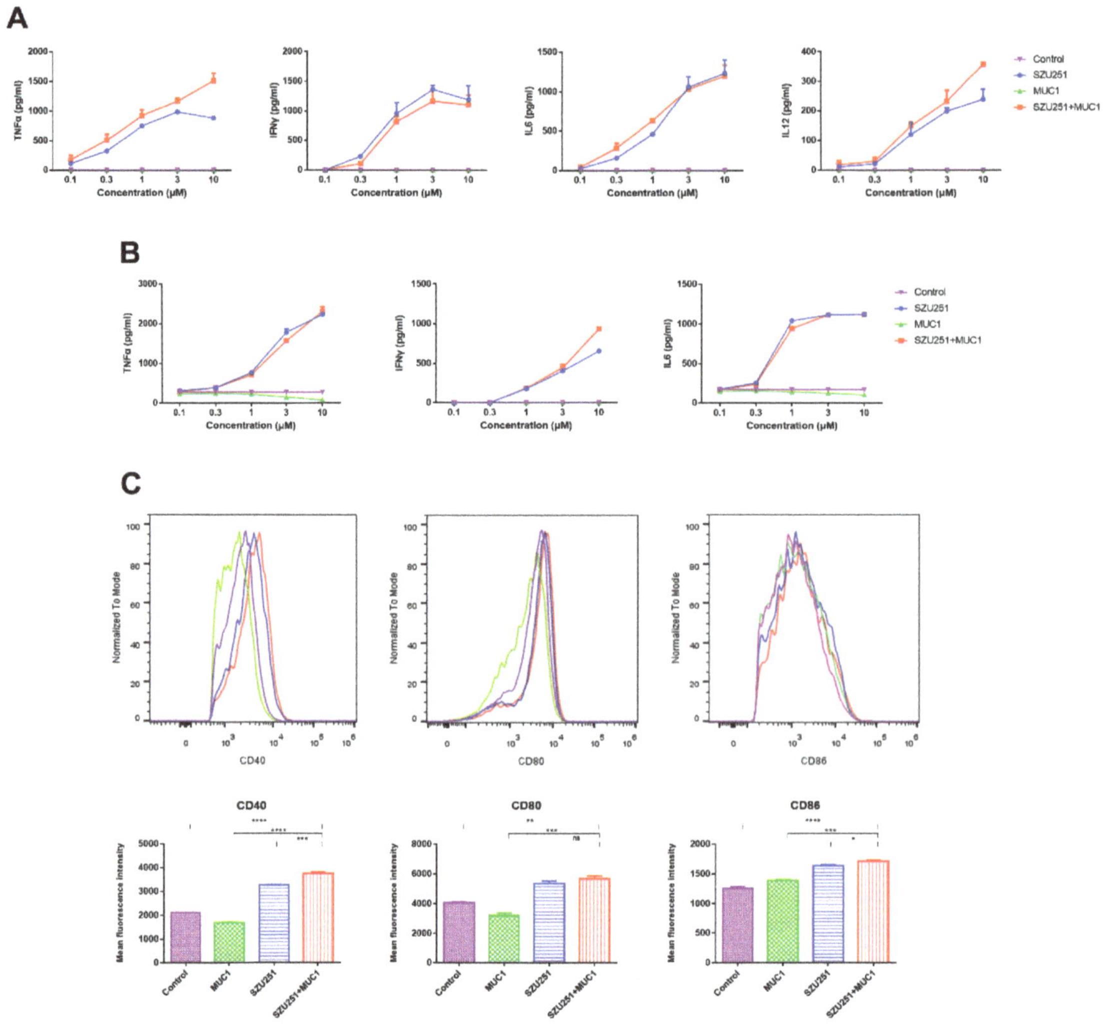

Figure 2. In vitro immunostimulatory responses of the vaccine candidates. (**A**) Induction of TNFα, IFNγ, IL6 and IL12 in mouse spleen lymphocytes after 24 h incubation. (**B**) Induction of TNFα, IFNγ and IL6 in mouse BMDCs after 24 h incubation. (**C**) Maturation of BMDCs was evaluated by measuring the expression of the surface molecules CD40, CD80 and CD86 by flow cytometry after 24 h treatment with 3 μM vaccine candidates. * $p < 0.05$, ** $p < 0.01$, *** $p < 0.001$, **** $p < 0.0001$, "ns" not significant. TNF, tumor necrosis factor; IFN, interferon; IL, interleukin. Data are presented as the mean ± SE; $n = 3$.

2.3. SZU251 + MUC1 + Al Inhibited 4T1 Mouse Breast Tumor Growth in the Prophylactic Schedule

The antitumor effects of the vaccine candidates were further examined in a mouse breast cancer model with the prophylactic schedule. In brief, each mouse was intramuscularly administered with vaccines every two weeks for a total of three times and then subcutaneously challenged with 4T1 cells. Twenty-one days after the tumor inoculation, the mice were sacrificed and the tumors were collected (Figure 3A). The tumor volumes, tumor weights and representative images of the tumors are illustrated in Figures 3B,C and

S5. All of the vaccine candidates displayed antitumor effects compared to the saline control (NaCl), except the aluminum adjuvant alone (Al). Notably, the tumor volumes and tumor weights when treated with SZU251 + MUC1 + Al were much lower than when treated with SZU251 + Al or MUC1 + Al. The results also showed that SZU251 + MUC1 + Al markedly improved the long-term survival rate compared to the other groups during a 70-day observation period (Figure 3D).

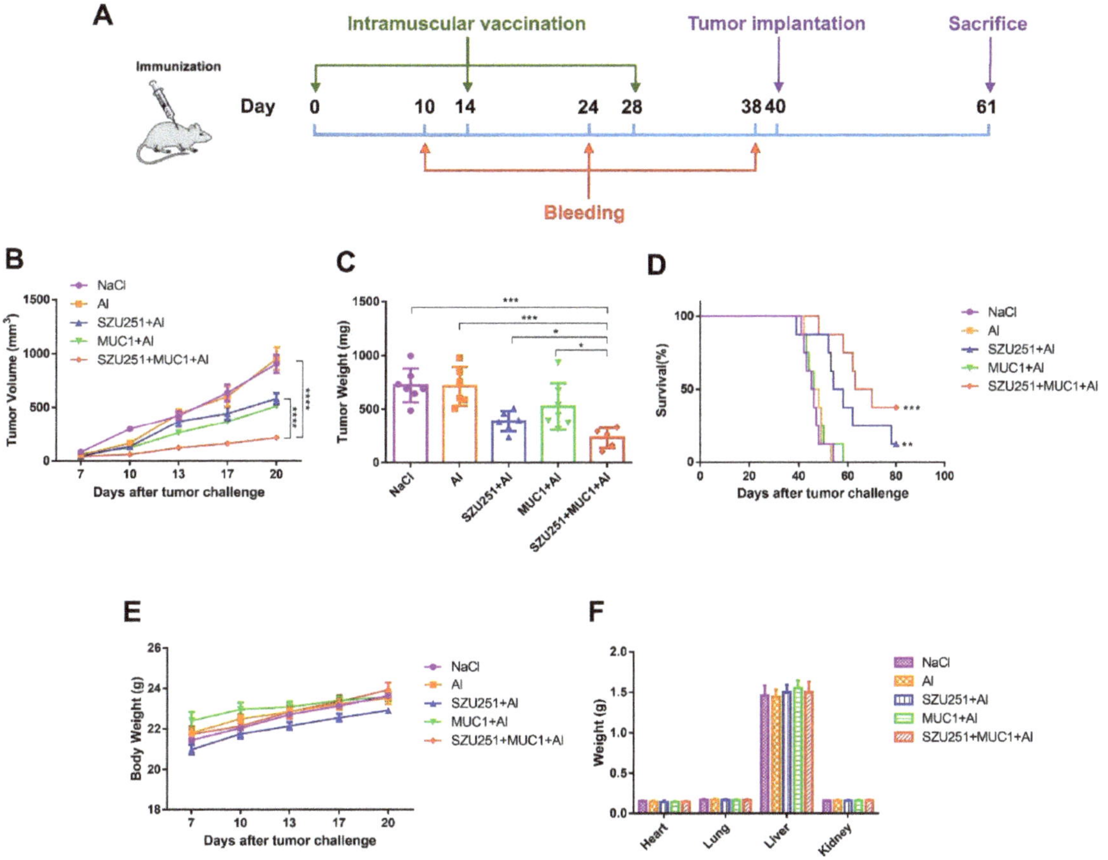

Figure 3. SZU251 + MUC1 + Al inhibited 4T1 mouse breast tumor growth after three immunizations in the prophylactic schedule. (**A**) Schematic diagram illustrating the prophylactic vaccination. (**B**) Tumor growth curves were measured twice a week until day 21 after tumor implantation. (**C**) Tumor weights were determined at the time of sacrifice. (**D**) Survival curves of the tumor-bearing mice. (**E**) Body weights were measured twice a week until day 21 after tumor implantation. (**F**) Weights of the major organs (hearts, lungs, livers and kidneys) were determined at the time of sacrifice. Data are presented as the mean ± SE; n = 5–8 mice/group. * $p < 0.05$, ** $p < 0.01$, *** $p < 0.001$, **** $p < 0.0001$.

During the experiment, all of the mice appeared healthy with no visible signs of pain, distress or discomfort. No significant differences in the body weights or the weights of the major organs could be detected between the groups (Figure 3E,F), indicating that the vaccines were well tolerated at these doses. HE staining of the major organs was carried out, and no serious structural and pathological changes were observed in all of the groups (Figure S6). Blood tests were performed, and the results are shown in Figure S7 and Table 1. There were some individual slight decreases in the WBC and PLT counts in the vaccine groups that were still considered within the normal range, while other parameters (RBC, HGB, HCT and MCV) displayed no significant abnormality.

Table 1. Blood routine examination of the mice in the prophylactic schedule.

Group	NaCl	Al	SZU251 + Al	MUC1 + Al	SZU251 + MUC1 + Al
WBC (10^9/L)	30.05 ± 4.96	28.70 ± 9.49	27.68 ± 7.59	18.63 ± 5.74	13.85 ± 3.24
RBC (10^{12}/L)	11.17 ± 0.56	11.64 ± 0.95	11.74 ± 0.69	11.07 ± 0.71	11.40 ± 0.49
HGB (g/L)	175.20 ± 7.73	186.50 ± 9.33	181.25 ± 5.56	171.20 ± 10.69	169.00 ± 8.69
HCT (%)	55.40 ± 2.42	57.45 ± 4.62	55.24 ± 3.23	55.36 ± 3.23	54.12 ± 2.91
MCV (fL)	49.66 ± 0.47	49.40 ± 0.86	47.11 ± 0.98	50.08 ± 1.00	47.52 ± 1.08
PLT (10^9/L)	1022.67 ± 168.50	1045.50 ± 258.30	957.50 ± 208.63	795.50 ± 138.85	809.00 ± 43.70

Abbreviations: WBC, white blood cell; RBC, red blood cell; HGB, hemoglobin; HCT, hematocrit; MCV, mean corpuscular volume; PLT, platelet.

2.4. SZU251 + MUC1 + Al Induced Tumor-Specific Immune Responses in the Prophylactic Schedule

Serum antibody titers against MUC1 were judged by the ELISA method to prove the impacts of the vaccinations on humoral immunity (Table 2). As expected, no antibody could be found in the vaccine groups without the MUC1 component (NaCl, Al and SZU251 + Al), while MUC1 + Al elicited total IgG, IgG1, IgG2a and IgM responses. SZU251 + MUC1 + Al prominently elevated the antibody titers to much higher levels than MUC1 + Al did. Meanwhile, the more times vaccinations were administered, the more potent antibody responses we could detect. In addition, no significant changes of IgG2a and IgM could be found between day 24 and day 38 in the SZU251 + MUC1 + Al group. It is supposed that these types of antibodies reached a plateau after two vaccinations. To validate whether the serum samples could recognize the native MUC1 antigen on cancer cells, binding of the serum samples to MUC1-expressing 4T1 cells was examined by flow cytometry (Figure 4A). The serum samples from SZU251 + MUC1 + Al reacted much more strongly with 4T1 cells than the other vaccine candidates did. B16 cells, in which the expression of MUC1 antigen is relatively low, were used as a negative control, and no obvious binding was observed.

Table 2. Anti-MUC1 antibody titers determined by ELISA after three immunizations with vaccine candidates in the prophylactic schedule.

Antibody		Titer				
		NaCl	Al	SZU251 + Al	MUC1 + Al	SZU251 + MUC1 + Al
IgG	Day 10	0	0	0	1000	16,000
	Day 24	0	0	0	32,000	1,024,000
	Day 38	0	0	0	256,000	2,048,000
IgG1	Day 10	0	0	0	0	0
	Day 24	0	0	0	8000	64,000
	Day 38	0	0	0	8000	128,000
IgG2a	Day 10	0	0	0	0	32,000
	Day 24	0	0	0	16,000	2,048,000
	Day 38	0	0	0	16,000	2,048,000
IgM	Day 10	0	0	0	0	16,000
	Day 24	0	0	0	4000	128,000
	Day 38	0	0	0	16,000	128,000

Antibody titer was defined as the maximum serum dilution when the absorption ratio of the experimental group to the negative control group was ≥2.0.

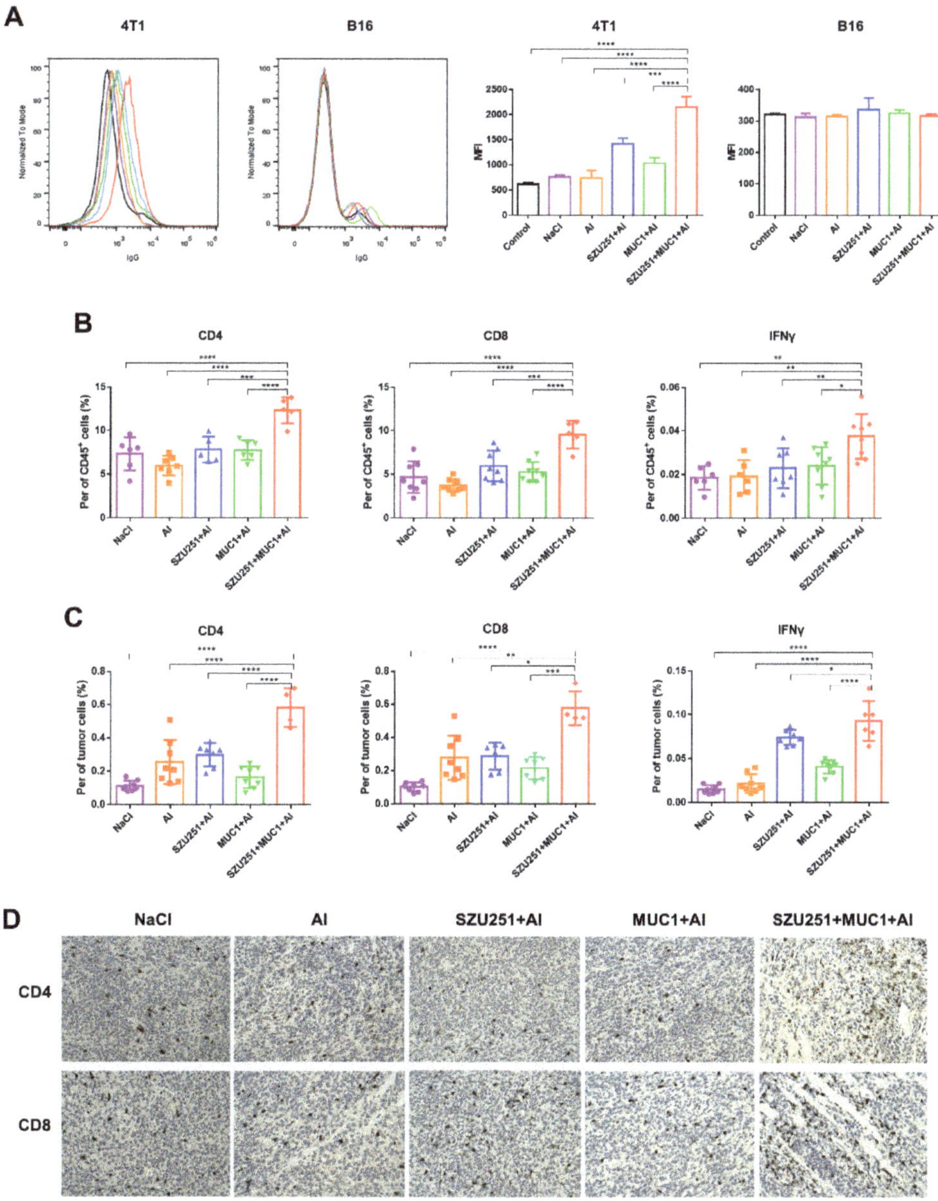

Figure 4. SZU251 + MUC1 + Al induced tumor-specific immune responses in the prophylactic schedule. (**A**) Specific anti-MUC1 antibody of the serum samples was tested on 4T1 and B16 cells using flow cytometry. NaCl (purple), Al (orange), SZU251 + Al (blue), MUC1 + Al (green) and SZU251 + MUC1 + Al (red). (**B**) Percentages of $CD3^+$ $CD4^+$ T cells, $CD3^+$ $CD8^+$ T cells and $CD3^+$ $CD8^+$ $IFN\gamma^+$ T cells in splenocytes were measured using flow cytometry. (**C**) Percentages of $CD3^+$ $CD4^+$ T cells, $CD3^+$ $CD8^+$ T cells and $CD3^+$ $CD8^+$ $IFN\gamma^+$ T cells in TILs were measured using flow cytometry. (**D**) Representative IHC staining for CD4 and CD8 of the tumor tissues in the prophylactic schedule (200× magnification). Data are presented as the mean ± SE; n = 5–8 mice/group. * $p < 0.05$, ** $p < 0.01$, *** $p < 0.001$, **** $p < 0.0001$.

Cellular immune responses were determined by analyzing $CD4^+$ and $CD8^+$ T cells in the spleens, while activated $CD8^+$ T cells were further identified by intracellular IFNγ staining. As shown in Figure 4B, SZU251 + MUC1 + Al boosted the percentages of total $CD4^+$, total $CD8^+$ and $CD8^+$ $IFN\gamma^+$ cells more effectively than any other vaccines did. T cells were also evaluated in the tumors to study whether tumor-specific T cells were recruited to the tumor sites. SZU251 + MUC1 + Al increased the percentages of total $CD4^+$, total $CD8^+$ and $CD8^+$ $IFN\gamma^+$ cells to the highest levels, which is consistent with the results in the spleens (Figure 4C). IHC staining was performed in the tumor tissues, and the results are shown in Figure 4D. The differences in $CD4^+$ and $CD8^+$ T cells were negligible in the NaCl, Al, SZU251 + Al and MUC1 + Al groups, while $CD4^+$ and $CD8^+$ T cells infiltration was significantly promoted by SZU251 + MUC1 + Al.

2.5. SZU251 + MUC1 + Al Inhibited 4T1 Mouse Breast Tumor Growth in the Therapeutic Schedule

The antitumor effects of the vaccine candidates were also examined in a mouse breast cancer model with a therapeutic schedule. In brief, each mouse was first subcutaneously challenged with 4T1 cells and then intramuscularly administered with vaccines once a week for a total of three times. Twenty-one days after the tumor inoculation, the mice were sacrificed and the tumors were collected (Figure 5A). The tumor volumes, tumor weights and representative images of the tumors are illustrated in Figures 5B,C and S8. Long-term survival curves were recorded as well, as shown in Figure 5D. SZU251 + MUC1 + Al evidently suppressed tumor growth and promoted the survival rate compared to the other vaccine groups. In addition, as shown in Figure 5B,D, MUC1 + Al was effective against the primary tumor growth but not the death of the mice, which might be due to the lung metastasis in the 4T1 model. Hence, SZU251 probably enhanced the antitumor effects of MUC1 + Al.

Toxicity was thoroughly determined, and the vaccines were well tolerated at these doses in the therapeutic schedule. There were no significant differences of the body weights or the weights of the major organs, and no structural or pathological changes of the major organs in all of the vaccine groups (Figure 5E,F and S9). No significant abnormality could be detected in all of the parameters (WBC, RBC, HGB, HCT, MCV and PLT) in blood tests (Figure S10 and Table 3).

Table 3. Blood routine examination of the mice in the therapeutic schedule.

Group	NaCl	Al	SZU251 + Al	MUC1 + Al	SZU251 + MUC1 + Al
WBC (10^9/L)	13.33 ± 0.78	9.68 ± 5.12	15.48 ± 5.31	12.60 ± 2.52	15.30 ± 3.48
RBC (10^{12}/L)	9.85 ± 0.40	10.00 ± 0.94	9.57 ± 0.33	8.70 ± 0.46	9.34 ± 0.81
HGB (g/L)	144.80 ± 4.15	142.00 ± 7.00	144.40 ± 7.54	134.00 ± 5.29	140.20 ± 9.40
HCT (%)	44.83 ± 1.35	45.93 ± 3.95	43.86 ± 1.95	40.56 ± 1.70	43.34 ± 3.67
MCV (fL)	46.92 ± 0.47	46.03 ± 0.76	45.77 ± 0.21	46.98 ± 0.89	45.87 ± 0.35
PLT (10^9/L)	739.40 ± 128.80	633.80 ± 288.80	1056.00 ± 214.99	933.40 ± 188.03	956.75 ± 231.80

Abbreviations: WBC, white blood cell; RBC, red blood cell; HGB, hemoglobin; HCT, hematocrit; MCV, mean corpuscular volume; PLT, platelet.

2.6. SZU251 + MUC1 + Al Induced Tumor-Specific Immune Responses in the Therapeutic Schedule

Humoral immune responses of the vaccinations are displayed as serum antibody titers against MUC1 judged by the ELISA method (Table 4). MUC1 + Al and SZU251 + MUC1 + Al remarkably elicited total IgG, IgG1, IgG2a and IgM responses, while SZU251 + MUC1 + Al was much more potent than MUC1 + Al. The serum antibody titers from 20 days after the first immunization were higher than those from 10 days after the first immunization. Binding of the serum samples to the tumor cells was examined by flow cytometry (Figure 6A). The serum samples from SZU251 + MUC1 + Al reacted strongly to 4T1 cells with high MUC1 expression but not B16 cells with low MUC1 expression.

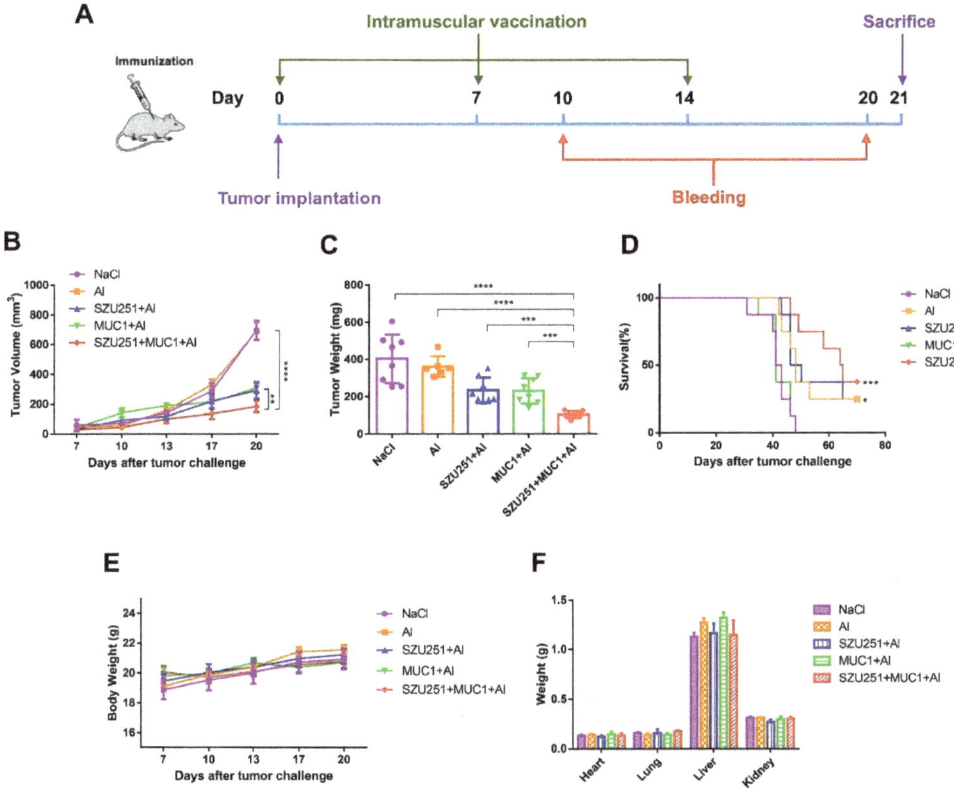

Figure 5. SZU251 + MUC1 + Al inhibited 4T1 mouse breast tumor growth after three immunizations in the therapeutic schedule. (**A**) Schematic diagram illustrating the therapeutic vaccination. (**B**) Tumor growth curves were measured twice a week until day 21 after tumor implantation. (**C**) Tumor weights were determined at the time of sacrifice. (**D**) Survival curves of the tumor-bearing mice. (**E**) Body weights were measured twice a week until day 21 after tumor implantation. (**F**) Weights of the major organs (hearts, lungs, livers and kidneys) were determined at the time of sacrifice. Data are presented as the mean ± SE; n = 5–8 mice/group. * $p < 0.05$, ** $p < 0.01$, *** $p < 0.001$, **** $p < 0.0001$.

Table 4. Anti-MUC1 antibody titers determined by ELISA after three immunizations with vaccine candidates in the therapeutic schedule.

Antibody		Titer				
		NaCl	Al	SZU251 + Al	MUC1 + Al	SZU251 + MUC1 + Al
IgG	Day 10	0	0	0	32,000	64,000
	Day 20	0	0	0	128,000	1,024,000
IgG1	Day 10	0	0	0	0	0
	Day 20	0	0	0	32,000	32,000
IgG2a	Day 10	0	0	0	32,000	64,000
	Day 20	0	0	0	32,000	1,024,000
IgM	Day 10	0	0	0	8000	128,000
	Day 20	0	0	0	32,000	256,000

Antibody titer was defined as the maximum serum dilution when the absorption ratio of the experimental group to the negative control group was ≥2.0.

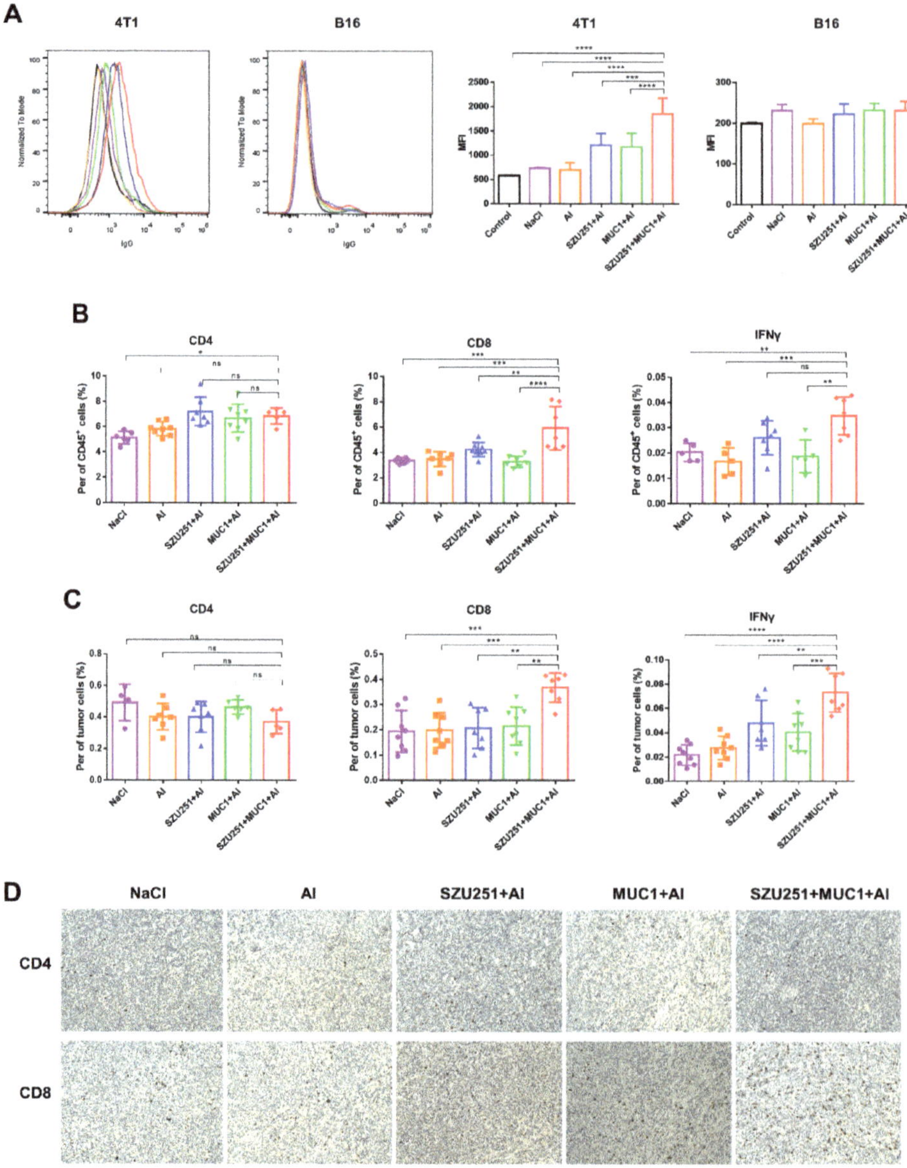

Figure 6. SZU251 + MUC1 + Al induced tumor-specific immune responses in the therapeutic schedule. (**A**) Specific anti-MUC1 antibody of the serum samples was tested on 4T1 and B16 cells using flow cytometry. NaCl (purple), Al (orange), SZU251 + Al (blue), MUC1 + Al (green) and SZU251 + MUC1 + Al (red). (**B**) Percentages of $CD3^+$ $CD4^+$ T cells, $CD3^+$ $CD8^+$ T cells and $CD3^+$ $CD8^+$ $IFN\gamma^+$ T cells in splenocytes were measured using flow cytometry. (**C**) Percentages of $CD3^+$ $CD4^+$ T cells, $CD3^+$ $CD8^+$ T cells and $CD3^+$ $CD8^+$ $IFN\gamma^+$ T cells in TILs were measured using flow cytometry. (**D**) Representative IHC staining for CD4 and CD8 of the tumor tissues in the therapeutic schedule (200× magnification). Data are presented as the mean ± SE; n = 5–8 mice/group. * $p < 0.05$, ** $p < 0.01$, *** $p < 0.001$, **** $p < 0.0001$.

Cellular immune responses to the vaccinations were determined by analyzing $CD4^+$ and $CD8^+$ T cells in the spleens and the tumors, while activated $CD8^+$ T cells were further identified by intracellular IFNγ staining (Figure 6B,C). SZU251 + MUC1 + Al boosted the percentages of total $CD8^+$ and $CD8^+$ $IFN\gamma^+$ cells both in the spleens and the tumors more effectively than any other vaccines. However, $CD4^+$ T cells were not impacted as much as $CD8^+$ T cells, where SZU251 + MUC1 + Al exhibited higher percentages of total $CD4^+$ T cells only in the spleens compared to the saline control (NaCl). IHC staining of the tumor tissues proved that the infiltration of $CD8^+$ T cells, but not $CD4^+$ T cells, was significantly promoted by SZU251 + MUC1 + Al (Figure 6D).

3. Discussion

Breast cancer, especially triple-negative breast cancer (TNBC), is a biologically and clinically heterogeneous disease which has been considered a major cause of morbidity and mortality due to its aggressive behavior and poor prognosis [15]. Traditional treatments, including surgery and chemotherapy, are not satisfactory enough for all of the patients. Decades of investigations on breast cancer have led to the development of new therapeutic options in recent years, such as immune checkpoint inhibitors, antibody–drug conjugates, tumor vaccines and other promising drug combinations.

Tumor vaccines come into effect by training the immune system to attack the cells that contain the tumor-associated antigens (TAAs) in the vaccines. A protein with an abnormal structure produced by tumor cells due to mutation can be an effective TAA, and abnormal glycosylation is an example of the mutation. MUC1 is a typical glycosylated tumor antigen expressed on the surface of cancer cells. Compared to normal tissues, the expression levels and glycosylation patterns of MUC1 are highly dissimilar in tumor tissues [16]. In a phase 2b/3 trial for advanced non-small-cell lung cancer, TG4010 (a modified vaccinia Ankara expressing MUC1) plus chemotherapy showed a significant improvement in progression-free survival compared to a placebo plus chemotherapy [17]. MUC1-specific T cells and antibodies are also identified in breast cancer patients, and MUC1-specific immunity is beneficial in the treatment of breast cancer [18,19]. It was reported that an MUC1 mRNA nano-vaccine induced a strong, antigen-specific, in vivo cytotoxic T lymphocyte response against TNBC cells [20]. However, the MUC1 antigen itself does not have enough immunogenicity to arouse efficient immune responses. Therefore, we introduced a small-molecule TLR7 agonist as a promising adjuvant for the activation of a broad spectrum of APCs, including different types of DCs [21]. The novel TLR7 agonist, SZU251, was synthesized by our group for the first time and exerted stronger effects on TLR7 activation than imiquimod (Figures 1B and S3). Moreover, after mixing with MUC1, SZU251 + MUC1 maintained similar immunostimulatory effects as SZU251. In agreement with the previous studies, SZU251 + MUC1 caused a rapid induction of common cytokines such as IL6, TNFα and IFNγ in mouse BMDCs and spleen lymphocytes in vitro (Figure 2A,B), favoring Th1-mediated immune responses and cytotoxic T cells acting on cancer cells [22]. SZU251 + MUC1 also promoted the expression of costimulatory molecules (CD40, CD80 and CD86) in mouse BMDCs in a concentration-dependent manner (Figures 2C and S4), inferring the maturation of BMDCs and the activation of T cells [23]. Additionally, SZU251 is an isothiocyanate (-N=C=S) derivative of our previous TLR7 agonist, SZU101, which could be easily coupled to the amino groups of the proteins. In future studies, we will conjugate the TLR7 agonists to other tumor-associated antigens, especially macromolecular protein antigens, where SZU251 could be more convenient for conjugation than SZU101.

An outstanding tumor vaccine should trigger cellular and humoral immunity simultaneously to elicit robust and long-lasting immune responses [24]. We have constructed several tumor vaccines by conjugating TLR7 agonists and TAAs and have proven the anti-tumor effects of the vaccines in previous studies. Nevertheless, the humoral responses of the vaccines are still not satisfactory, even with the introduction of our TLR7 agonists [8,14]. The aluminum adjuvant is well known for the promotion of humoral immune responses by adsorbing antigens and altering their immunological properties [25]. Therefore, we constructed a novel vaccine (SZU251 + MUC1 + Al) with three components in this study. SZU251 and MUC1 were first mixed at a molar ratio of 3:1 to form SZU251 + MUC1, and then, SZU251 + MUC1 was absorbed on the aluminum adjuvant to form SZU251 + MUC1 + Al (Figure 1C). Almost all of the experiments in this study displayed that the aluminum adjuvant (Al) alone exerted no effects on tumor inhibition or immune stimulation compared to the control. Therefore, we only included the SZU251 + MUC1 + Al group and not the SZU251 + MUC1 group, and the efficacy of SZU251 + MUC1 + Al could be determined compared to either the control or Al alone. After three immunizations, a tumor challenge model was generated in BALB/c mice via the implantation of mouse 4T1 TNBC cells (Figure 3A). SZU251 + MUC1 + Al showed the best prophylactic effects with suppression of the tumor growth and extension of the survival time of the mice (Figures 3B,D and S5). It is reported that functional activation of TLR7 might lead to the loss of body weight and a decrease in platelet counts, and chronic TLR7 signaling drives anemia through the differentiation of specialized hemophagocytes [26]. In our toxicity studies, except for a slight decrease in the WBC count, no significant differences could be detected between the groups in the body weights of the mice, the weights and HE staining of the major organs and the results of the blood tests (Figures 3E,F, S6 and S7). Thus, the safety of SZU251 + MUC1 + Al was displayed in the prophylactic schedule.

As for the humoral responses, SZU251 + MUC1 + Al induced a substantial increase in antibody titers for MUC1-specific IgG, IgG1, IgG2a and IgM in a time-dependent manner (Table 2). IgG2a and IgG1 represent immune Th1 and Th2 tendencies, respectively. Th1 is the most important helper cell type in cancer immunity, involved in tumor cell killing by secreting cytokines that activate tumor cell surface death receptors and induce epitope spreading [27,28]. Our results demonstrated that after three immunizations, the ratio of IgG2a to IgG1 of SZU251 + MUC1 + Al was the highest compared to the other groups, indicating the tendency of Th1. IgM is also part of the first line of immunological defense. Natural IgM is associated with the recognition and removal of cancerous cells, and some IgM antibodies are used in the diagnosis and treatment of breast cancer [29]. Furthermore, the recognition of native MUC1 antigen on 4T1 cells by the antibodies was verified by flow cytometry (Figure 4A).

As for the cellular responses, lymphocytes in the spleens and the tumors were analyzed by flow cytometry. The main types of lymphocytes in cell-mediated immunity are $CD4^+$ and $CD8^+$ T cells, which play a critical role in the induction of an effective immune response against tumors [30]. As shown in Figure 4B, significant increases in $CD4^+$, $CD8^+$ and activated $CD8^+$ T cells were detected in the spleens treated with SZU251 + MUC1 + Al. $CD4^+$, $CD8^+$ and activated $CD8^+$ T cell infiltration to the tumor microenvironment was also demonstrated by flow cytometry and IHC staining (Figure 4C,D).

In addition to the above results of the vaccines in the prophylactic schedule, we explored their benefits in the therapeutic schedule. BALB/c mice were first implanted with mouse 4T1 TNBC cells and then treated with the vaccines three times. As shown in Figure 5 and Figures S8–S10, SZU251 + MUC1 + Al displayed high efficacy and low toxicity in the treatment of breast cancer. Humoral and cellular immune responses were determined in the therapeutic schedule, with similar effects as the prophylactic schedule (Figure 6 and Table 4). Only the promotion of $CD4^+$ T cells by SZU251 + MUC1 + Al was weak, which may be due to the limited time of drug administration with low recognition of epitopes [31].

In conclusion, a novel vaccine for breast cancer was developed with three components: a small-molecule TLR7 agonist (SZU251), an MUC1-related peptide (MUC1) and an aluminum adjuvant (Al). Our results demonstrated that SZU251 + MUC1 + Al was effective and safe against a tumor expressing the MUC1 antigen in both prophylactic and therapeutic schedules. Incorporation of the TLR7 agonist and the aluminum adjuvant induced non-specific immune responses and substantially enhanced the specific humoral and cellular immune responses to the MUC1 antigen in the mice. Therefore, our vaccine candidate may have beneficial effects in the prevention and treatment of breast cancer patients.

4. Materials and Methods

4.1. Compounds

The MUC1 peptide used in the present study was obtained by the solid phase method using an Fmoc strategy (ChinaPeptides Co., Ltd., Shanghai, China).

SZU251 was synthesized by our laboratory with the following three steps. First, 1 g SZU101, 540 mg 2-(2-(2-(2-azidoethoxy)ethoxy)ethoxy)ethan-1-amine, 460 mg HOBT, 660 mg EDC and 1200 μL DIPEA were dissolved in 3 mL DMF, and the reaction was monitored by LC-MS at room temperature overnight. The reaction was purified by HPLC to yield 700 mg A1 as a white solid with 48.3% yield. Second, 500 mg A1 and 50 mg Pd/C (10%) were dissolved in 5 mL of MeOH, evacuated and passed through H_2, and the reaction was monitored by LC-MS for 2 h at room temperature. After the reaction, Pd/C was removed by filtration, and the product A2 was obtained by spin-drying with MeOH. Third, 500 mg A2, 130 mg CS_2 and 230 μL TEA were dissolved in 2 mL DMF and reacted overnight at room temperature; then, 310 mg TsCl was added, and the reaction was monitored by LC-MS for 5 h at room temperature. After the reaction, it was purified by HPLC, and 320 mg of white solid (SZU251) was obtained at 60% yield.

SZU251: ^{1}H NMR (500 MHz, DMSO-d6) δ 10.08 (s, 1H), 8.28 (t, J = 5.8 Hz, 1H), 7.86 (t, J = 5.5 Hz, 1H), 7.23 (d, J = 8.1 Hz, 2H), 7.18 (d, J = 8.1 Hz, 2H), 6.50 (s, 2H), 4.82 (s, 2H), 4.29–4.23 (d, J = 5.0 Hz, 2H), 4.20 (d, J = 5.8 Hz, 2H), 3.80 (t, J = 5.1 Hz, 2H), 3.61 (t, J = 5.1 Hz, 2H), 3.60–3.55 (m, 4H), 3.55–3.46 (m, 7H), 3.38 (t, J = 5.9 Hz, 2H), 3.27 (s, 3H), 3.17 (q, J = 5.8 Hz, 2H), 2.33 (s, 4H). ^{13}C NMR (125 MHz, DMSO-d6) δ 171.36, 171.28, 159.81, 152.22, 149.13, 147.72, 138.78, 135.60, 127.48 (2C), 127.34 (2C), 98.33, 70.23, 69.77, 69.75, 69.63, 69.57, 69.10, 68.42, 65.26, 58.07, 45.04, 42.13, 41.75, 40.11, 38.54, 30.73, 30.71.

An aluminum adjuvant (Alhydrogel® adjuvant 2%, InvivoGen, San Diego, CA, USA) containing 8% of the total volume was added to the antigen solution (SZU251, MUC1 or SZU251 + MUC1) and mixed well at 1000 rpm at room temperature for 2 h. Here, for SZU251 + MUC1 + Al, SZU251 and MUC1 were first mixed at a molar ratio of 3:1 and then mixed thoroughly with the aluminum adjuvant.

4.2. Mice and Cell Lines

For the experiments, 4T1 mouse breast cancer cells, mouse BMDCs and mouse spleen lymphocytes were cultured in RPMI-1640 medium supplemented with 10% FBS, 100 μg/mL penicillin and 100 μg/mL streptomycin (Gibco, Waltham, MA, USA) at 37 °C in a humidified atmosphere with 5% CO_2. All experiments were performed using mycoplasma-free cells.

Female 4-week-old BALB/c mice (weight 15–20 g) were purchased from the Guangdong Medical Laboratory Animal Center, China. All mice were housed under constant specific pathogen-free laboratory conditions at 18–22 °C, 50–60% humidity, 12 h of light/dark cycling and with ad libitum access to water and food. The protocol for animal experiments was approved by the Ethics Committee of Experimental Animals of Shenzhen University, Shenzhen, China.

4.3. HEK-Blue Assay

HEK-Blue hTLR7 and hTLR8 cells (InvivoGen) were cultured in DMEM supplemented with 10% FBS, blasticidin, zeocin and normocin (InvivoGen). The cells (2.5×10^4/well) were seeded into 96-well plates and then treated with the compounds in HEK-Blue Detection medium (InvivoGen) for 14 h. The activation of TLR7 and TLR8 was assessed by a microplate reader (BioTek, Winooski, VT, USA).

4.4. BMDCs Preparation

All femurs and tibias of the mice were surgically removed. The ends of the bones were cut off with scissors, and the bones were repeatedly rinsed out into the culture dish. The bone marrow suspension was collected, filtered with a 200-micron nylon mesh and centrifuged at 1200 rpm for 5 min. Then, the obtained mouse bone marrow cells were adjusted to 2×10^5 cells/mL and seeded into 100 mm bacterial culture dishes (Petri dish), and recombinant mouse GM-CSF (20 ng/mL) was added. On day 3, 10 mL of complete culture medium containing 20 ng/mL GM-CSF was added to the culture dish. On day 6 and day 8, the old culture medium was collected in half volume, centrifuged and resuspended with complete culture medium containing 20 ng/mL GM-CSF; then, the cell suspension was put back into the original dish. On day 10, the cells could be collected as BMDCs.

4.5. BMDCs Maturation

The prepared BMDCs (1×10^6 cells/mL/well) were cultured with the compounds for 24 h at 37 °C and stained with anti-mouse CD11c-PerCP/Cy5.5, CD40-APC, CD80-FITC and CD86-BV650 antibodies (BioLegend, San Diego, CA, USA) at room temperature for 30 min in the dark. The cells were analyzed using an Attune NxT Flow Cytometer (Invitrogen, Waltham, MA, USA).

4.6. Spleen Lymphocytes Preparation

The spleens of the mice were surgically removed and ground with 4 mL of mouse lymphocyte separation solution. After centrifugation, the middle layer of the liquid was harvested and washed once by centrifugation. Then, 1 mL of lysis solution was used to lyse the erythrocytes for 1 min, and the remaining cells were collected as mouse spleen lymphocytes after washing with PBS.

4.7. Analysis of Cytokine Levels by ELISA

The prepared spleen lymphocytes and BMDCs (1×10^6 cells/mL/well) were cultured with the compounds for 24 h. The total cytokine levels generated by the vaccine candidates were evaluated using ELISA kits (IFNγ, TNFα, IL6 and IL12) (Invitrogen) as per the manufacturer's instructions. Briefly, high-binding 96-well plates were coated with capture antibodies at 4 °C overnight. Then, the coated plates were blocked by diluent buffer, added with the samples and incubated for 2 h at room temperature. Following incubation with the detection antibodies and Avidin-HRP for 1 h at room temperature, the plates were incubated with 3,3′,5,5′-tetramethylbenzidine (TMB) substrate solution for 15–20 min and then sulfuric acid. Absorbance was measured at 450 nm with a microplate reader (BioTek, Winooski, VT, USA).

4.8. Immunization of Mice

Five groups of female BALB/c mice were examined with different vaccine candidates: (i) normal saline (NaCl); (ii) Alhydrogel adjuvant 2% (Al); (iii) SZU251 (57 μg) with Alhydrogel adjuvant (SZU251 + Al); (iv) MUC1 (100 μg) with Alhydrogel adjuvant (MUC1 + Al); (v) SZU251 + MUC1 (157 μg) with Alhydrogel adjuvant (SZU251 + MUC1 + Al). In the prophylactic schedule, mice were immunized by intramuscular injection into the right hind leg on day 0, day 14 and day 28. The mice were bled on day 10, day 24 and day 38 after boost immunizations. In the therapeutic schedule, mice were immunized by intramuscular injection into the right hind leg on day 0, day 7 and day 14. The mice were bled on day 10 and day 20 after boost immunizations. Mouse blood samples were placed at 4 °C overnight and centrifuged at 3000 rpm for 10 min. The supernatant was collected as serum and stored at −80 °C before use.

4.9. Analysis of Antibody Titers and Subtypes by ELISA

The antibody titers and antibody isotypes generated by the vaccine candidates were measured by ELISA. The MUC1 peptide as an antigen was dissolved in a $NaHCO_3/Na_2CO_3$

buffer with a final concentration of 1 μg/mL. Next, 96-well plates were coated with MUC1 and incubated at 4 °C overnight. Then, the coated plates were blocked with dilution buffer and incubated with the serially diluted serum samples at 37 °C for 2 h. Following incubation with either HRP-linked goat anti-mouse antibody IgG, IgM, IgG1 or IgG2a (Immunoway, Plano, TX, USA) (1:5000 dilution) at 37 °C for 1 h, the plates were incubated with TMB substrate solution for 15 min and then sulfuric acid. Absorbance was recorded at 450 nm with a microplate reader (BioTek). Antibody titer was defined as the maximum serum dilution when the absorption ratio of the experimental group to the negative control group was $\geq$2.0.

4.10. Analysis of Antigen Recognition Ability

Serum samples collected after immunizations were diluted 10-fold and incubated with a 4T1 or B16 single cell suspension on ice for 1 h in the dark. Next, the cells were washed and incubated with goat anti-mouse IgG γ-chain-specific antibody conjugated to fluorescein isothiocyanate (IgG-FITC; Biolegend) on ice for 1 h in the dark. Cells were analyzed using an Attune NxT Flow Cytometer (Invitrogen) and FlowJo X 10.0.7 R2 software (BD Biosciences, Franklin Lakes, NJ, USA).

4.11. Evaluation of Antitumor Effects in Vaccinated Mice

For the prophylactic schedule, on day 40, mice were subcutaneously implanted with 4T1 cell suspension (3×10^5 cells). Tumor size was measured twice a week, and tumor volume was calculated according to the following equation: Volume = $L \times W^2/2$, where L (length) and W (width) are the long and short axes of the tumor, respectively. On day 61, the mice were sacrificed with carbon dioxide inhalation to reduce animal pain, and tumors and spleens were surgically dissected, weighed and photographed. For the therapeutic schedule, mice were implanted with 4T1 cell suspension on day 0, and tumors and spleens were removed on day 21. In both sets of experiments, an additional 40 mice were used to assess long-term survival until their natural death or until the tumor size reached 20 mm in diameter.

4.12. Detection of T Lymphocytes

Spleens and tumors were harvested at the time of sacrifice. Spleens were ground and red blood cells were lysed. Tumors were ground and digested with collagenase A and DNase I at 37 °C for 40 min. Single cell suspensions of splenocytes and tumor cells (2×10^6 cells/spleen, 5×10^6 cells/tumor) were stained with anti-mouse CD45-BV605, CD3-FITC, CD4-PE and CD8-PerCP/Cy5.5 antibody (cell surface antibody) (BioLegend), protecting from light at 4 °C for 30 min. The cells were fixed and permeabilized for intracellular staining (Fixation/Permeabilization Solution kit, eBioscience, San Diego, CA, USA) and stained with anti-mouse IFNγ-BV421 antibody (intracellular antibody) (BioLegend) at room temperature for 30 min in the dark. Subsequently, cells were analyzed using an Attune NxT Flow Cytometer (Invitrogen) and FlowJo v10 software.

4.13. Immunohistochemistry (IHC) Staining

Tumor tissues were fixed in 4% paraformaldehyde, embedded in paraffin and transversely cut. Sections were stained with anti-mouse CD4 and CD8 primary antibody (Abcam, Cambridge, UK) at 4 °C overnight and HRP-conjugated secondary antibody (Abcam) at room temperature for 1 h. Additionally, 3′-3-diaminobenzidine (DAB) substrate was used for color development. Cell nuclei were further stained with hematoxylin. The sections were dehydrated, sealed and photographed with an EVOS fluorescence microscope (Invitrogen) at 200× magnification.

4.14. Toxicity Studies

After tumor implantation, the body weights of the mice were measured twice a week. At the time of sacrifice, blood samples and major organs (hearts, livers, lungs and kidneys) were collected. The weights of the major organs were measured, and pathological changes of the major organs were observed by HE staining. Routine blood examination was performed using a blood test instrument (Mindray, Shenzhen, China), where the changes in white blood cell (WBC) count, red blood cell (RBC) count, hemoglobin (HGB), hematocrit (HCT), mean corpuscular volume (MCV) and platelet (PLT) were recorded.

4.15. Statistical Analysis

Statistical analysis was performed using the GraphPad Prism5 software. A two-tailed *t*-test and one-way ANOVA test were performed for comparisons between two and more groups, respectively. Statistical analysis of the data from the tumor growth curves was performed using a two-way ANOVA test. Data are presented as the mean ± SE. Differences were considered statistically significant when $p < 0.05$.

Supplementary Materials: The following supporting information can be downloaded at: https://www.mdpi.com/article/10.3390/ijms232315130/s1.

Author Contributions: Conceptualization, X.W. and G.J.; methodology, S.Z., Y.L., J.Z. and J.W.; software, S.Z.; validation, S.Z., Y.L., J.Z. and J.W.; formal analysis, S.Z., Y.L. and J.Z.; investigation, S.Z., Y.L., J.Z. and J.W.; resources, X.W.; data curation, S.Z.; writing—original draft preparation, S.Z.; writing—review and editing, X.W.; visualization, X.W. and S.Z.; supervision, X.W.; funding acquisition, X.W. All authors have read and agreed to the published version of the manuscript.

Funding: This research was funded by the Guangdong Science and Technology Department, grant number 2022A1515012219, the Shenzhen Science and Technology Innovation Commission, grant number JCYJ20220531101605011, and the Shenzhen Key Medical Discipline Construction Fund, grant number SZXK062.

Institutional Review Board Statement: The study was conducted in accordance with the Declaration of Helsinki and approved by the Ethics Committee of the 3rd Affiliated Hospital of Shenzhen University (protocol 2019-SZLH-LW-009, approval on 15 October 2019).

Informed Consent Statement: Not applicable.

Data Availability Statement: The data needed to evaluate the conclusions are present in the paper and/or the Supplementary Materials.

Conflicts of Interest: The authors declare no conflict of interest.

References

1. Mokhtari, Y.; Pourbagheri-Sigaroodi, A.; Zafari, P.; Bagheri, N.; Ghaffari, S.H.; Bashash, D. Toll-like receptors (TLRs): An old family of immune receptors with a new face in cancer pathogenesis. *J. Cell. Mol. Med.* **2021**, *25*, 639–651. [CrossRef] [PubMed]
2. Petes, C.; Odoardi, N.; Gee, K. The Toll for Trafficking: Toll-Like Receptor 7 Delivery to the Endosome. *Front. Immunol.* **2017**, *8*, 1075. [CrossRef] [PubMed]
3. Ding, R.; Jiao, A.; Zhang, B. Targeting toll-like receptors on T cells as a therapeutic strategy against tumors. *Int. Immunopharmacol.* **2022**, *107*, 108708. [CrossRef] [PubMed]
4. Wan, D.; Que, H.; Chen, L.; Lan, T.; Hong, W.; He, C.; Yang, J.; Wei, Y.; Wei, X. Lymph-Node-Targeted Cholesterolized TLR7 Agonist Liposomes Provoke a Safe and Durable Antitumor Response. *Nano Lett.* **2021**, *21*, 7960–7969. [CrossRef]
5. Zhang, H.; Tang, W.L.; Kheirolomoom, A.; Fite, B.Z.; Wu, B.; Lau, K.; Baikoghli, M.; Raie, M.N.; Tumbale, S.K.; Foiret, J.; et al. Development of thermosensitive resiquimod-loaded liposomes for enhanced cancer immunotherapy. *J. Control. Release Off. J. Control. Release Soc.* **2021**, *330*, 1080–1094. [CrossRef]
6. Narayanan, J.S.S.; Ray, P.; Hayashi, T.; Whisenant, T.C.; Vicente, D.; Carson, D.A.; Miller, A.M.; Schoenberger, S.P.; White, R.R. Irreversible Electroporation Combined with Checkpoint Blockade and TLR7 Stimulation Induces Antitumor Immunity in a Murine Pancreatic Cancer Model. *Cancer Immunol. Res.* **2019**, *7*, 1714–1726. [CrossRef]
7. Wang, X.; Yu, B.; Cao, B.; Zhou, J.; Deng, Y.; Wang, Z.; Jin, G. A chemical conjugation of JQ-1 and a TLR7 agonist induces tumoricidal effects in a murine model of melanoma via enhanced immunomodulation. *Int. J. Cancer* **2021**, *148*, 437–447. [CrossRef]
8. Wang, X.; Liu, Y.; Diao, Y.; Gao, N.; Wan, Y.; Zhong, J.; Zheng, H.; Wang, Z.; Jin, G. Gastric cancer vaccines synthesized using a TLR7 agonist and their synergistic antitumor effects with 5-fluorouracil. *J. Transl. Med.* **2018**, *16*, 120. [CrossRef]
9. Beatson, R.E.; Taylor-Papadimitriou, J.; Burchell, J.M. MUC1 immunotherapy. *Immunotherapy* **2010**, *2*, 305–327. [CrossRef]

10. Rashidijahanabad, Z.; Huang, X. Recent advances in tumor associated carbohydrate antigen based chimeric antigen receptor T cells and bispecific antibodies for anti-cancer immunotherapy. *Semin. Immunol.* **2020**, *47*, 101390. [CrossRef]
11. Stergiou, N.; Urschbach, M.; Gabba, A.; Schmitt, E.; Kunz, H.; Besenius, P. The Development of Vaccines from Synthetic Tumor-Associated Mucin Glycopeptides and their Glycosylation-Dependent Immune Response. *Chem. Rec.* **2021**, *21*, 3313–3331. [CrossRef] [PubMed]
12. Zhou, S.H.; Li, Y.T.; Zhang, R.Y.; Liu, Y.L.; You, Z.W.; Bian, M.M.; Wen, Y.; Wang, J.; Du, J.J.; Guo, J. Alum Adjuvant and Built-in TLR7 Agonist Synergistically Enhance Anti-MUC1 Immune Responses for Cancer Vaccine. *Front. Immunol.* **2022**, *13*, 857779. [CrossRef] [PubMed]
13. Lakshminarayanan, V.; Thompson, P.; Wolfert, M.A.; Buskas, T.; Bradley, J.M.; Pathangey, L.B.; Madsen, C.S.; Cohen, P.A.; Gendler, S.J.; Boons, G.J. Immune recognition of tumor-associated mucin MUC1 is achieved by a fully synthetic aberrantly glycosylated MUC1 tripartite vaccine. *Proc. Natl. Acad. Sci. USA* **2012**, *109*, 261–266. [CrossRef] [PubMed]
14. Liu, Y.; Tang, L.; Gao, N.; Diao, Y.; Zhong, J.; Deng, Y.; Wang, Z.; Jin, G.; Wang, X. Synthetic MUC1 breast cancer vaccine containing a Toll-like receptor 7 agonist exerts antitumor effects. *Oncol. Lett.* **2020**, *20*, 2369–2377. [CrossRef] [PubMed]
15. Bianchini, G.; Balko, J.M.; Mayer, I.A.; Sanders, M.E.; Gianni, L. Triple-negative breast cancer: Challenges and opportunities of a heterogeneous disease. *Nat. Rev. Clin. Oncol.* **2016**, *13*, 674–690. [CrossRef]
16. Beckwith, D.M.; Cudic, M. Tumor-associated O-glycans of MUC1: Carriers of the glyco-code and targets for cancer vaccine design. *Semin. Immunol.* **2020**, *47*, 101389. [CrossRef]
17. Quoix, E.; Lena, H.; Losonczy, G.; Forget, F.; Chouaid, C.; Papai, Z.; Gervais, R.; Ottensmeier, C.; Szczesna, A.; Kazarnowicz, A.; et al. TG4010 immunotherapy and first-line chemotherapy for advanced non-small-cell lung cancer (TIME): Results from the phase 2b part of a randomised, double-blind, placebo-controlled, phase 2b/3 trial. *Lancet Oncol.* **2016**, *17*, 212–223. [CrossRef]
18. Guckel, B.; Rentzsch, C.; Nastke, M.D.; Marme, A.; Gruber, I.; Stevanovic, S.; Kayser, S.; Wallwiener, D. Pre-existing T-cell immunity against mucin-1 in breast cancer patients and healthy volunteers. *J. Cancer Res. Clin. Oncol.* **2006**, *132*, 265–274. [CrossRef]
19. Fremd, C.; Stefanovic, S.; Beckhove, P.; Pritsch, M.; Lim, H.; Wallwiener, M.; Heil, J.; Golatta, M.; Rom, J.; Sohn, C.; et al. Mucin 1-specific B cell immune responses and their impact on overall survival in breast cancer patients. *Oncoimmunology* **2016**, *5*, e1057387. [CrossRef]
20. Liu, L.; Wang, Y.; Miao, L.; Liu, Q.; Musetti, S.; Li, J.; Huang, L. Combination Immunotherapy of MUC1 mRNA Nano-vaccine and CTLA-4 Blockade Effectively Inhibits Growth of Triple Negative Breast Cancer. *Mol. Ther. J. Am. Soc. Gene Ther.* **2018**, *26*, 45–55. [CrossRef]
21. Beesu, M.; Salyer, A.C.; Brush, M.J.; Trautman, K.L.; Hill, J.K.; David, S.A. Identification of High-Potency Human TLR8 and Dual TLR7/TLR8 Agonists in Pyrimidine-2,4-diamines. *J. Med. Chem.* **2017**, *60*, 2084–2098. [CrossRef] [PubMed]
22. Zitvogel, L.; Galluzzi, L.; Kepp, O.; Smyth, M.J.; Kroemer, G. Type I interferons in anticancer immunity. *Nat. Rev. Immunol.* **2015**, *15*, 405–414. [CrossRef] [PubMed]
23. Van Gool, S.W.; Vandenberghe, P.; de Boer, M.; Ceuppens, J.L. CD80, CD86 and CD40 provide accessory signals in a multiple-step T-cell activation model. *Immunol. Rev.* **1996**, *153*, 47–83. [CrossRef] [PubMed]
24. Guy, B. The perfect mix: Recent progress in adjuvant research. *Nat. Rev. Microbiol.* **2007**, *5*, 505–517. [CrossRef]
25. He, P.; Zou, Y.; Hu, Z. Advances in aluminum hydroxide-based adjuvant research and its mechanism. *Hum. Vaccines Immunother.* **2015**, *11*, 477–488. [CrossRef]
26. Akilesh, H.M.; Buechler, M.B.; Duggan, J.M.; Hahn, W.O.; Matta, B.; Sun, X.; Gessay, G.; Whalen, E.; Mason, M.; Presnell, S.R.; et al. Chronic TLR7 and TLR9 signaling drives anemia via differentiation of specialized hemophagocytes. *Science* **2019**, *363*, eaao5213. [CrossRef]
27. Mailliard, R.B.; Egawa, S.; Cai, Q.; Kalinska, A.; Bykovskaya, S.N.; Lotze, M.T.; Kapsenberg, M.L.; Storkus, W.J.; Kalinski, P. Complementary dendritic cell-activating function of CD8+ and CD4+ T cells: Helper role of CD8+ T cells in the development of T helper type 1 responses. *J. Exp. Med.* **2002**, *195*, 473–483. [CrossRef]
28. Lin, W.W.; Karin, M. A cytokine-mediated link between innate immunity, inflammation, and cancer. *J. Clin. Invest.* **2007**, *117*, 1175–1183. [CrossRef]
29. Díaz-Zaragoza, M.; Hernández-Ávila, R.; Viedma-Rodríguez, R.; Arenas-Aranda, D.; Ostoa-Saloma, P. Natural and adaptive IgM antibodies in the recognition of tumor-associated antigens of breast cancer (Review). *Oncol. Rep.* **2015**, *34*, 1106–1114. [CrossRef]
30. Sato-Kaneko, F.; Yao, S.; Ahmadi, A.; Zhang, S.S.; Hosoya, T.; Kaneda, M.M.; Varner, J.A.; Pu, M.; Messer, K.S.; Guiducci, C.; et al. Combination immunotherapy with TLR agonists and checkpoint inhibitors suppresses head and neck cancer. *JCI Insight* **2017**, *2*, e93397. [CrossRef]
31. Stephens, A.J.; Burgess-Brown, N.A.; Jiang, S. Beyond Just Peptide Antigens: The Complex World of Peptide-Based Cancer Vaccines. *Front. Immunol.* **2021**, *12*, 696791. [CrossRef] [PubMed]

MDPI
St. Alban-Anlage 66
4052 Basel
Switzerland
Tel. +41 61 683 77 34
Fax +41 61 302 89 18
www.mdpi.com

International Journal of Molecular Sciences Editorial Office
E-mail: ijms@mdpi.com
www.mdpi.com/journal/ijms

www.ingramcontent.com/pod-product-compliance
Lightning Source LLC
LaVergne TN
LVHW070425100526
838202LV00014B/1529